RANDOM VIBRATIONS

RANDOM VIBRATIONS

Theory and Practice

PAUL H. WIRSCHING
The University of Arizona
Tucson, Arizona

THOMAS L. PAEZ
Sandia National Laboratories
Albuquerque, New Mexico

KEITH ORTIZ
Sandia National Laboratories
Albuquerque, New Mexico

A Wiley-Interscience Publication
JOHN WILEY & SONS, INC.
New York • Chichester • Brisbane • Toronto • Singapore

Library of Congress Cataloging in Publication Data:
Wirsching, Paul H.
    Random vibrations: theory and practice / by Paul H. Wirsching,
Thomas L. Paez, Keith Ortiz.
        p.   cm.
    ISBN 0-471-58579-3 (cloth : alk. paper)
    1. Random vibration.   I. Paez, Thomas L., 1949–   . II. Ortiz,
Keith. III. Title.
TA355.W57   1995
620.3—dc20   95-32229

Printed in the United States of America

10 9 8 7 6 5 4 3 2 1

To our families who have given us their support.

To Jeanne

P.H.W.

To Eileen

T.L.P.

To Karen, Keary, and Kaley

K.O.

CONTENTS

PREFACE

At the time of this writing, the field of random vibrations is making significant breakthroughs in its practice. Several factors are converging to make this happen. First, there is a proliferation of commercial, portable data acquisition and test equipment, made possible by low-cost personal computers. Second, the techniques for performing system identification and modal analysis are becoming established and widespread. Third, random vibration testing standards are being adopted for military hardware. Random vibrations is no longer only a specialist's field.

Even if it were not important in its own right, the study of random vibrations would be important because it is an excellent instructional vehicle for the tools of modern vibration analysis. These tools include random processes, Fourier transforms and spectral analysis, system identification and modal analysis, probabilistic (or reliability-based) design, and data acquisition and analysis.

The authors have been teaching, individually, random vibrations in graduate courses for many years and, together, a successful short course since 1988. In that time, we have been continuously frustrated by the lack of a text to recommend to our students as a single reference on the subject. We each have favorites for particular topics, but there is no one-stop shopping. For example, many of the "best" books on the subject assume much more familiarity and sophistication with the subject of probability than the majority of our students have. Others are so "practical" in their explanations that the student is not well prepared to do research. There seems to be nothing written for the middle ground. Furthermore, from our short-course experience, we wished to have a textbook that our students could take home and study after leaving our lectures to obtain greater depth of understanding than otherwise possible. This text is intended to fill these needs.

Hence, this book is written with the following expectations: (1) The student is expected to have had a course in probability and a course in vibrations, at least at the undergraduate elective level. But this will be the student's first exposure to random processes and, possibly, to multiple-degree-of-freedom dynamic systems. (2) Although this text contains enough material for a full year of graduate study, most college offerings are one semester or one quarter in length. So we expect that much of the material might be covered through independent study. In the case of practicing engineers, all of the material might be covered through independent study. Hence, the book makes liberal use of examples and illustrations throughout, and each chapter has a set of problems designed to facilitate this style of learning. (3) At the end, a graduate student should have a solid foundation in stationary, Gaussian random vibrations of linear systems and should be prepared to begin masters-level research in random vibrations, for example, on nonstationary or non-Gaussian random processes or on nonlinear systems. An industrial student should be equipped to understand design and test specifications and be able to work competently with data. Whatever the starting level, our hope is to increment each student's knowledge and understanding.

The scope of our presentation is limited by the allowable length of the book and by our intention to make a solid presentation of the fundamentals. Hence, we deliberately have not included material on the response of nonlinear systems and the tools normally associated with nonlinear random vibrations, such as equivalent linearization, Markov processes, or the Fokker–Planck equation. Likewise, we have minimized the material on nonstationary and non-Gaussian random processes. These topics are important, but there is no universal approach to solving these problems, and we could not hope to cover all possible solution methods. Therefore, some instructors may wish to supplement this text with specialized, advanced material as appropriate.

The book is conceptually divided into three parts. Chapters 1–6 review the theory of probability and present the theory of random processes in the time and frequency domains. Chapters 7–10 review the deterministic vibration of single- and multiple-degree-of-freedom linear systems, develop the response relations for random loadings, and discuss design for random vibrations. Chapters 11–14 review the theory of statistical estimation and present the theory of statistical signal processing. The appendices include material on the convergence of random processes, sample programs for the fast Fourier transform, some useful mathematical formulas, and pertinent probability tables.

There are many people to be thanked for the realization of this book in print. The authors especially wish to acknowledge Linda Reuter for typing the manuscript and Frank Cerra, formerly of John Wiley, and Minna Panfili, of John Wiley, for their encouragement. We also extend our thanks to our many colleagues who read sections of the book and recommended improvements and to our students, who in their efforts to understand random vibrations asked questions that led to improvements in the book. Finally, any passages that lack clarity or include errors are the sole responsibility of the authors.

1

INTRODUCTION

1.1 HISTORY AND MOTIVATION

The accurate analysis, design, and assessment of mechanical and structural systems, subjected to realistic dynamic environments, must consider the potential for random loads and randomness in structural and material properties. The engineering field that deals with these issues is known as random vibrations. The modern theory of random vibrations is the product of generations of work in the fields of deterministic structural vibrations, probabilistic analysis of mechanical system response, and random signal analysis.

The deterministic theory of structural vibrations has its roots at least as early as Lord Rayleigh's *The Theory of Sound* (1877). Rayleigh's book concerns itself with how sound is created and perceived and thus with the vibration and resonance of elastic solids and gases and with the propagation of acoustic waves in materials. His methods were adopted around the turn of the century by mechanical engineers who were concerned with the vibration of engineered structures.

By the time Stephen Timoshenko's *Vibration Problems in Engineering* appeared in 1928, a rigorous theory of structural vibrations had been established. The theory was motivated by the dynamic loadings resulting from the operation of machines, especially rotating machinery. Timoshenko gives many examples of vibration problems facing engineers of the day. All involve dynamic loads that are periodic with known frequencies. When transient loads arise, they are represented in analytical forms. The physical parameters of the system are known and the loading is deterministic. The study of this sort of vibration forms the basic material of most modern courses and texts in deterministic structural vibrations.

From these beginnings, structural vibration theory has evolved in the direction of greater complexity and realism, and technology has accelerated the rate of change. The digital computer makes possible the analysis of complex structures using the finite-element method and the computation of the response to an arbitrary load history using numerical integration.

However, many important modern vibration problems are not sinusoidal or even periodic. In fact, the loadings are nondeterministic, that is, they are random or stochastic. This requires the further extension of structural vibration theory to include techniques to characterize and predict random vibrations. The necessary extensions have their roots as early as the work of Albert Einstein.

The probabilistic theory of mechanical system behavior arose from theoretical investigations into the motion of particles suspended in a fluid, known as Brownian motion. Einstein's (1905) doctoral research established the first mathematical treatment of Brownian motion, showing that the probabilistic behavior of randomly excited particles suspended in a fluid medium is governed by the diffusion equation, a parabolic partial differential equation. A generation of researchers generalized Einstein's results using the diffusion equation framework. They developed, among other things, probabilistic characterizations for the motion of a linear, single-degree-of-freedom system excited by white noise (a temporally steady, random input) and the motion of multiple-degree-of-freedom systems excited by white noise. Several of the classical papers that made substantial advances in the stochastic analysis of system response are compiled in a volume entitled *Selected Papers on Noise and Stochastic Processes*, edited by Nelson Wax (1954).

Other developments in the mathematical theory of random processes in the 1920s, 1930s, and 1940s provided new methods for the characterization of random processes including Fourier analysis, autocorrelation functions, and spectral density functions. These spectral methods deal directly with the governing equations of motion and may be more practically applied to complex systems than diffusion-equation-based methods. Hence, the spectral approach is more predominant today, particularly for the analysis of linear systems. Much of the original development appears in the communications, electrical engineering, and mathematics literature. Some classical papers in this line of development are also contained in *Selected Papers on Noise and Stochastic Processes*.

By the late 1950s, in response to the demands of military and civilian aerospace programs, the need for a coherent theory of random vibration of mechanical systems was apparent. Rockets, jet engines, and turbulence are classic sources of random stimulation leading to random vibration. A symposium in 1958 and the publication of the collected papers under the title *Random Vibration*, edited by Stephen Crandall, signaled the start of the modern study of random vibration of mechanical systems and structures. The articles in this volume cover topics ranging from random vibration analysis to signal measurement, parameter estimation, testing, and design of mechanical systems for random environments.

The first textbook to be devoted to random vibrations is *Ran*
in Mechanical Systems by Stephen Crandall and William Ma
book is sponsored by the American Society of Mechanical En
major motivation is to "draw together the fundamental facts anu ....
random vibration in a form particularly suited to mechanical engineers." The
book is a presentation of classical results in a mechanical setting, and it con-
centrates on stationary random processes and on linear time-invariant systems
of one and two degrees of freedom. It is of an introductory nature, but it remains
a classic in the field.

During the 1960s and beyond, random vibrations theory was extended to
other types of dynamic systems. Several texts appeared, and early among these
is the classic text by Y. K. Lin, entitled *Probabilistic Theory of Structural
Dynamics* (1967). It develops, in detail, the mathematical theory of random
vibrations. Among the many texts that cover the linear and nonlinear theories
of random vibration of many different types of mechanical systems and structures are those by Augusti, Baratta, and Casciati (1984); Bolotin (1984); Crandall (1963); Ibrahim (1985); Newland (1984); Nigam (1983); Roberts and
Spanos (1990); Robson (1964); Schueller and Shinozuka (1987); Soong and
Grigoriu (1993); and Yang (1986).

While the theory of dynamic response to random loadings was being developed, there were simultaneous developments in the measurement of environments and the statistical estimation of the parameters of random processes.
An early discussion of the important topic of spectral density estimation is
given by Rona (1958). A text considered to be a classic among mechanical
and structural system analysts is *Measurement and Analysis of Random Data*
by Julius Bendat and Allan Piersol, first published in 1966. This book and its
subsequent revisions (Bendat and Piersol, 1971, 1986) became the standard
references for estimation of the parameters of random sources. The practical
breakthrough in the estimation of signal and system parameters came with the
development of the fast Fourier transform by Cooley and Tukey (1965). When
implemented on digital computers, it made possible the rapid assessment of
physical system behavior.

Today, the practice of random vibrations is widespread in industry. One
reason for this is the availability of commercial, portable data acquisition and
test equipment having the capability to analyze or synthesize random vibrations
automatically. Another reason is the recent development of modal analysis as
an analytical and design tool and the use of random forcing functions as input
to these procedures. Other reasons include the adoption of random vibration
testing standards for military work and interest in using random vibrations for
environmental stress screening.

Some typical applications of random vibrations today are as follows:

- Vibration of a bridge or a tall building in the wind

- Vibration of a building in an earthquake

- Vibration of a ship's hull or an offshore oil platform in stormy ocean waves
- Vibration of an automobile due to engine and drive train vibration, pavement roughness and wind, and resulting noise inside its passenger compartment
- Vibration of aerospace vehicles due to turbulence, acoustic pressure, or rough burning engines
- Vibration of nonstructural elements, such as electronic components, due to support motions

The methods developed in this book will prepare the student or practitioner to work on such problems.

1.2 EXAMPLE

We are motivated to pursue the random vibration analysis of systems subjected to random inputs for the following reason: When the response of a mechanical system is forced by a random vibration excitation, the specific response to a future event cannot be computed because the dynamic environments to be realized in the future cannot be precisely predicted.

Consider Figures 1.1 and 1.2, for example. These figures show two segments of time histories (Figures 1.1a and 1.2a) of an excitation from a single random vibration source that occurs in connection with an aerospace environment. The acceleration responses that correspond to the individual inputs are shown in Figures 1.1b and 1.2b. These occur at a point on an aerospace shell structure. Though the excitations come from the same nominal source, they are not identical. For example, the responses have the same general character but differ at almost all times and their peak values differ. Any future excitation and the response it excites will differ from those measured in the past. Structural responses to random excitations are random processes.

The major objectives of random vibrations and random signal analysis investigations are the analysis of random responses given the characteristics of random inputs, the design of mechanical systems for random environments, the characterization of random inputs and responses, and the testing of mechanical systems subjected to random inputs.

For example, the following tasks might be associated with the excitations and responses shown in Figures 1.1 and 1.2:

- Acquiring the data shown
- Estimating parameters of random excitation, random response, or a structural system, based on measured excitations and responses
- Analyzing random response based on a description of the structure and a specification of the random excitation

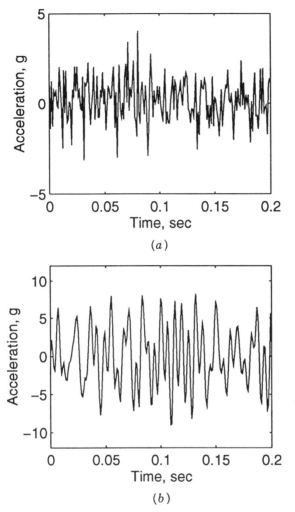

Figure 1.1 (*a*) Segment of random excitation time history from aerospace vehicle. (*b*) Random vibration response excited by input in part (*a*).

- Designing structural and mechanical systems to survive random environments
- Modifying a structure to diminish its response to a random environment
- Judging the health of a structure based on measured excitations and responses and system parameters
- Performing laboratory tests on structures and components that simulate random field environments
- Controlling the motion of a structure through variation of its parameters or application of control forces

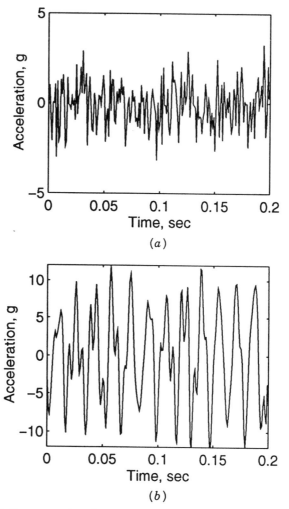

Figure 1.2 (*a*) Segment of random excitation time history measured at same point and during same environment as signal in Figure 1.1*a*. (*b*) Random vibration response excited by input in part (*a*).

1.3 SCOPE AND ORGANIZATION

The purpose of this text is to introduce the student to the theory of random vibrations. The authors wish to make the material accessible to advanced undergraduates and first-year graduate students. We wish to provide a solid foundation in preparation for research, but without the burden of advanced material. Hence, the emphasis of this book is on stationary, ergodic, and Gaussian random processes and on linear time-invariant dynamic systems. The mathematical developments are rigorous and, generally, not limited to the specializations above. However, the most useful, specialized results are quickly determined.

The field continues to advance in many directions simultaneously. The authors could not hope to do justice to advanced topics, such as transient processes (e.g., shock), nonlinear response and the tools for analyzing it (e.g., Markov processes and the Fokker–Planck equation), simulation, and structures with random system properties (known as stochastic differential equations, or probabilistic design). So, rather than trying to be all things to all people, we decided to only cover "the basics." We expect that the instructor who wishes to cover special, advanced topics will supplement the course with pertinent material.

The beginning student may ask why we chose to approach the subject by first presenting the theory of random processes. After all, the spectral density and transfer functions could be developed from the theory of Fourier transforms with which many readers are familiar. This alternate approach would indeed provide us with the correct excitation–response relations. However, we would not be able to answer fundamental, real-world problems like: How should random vibrations be measured (e.g., how long a measurement, how many measurements, how fast to sample, how good are the resulting estimates)? Given data already in hand, how should it be analyzed and what are the trade-offs?

Hence, we have conceptually divided this book into three major parts. In the first part, we present the mathematical foundations of our study, random processes. In the second part, we apply the mathematical foundations to the analysis and design problems. In the third part, we explain the essential elements of random signal analysis.

The first part of the text begins with Chapter 2, which presents basic elements of the theory of probability, emphasizing the significance of correlation between two random variables. Chapter 3 introduces random processes and discusses the autocorrelation and cross-correlation functions, which characterize a random process in the time domain. In Chapter 4, Fourier transforms are presented so they can be applied in Chapter 5, which presents the frequency-domain characterization of a random process using spectral density functions. Chapter 6 presents some probabilistic measures of a stationary random process that are useful in the design problem, including level crossings, envelopes, and peaks.

The second part of the text begins with Chapter 7, which reviews the theory of single-degree-of-freedom systems in the time and frequency domains. Chapter 8 derives the random vibration response of a single-degree-of-freedom system to a random excitation environment. Since most vibration analysis today is done with finite elements, Chapter 9 presents the matrix analysis of multiple-degree-of-freedom systems, including modal decomposition, and the chapter presents the response of each mode and the combination of the modal responses. Chapter 10 presents the basic design problem and contains an extended discussion of metal fatigue due to random vibrations.

The third part of the text starts with Chapter 11 and an examination of the desirable qualities of statistical estimators of the parameters of a random process. In Chapter 12, the estimation of the correlation function in the time domain is discussed. Chapter 13 introduces the discrete Fourier transform and some of the techniques associated with the analysis of data using windows. In Chapter 14, the estimation of spectral density functions, frequency response functions, and related quantities is discussed.

The appendices contain a discussion of some of the fine points associated with differentiation and integration of random processes as well as the standard tables and formulas usually contained in a book of this nature.

2

RANDOM VARIABLES

An understanding of the theory of probability is essential to our development of the theory of random vibrations. This chapter reviews the basics of the theory of probability, random variables, and their probability distributions. Due to its importance in subsequent material, special attention is paid to the correlation of jointly distributed random variables. The treatment here is necessarily brief; for more detailed treatments, the reader is referred to standard texts devoted entirely to this material, for example, Feller (1957), Benjamin and Cornell (1970), Ang and Tang (1975, 1984), and Hines and Montgomery (1990).

2.1 ESSENTIAL PROBABILITY

The systematic mathematical study of probability dates back to the sixteenth and seventeenth centuries, when Gerolamo Cardano (1501–1576), Pierre de Fermat (1601–1665), Blaise Pascal (1623–1662), and Christiaan Huygens (1629–1695) studied games of chance involving cards and dice. In such problems, the probability of an event occurring may be known in advance of any experimentation, that is, a priori, being fixed by the number of cards in the deck or the number of faces on each die. Jakob Bernoulli (1655–1705) extended the mathematical analysis to problems in which the probability of an event is estimated as the relative frequency of the event in many observations, that is, a posteriori. The modern mathematical theory of probability is attributed to Andrey Nikolayevich Kolmogorov (1903–1987), who in 1933 placed it upon an axiomatic foundation, building upon the theory of measure developed by Henri-Léon Lebesgue (1875–1941) in 1901 and by Émile Borel (1871–1956) in 1909. This section presents the essential elements of modern probability theory: set theory, Kolmogorov's axioms, and conditional probability.

2.1.1 Set Theory

A set is a collection of elements, such as $A = \{$the even faces on a die$\} = \{2, 4, 6\}$ or $B = \{$the positive velocities of a mass $M\} = \{V_M > 0\}$. The set that contains all possible elements is called the universal set, designated U. The set that contains no elements is called the empty or null set, designated $\varnothing$. The complement of a set, written as $\overline{A}$ (or A' or $\neg\ A$) and read as "not A," is the set containing all elements of the universe that are not in A. Thus, the empty set is the complement of the universal set.

There are two important binary operations on sets, union and intersection. The union of sets A and B is the set of elements that are either in A or in B or in both. This is written as $A \cup B$, where $\cup$ is read as "union" or as "or." The intersection of sets A and B is the set of elements that are in both A and B. This is written as $A \cap B$, where $\cap$ is read as "intersection" or as "and." If the intersection of A and B is the empty set, i.e., $A \cap B = \varnothing$, then A and B are called mutually exclusive or disjoint. These concepts are illustrated using Venn diagrams in Figure 2.1.

Let E be a random experiment, that is, one with a random or unpredictable outcome. For example, we might toss a die and observe the number of spots on its top face. The set of all possible outcomes of a random experiment is called its sample space S. A random event is defined to be a set of sample outcomes, or sample realizations, of E. In this example, $S = \{1, 2, 3, 4, 5,$

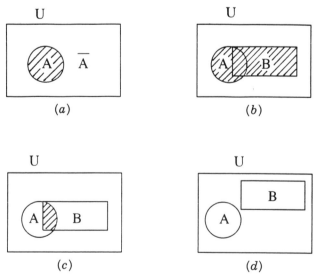

Figure 2.1 Venn diagrams: (a) A and its complement; (b) $A \cup B$; (c) $A \cap B$; (d) $A \cap B = \varnothing$.

6} and we might be interested in the event that the two-spot face is up, {2}, or that a face with an even number of spots is up, {2, 4, 6}.

The careful definition of a random event is essential to the determination of its probability of occurrence. Notice that in E above we are interested in only that face of the die that is on top; the path the die took, the orientation of the die otherwise, and any other specifics of the mechanics of the experiment are not of interest. If these other things were of interest to us, we would need to redefine the experiment.

Example 2.1: Peak Values of Motion of Oscillating Mass. Consider an experiment in which we observe the motion of a mass M that is suspended by a spring and oscillates due to some forcing function. Its position at rest is designated $u_0 = 0$. The event of interest, call it E, is to observe the displacement of the mass whenever it is at a maximum, or peak. Let u, v, and w be the observed displacement, velocity, and acceleration, respectively. How do we describe this event in terms of u, v, and w?

There are two conditions that must exist when M is at a maximum: Its velocity must be equal to zero and its acceleration must be negative. We write these conditions individually as $\{v = 0\}$ and $\{w < 0\}$. There are no restrictions on the displacement, u, that is, it could range from $-\infty$ to $+\infty$. Thus, the event of interest is the intersection of these conditions, namely:

$$E = \{-\infty < u < +\infty \cap v = 0 \cap w < 0\}$$

As long as there is no confusion, it is customary to rewrite this kind of statement by replacing the intersection symbols with commas, as follows:

$$E = \{-\infty < u < +\infty, v = 0, w < 0\}$$

2.1.2 Axioms of Probability

If E is described by a statement, such as in the previous example, the probability of E may be written as $P(E)$, for example, $P(-\infty < u < +\infty, v = 0, w < 0)$; p_E may also be used.

The mathematics of probability is founded upon three intuitive axioms. The first two axioms serve to normalize the measure.

Axiom I. Let A be a random event and let $P(A)$ be the probability that event A occurs, that is, that the conditions for A are true; then, $P(A)$ is a number between 0 and 1, that is,

$$0 \leq P(A) \leq 1 \qquad (2.1)$$

Axiom II. Let S be the event that any event in the sample space occurs, that is, the certain event; then, $P(S)$ is equal to 1, that is,

$$P(S) = 1 \qquad (2.2)$$

Axiom III. This is known as the axiom of additivity: Let A and B be mutually exclusive; then the probability of either A or B occurring is equal to the sum of their probabilities, that is,

$$P(A \cup B) = P(A) + P(B) \text{ if and only if } A \cap B = \emptyset \qquad (2.3)$$

In general, the intersection of A and B is not empty and the probability of their union is equal to the sum of $P(A$ and not $B)$ plus $P(B$ and not $A)$ plus $P(A$ and $B)$. This may be shown to be

$$P(A \cup B) = P(A) + P(B) - P(A \cap B) \qquad (2.4)$$

Example 2.2: Probability of Two Peaks Above a Threshold. Continuing the experiment with the oscillating mass, let E_1 and E_2 be the events that the first and second peaks, respectively, are above a given threshold value, u_{th}. The probability that either the first or the second peak is above u_{th} (both may be above u_{th}) is given by

$$P(E_1 \cup E_2) = P(E_1) + P(E_2) - P(E_1 \cap E_2)$$

2.1.3 Conditional Probability and Independence

A probability is a measure of our uncertainty about an event. As such, the number we assign to the probability of an event depends on the state of our knowledge or information. If we gain new information, our uncertainty may decrease (or increase) and the probability may change. We say the probability is conditional upon our state of information. Thus, if A and B are random events in an experiment, the probability of one occurring may change if we know that the other has occurred. We write a conditional probability as follows, using a vertical bar, |, read as "given," to set off the conditioning event:

$$P(A|B) = P(\text{event } A \text{ occurs} \mid \text{we know event } B \text{ occurs}) \qquad (2.5)$$

Consider the Venn diagram shown in Figure 2.2. The probability of an event may be thought of as the relative area occupied by the event in the Venn

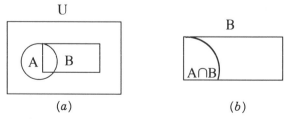

           (a)   (b)

Figure 2.2   Venn diagram: (a) A and B in original domain U; (b) $A \cap B$ in conditional domain given B.

diagram. The knowledge that B occurs has the effect of redefining the sample space to include only elements in set B. If A is also to occur, it must be part of the intersection of A and B, and its conditional probability is given by the ratio

$$P(A|B) = \frac{P(A \cap B)}{P(B)} \tag{2.6}$$

Mathematically, this is simply a renormalization so that, in the new sample space, the probability of B is equal to 1, that is, $P(B|B) = 1$.

The above definition of conditional probability may be rewritten as follows:

$$P(A \cap B) = P(A|B) \cdot P(B) \tag{2.7}$$

This form is known as the chain or multiplication rule. It provides a way to calculate the probability of the intersection of A and B that is always correct and is frequently the most logical way to calculate the intersection. (Note that one could also make A the conditioning event by making the obvious substitutions.)

If the knowledge that B occurs does not change the assessment of the probability of A, we say that A and B are independent, that is,

$$P(A|B) = P(A) \tag{2.8}$$

if and only if A and B are independent. It follows that the probability of $\{A \cap B\}$ is the product of $P(A)$ and $P(B)$:

$$P(A \cap B) = P(A) \cdot P(B) \tag{2.9}$$

if and only if A and B are independent. Equation (2.9) is sometimes used as the definition of independence.

Example 2.3: Probability of Two Independent Peaks Above a Threshold. Again continuing the experiment with the oscillating mass, suppose the probabilities of E_1 and E_2 are both equal to 0.01 and that they are independent. The probability that either the first or second peak is above u_{th} (both may be above u_{th}) is given by

$$P(E_1 \cup E_2) = P(E_1) + P(E_2) - P(E_1 \cap E_2)$$

$$= P(E_1) + P(E_2) - P(E_1) \cdot P(E_2)$$

$$= 0.01 + 0.01 - (0.01)(0.01)$$

$$= 0.0199$$

2.2 SINGLE RANDOM VARIABLES

If the outcome of a random experiment is represented by a number, we call
the number a random variable. This section presents the theory of single random
variables, discrete and continuous: their distributions, measures of their central
tendency and dispersion, and expected values. The Bernoulli, Poisson, uniform, and Rayleigh probability distributions are introduced as examples. Special attention is paid to Gaussian, or normal, random variables because of their
importance in our work.

2.2.1 Discrete Random Variables and Probability Mass Function

For our notation, we let a capital letter, say X, represent the random variable
in general, and we use the lowercase of that letter, in this case, x, to represent
a particular value that X may take. Our interest is in the probabilities of events
such as {a sample realization of random variable X has the value x}, that is,
$\{X = x\}$. The assignment of probabilities for all values of X is accomplished
with a function known as a probability distribution function. There are two
kinds of random variables, requiring slightly different treatment: discrete and
continuous.

A discrete random variable is one for which the sample space consists of
integer values. The sample space for our die-throwing example, {1, 2, 3, 4,
5, 6}, is an example. For a discrete random variable, the probability distribution
function is called a probability law or a probability mass function, designated
$p_X(x)$. Notice that the function is subscripted by the name of the random variable it describes.

From the first axiom of probability, we know that $p_X(x)$ must have values
between zero and one; that is,

$$0 \le p_X(x) \le 1 \tag{2.10}$$

From the second axiom, we know that the sum of $p_X(x)$ over all possible values
of x must be equal to 1, that is,

$$\sum_{\text{all } x} p_X(x) = 1 \tag{2.11}$$

Example 2.4: Bernoulli Distribution. A random variable, say K, with the
sample space $S = \{0, 1\}$ might be used to indicate whether a certain condition
is true or not. For instance, suppose we look at the oscillating mass of our
previous examples at some arbitrary time t; if it is above the threshold, that is,
$u(t) > u_{\text{th}}$, we would assign $k = 1$ to the outcome; otherwise we would give
it $k = 0$. Used in this way, the random variable is known as an indicator
variable. Let the probability of the event occurring (i.e., $k = 1$) be equal to
p. The probability distribution of K, $p_K(k)$, is given by

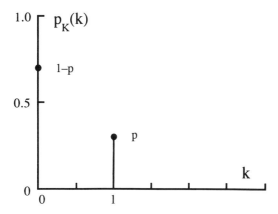

Figure 2.3 Bernoulli distribution.

$$p_K(k) = \begin{cases} p & k = 1 \\ 1 - p & k = 0 \\ 0 & \text{otherwise} \end{cases}$$

This simple distribution is known as the Bernoulli distribution, named for Jakob Bernoulli, and is shown in Figure 2.3.

Example 2.5: Poisson Distribution. A random variable with the sample space $S = \{0, 1, 2, 3, \ldots\}$ might be used to count the number of occurrences of some event, such as the number of times our oscillating mass crosses the threshold u_{th} in a time interval, say $(0, t)$. Under certain special conditions, the following distribution might describe the probability that the number of crossings in $(0, t)$ is equal to N:

$$p_N(n) = \begin{cases} \dfrac{(\lambda t)^n e^{-\lambda t}}{n!} & n = 0, 1, 2, \ldots \\ 0 & \text{otherwise} \end{cases}$$

This distribution is known as the Poisson distribution, named for Siméon-Denis Poisson (1781–1840), and is shown in Figure 2.4. To see that the sum of the probabilities is equal to 1, note that $e^x = \sum_{n=0}^{\infty} x^n/n!$.

2.2.2 Continuous Random Variables and Probability Density Function

A continuous random variable is one for which the sample space consists of a range of values on the real number line. Two examples are $S = \{$real numbers

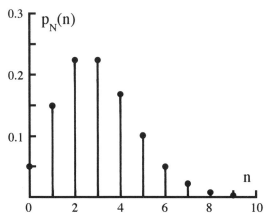

Figure 2.4 Poisson distribution, $\lambda t = 3$.

in the interval [0, 1]} and $S = \{$all real numbers greater than or equal to zero$\}$. (Since, in our studies, the random variable is usually physically realizable, e.g., a measure of displacement, velocity, or acceleration, there may be a practical upper limit to its range, but for our purposes the limit is often so high that it is conveniently treated as infinity.)

The difference between the treatments of discrete and continuous random variables is that a finite nonzero probability can be ascribed to a particular value of a discrete random variable, but not a continuous random variable. The number of points on the real number line between any two points is infinite. Thus, if we try to assign a nonzero probability to each value of a continuous random variable, the total probability would be infinite and the second axiom would be violated.

To solve this dilemma, we adopt the approach taken in describing the distribution of mass within a body: the infinitesimally small point at coordinates (x, y, z) has an associated mass density greater than or equal to zero; one calculates the mass of the body by integrating the density over the volume occupied by the body. We analogously define the probability density associated with a point in the sample space of a continuous random variable; one calculates the probability that a random outcome occurs within a region of the sample space by integrating the probability density over the region.

The function that describes the distribution of probability density over the sample space of the continuous random variable, X, is called the probability density function (PDF) and is designated $f_X(x)$. [It is from the mass analogy that $p_X(x)$ is called the probability mass function.] Thus, to find the probability of X occurring between a and b, $f_X(x)$ is integrated from a to b (Figure 2.5):

$$P(a < X \leq b) = \int_a^b f_X(x)\, dx \tag{2.12}$$

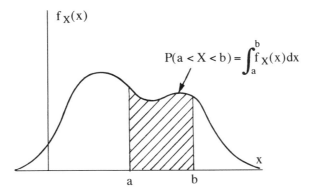

Figure 2.5 Probability density function.

To satisfy the first axiom of probability, $f_X(x)$ is positive valued:

$$f_X(x) \geq 0 \qquad (2.13)$$

To satisfy the second axiom of probability, the total area under $f_X(x)$ is equal to 1:

$$\int_{-\infty}^{+\infty} f_X(x) \, dx = 1 \qquad (2.14)$$

The following heuristic is useful. Consider a continuous random variable X. One may think of the probability that X is between x and $x + dx$, where dx is a differential element of infinitesimal width, as being the probability that X is approximately equal to x, that is, $P(x < X \leq x + dx) = P(X \approx x)$. From Eq. (2.12), the probability that X falls in this interval is given by

$$P(x < X \leq x + dx) = \int_{x}^{x+dx} f_X(u) \, du = f_X(x) \cdot dx \qquad (2.15)$$

The second step results from the mean-value theorem of calculus, which says that the integral is equal to the average height of the function times the width of the interval, that is, $f_X(x)$ times dx. This will be true as long as $f_X(x)$ is a smooth function without discontinuities. Thus, the probability that X is "approximately equal" to a specific value x is equal to $f_X(x)$ times dx:

$$P(X \approx x) = f_X(x) \cdot dx \qquad (2.16)$$

Just remember that this is a heuristic. The probability that a continuous random variable X takes on a specific value x is really zero.

We note that it is sometimes convenient to talk about point masses, that is, masses that are concentrated, or lumped, at infintesimally small points. It is possible to treat lumped and distributed masses in a unified manner using a tool known as the Dirac delta function, which is discussed in detail in Chapter 4. For now, we may consider the Dirac delta function to be a function that is the limiting case of a rectangle of width Δx and height $1/\Delta x$ as Δx tends to zero. So defined, it is evident that the area under the function remains constant and equal to 1. The Dirac delta function is denoted $\delta(x - a)$ and, for practical purposes, is zero everywhere except at $x = a$, where it is singular.

It is likewise possible to treat discrete and continuous random variables together in a unified manner using Dirac delta functions. If X takes on a discrete value, say a, with finite probability $p_X(a)$, then the product $p_X(a)$ times $\delta(x - a)$ is added to the PDF. The condition that the total area under the density function is equal to 1 still applies. Details of this approach are left to the reader.

Since a unified approach is possible, we will drop the use of the term *probability mass function* in the general discussion. "Probability density function" is interpreted as encompassing both. Also, unless the discrete formulas add value to the presentation, our subsequent mathematical developments will be given only in terms of continuous random variables.

Example 2.6: Uniform Distribution. Suppose the phase angle of some function is a random variable Θ and it is equally likely to fall in any interval of width $d\theta$ between 0 and 2π. We say that the phase angle is uniformly distributed over the interval. The probability density function $f_\Theta(\theta)$ is given by

$$f_\Theta(\theta) = \begin{cases} \dfrac{1}{2\pi} & 0 \leq \theta \leq 2\pi \\ 0 & \text{otherwise} \end{cases}$$

This distribution is shown in Figure 2.6. The total probability found by integrating $f_\Theta(\theta)$ over the entire sample space, that is, from 0 to 2π, is obviously equal to 1. More generally, if the limits were a and b, the reader should determine what the constant value of the density function would be.

Example 2.7: Rayleigh Distribution. Under certain conditions, the heights of the peak values, H, of a randomly oscillating mass is given by the Rayleigh probability density function, named after Lord Rayleigh (1842–1919). (The

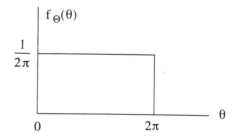

Figure 2.6 Uniform probability density function over $(0, 2\pi)$.

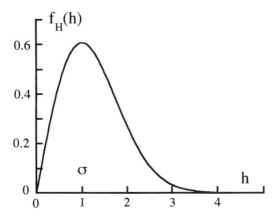

Figure 2.7 Rayleigh probability density function, $\sigma = 1$.

applicable conditions and the derivation of the distribution are discussed in Chapter 6.) The probability density function $f_H(h)$ is given by

$$f_H(h) = \frac{h}{\sigma^2} \exp\left[-\frac{1}{2}\left(\frac{h}{\sigma}\right)^2\right] \quad h > 0$$

This distribution is shown in Figure 2.7. The reader should verify that the integral of this function from zero to infinity is equal to 1. Note that we have omitted the statement that the function is equal to zero for h less than zero. For convenience, as long as there is no possibility for confusion, we may omit statements that the function is equal to zero for the ranges of the random variable not mentioned.

2.2.3 Cumulative Distribution Functions

The cumulative distribution function (CDF) $F_X(x)$ is an alternate way to describe the probability distribution for both discrete and continuous random variables. This function is defined for all values of a random variable X from $-\infty$ to $+\infty$ and is equal to the probability that X is less than or equal to a specific value x:

$$F_X(x) = P(X \leq x) \tag{2.17}$$

If X is discrete, $F_X(x)$ is calculated by summing the probabilities of all values of X less than or equal to x. If X is continuous, $F_X(x)$ is calculated by integrating the PDF for all values of X less than or equal to x:

$$F_X(x) = \int_{-\infty}^{x} f_X(u)\, du \tag{2.18}$$

Conversely, the PDF may be found from the CDF by differentiation:

$$f_X(x) = \frac{dF_X(x)}{dx} \tag{2.19}$$

The CDF is a nondecreasing function of x (its slope is always greater than or equal to zero) with lower and upper limits of 0 and 1, respectively.

Since the CDF and the PDF are uniquely related and contain equivalent information, we will refer to both functions as distribution functions. Which particular function we mean will be obvious from the context, if important.

Example 2.8: Cumulative Distributions for Example Distributions. Figure 2.8 shows the CDFs for our example discrete random variables, Bernoulli and Poisson. For discrete random variables, the CDF has the appearance of a staircase because there are discrete jumps at each value of x that has a nonzero probability; the heights of the steps are equal to the probabilities. (Note that

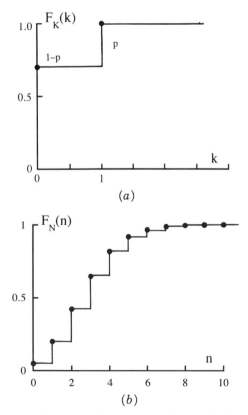

Figure 2.8 Cumulative distribution functions: (*a*) Bernoulli distribution; (*b*) Poisson distribution, $\lambda t = 3$.

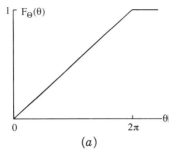 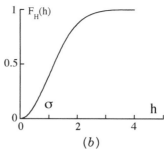

(a) (b)

Figure 2.9 Cumulative distribution functions: (a) uniform distribution $(0, 2\pi)$; (b) Rayleigh distribution, $\sigma = 1$.

the derivatives of these functions are singular at the discrete jumps. As mentioned in the previous section, the density function may be written using Dirac delta functions to represent these instantaneous jumps.)

Figure 2.9 shows the CDFs for our example continuous random variables. For continuous random variables, the CDF is smooth in appearance. For the uniformly distributed phase angle Θ, the equation for the CDF is

$$
F_\Theta(\theta) = \begin{cases} 0 & \theta < 0 \\ \dfrac{\theta}{2\pi} & 0 \le \theta \le 2\pi \\ 1 & \theta > 2\pi \end{cases}
$$

Note the need to provide different equations for the different intervals. For the Rayleigh distributed height of a peak, H, the equation is

$$
F_H(h) = \begin{cases} 0 & h < 0 \\ 1 - \exp\left[-\dfrac{1}{2}\left(\dfrac{h}{\sigma}\right)^2\right] & h \ge 0 \end{cases}
$$

2.2.4 Measures of Central Tendency and Dispersion

Some of the key questions we wish to answer with probability distributions have to do with characterizing the average behavior of a random variable X and estimating how frequently significant deviations from the average occur. A measure of central tendency is captured by the mean value μ_X of the probability distribution. The variation from the mean is captured, to first order, by the variance σ_X^2 and its relatives, the standard deviation σ_X and coefficient of variation C_X. (Notice that the convention of subscripting with the name of the random variable is continued.)

The mean value μ_X is simply the average value. It is also known as the expected value. If X is discrete, the mean is calculated by adding each possible value of X multiplied by its probability:

$$\mu_X = \sum_{\text{all } x_i} x_i \cdot p_X(x_i) \tag{2.20}$$

If X is continuous, the idea is the same: Heuristically, we wish to add each value of X multiplied by its probability. Summation is replaced by integration and $p_X(x_i)$ is replaced by $f_X(x) \, dx$:

$$\mu_X = \int_{-\infty}^{+\infty} x \cdot f_X(x) \, dx \tag{2.21}$$

Figure 2.10 illustrates the calculation of the mean value. From this figure, the mechanical interpretation of the mean as the centroid, or center of mass, of the PDF is evident. (Normally, to find the centroid, the additional step of division by the total area of the figure is required, but this is unnecessary here because the total area of the PDF is equal to 1.) The mean value may also be thought of as the first moment of the PDF.

The variance σ_X^2 is the second moment of the PDF about the mean:

$$\sigma_X^2 = \int_{-\infty}^{+\infty} (x - \mu_X)^2 \cdot f_X(x) \, dx \tag{2.22}$$

In the mechanical interpretation, the variance is the moment of inertia of the function measured about the centroid (Figure 2.10). Furthermore, since the

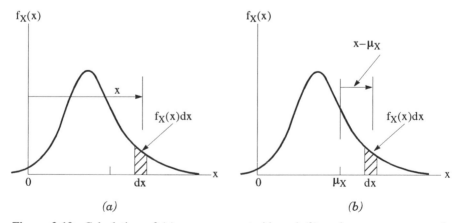

Figure 2.10 Calculation of (a) mean, or centroid, and (b) variance, or moment of inertia.

area of the function is equal to 1, it is also equal to the square of the radius of gyration about the mean.

The positive square root of the variance is known as the standard deviation σ_X:

$$\sigma_X = \sqrt{\sigma_X^2} \tag{2.23}$$

The standard deviation is often preferred over the variance as a measure of dispersion because it has the same units as X, whereas the variance has units of X^2.

The coefficient of variation C_X is a dimensionless measure of variation created by the ratio of the standard deviation to the mean (defined when the mean is nonzero):

$$C_X = \frac{\sigma_X}{\mu_X} \tag{2.24}$$

Some readers will recognize C_X as the inverse of the signal-to-noise ratio found in data acquisition. All three measures of variation, σ_X^2, σ_X, and C_X, increase with greater variation about the mean value.

Example 2.9: Mean and Variance of Example Distributions. The mean, variance, and coefficient of variation for the four example distributions are as follows:

Bernoulli variable K with parameter p: $\mu_K = p$, $\sigma_K^2 = p(1 - p)$, $C_K = \sqrt{1/p - 1}$

Poisson variable N with parameter λt: $\mu_N = \lambda t$, $\sigma_N^2 = \lambda t$, $C_N = 1/\sqrt{\lambda t}$

Uniform variable Θ with range $(0, 2\pi)$: $\mu_\Theta = \pi$, $\sigma_\Theta^2 = \frac{1}{3}\pi^2$, $C_\Theta = 1/\sqrt{3}$

Rayleigh variable H with parameter σ: $\mu_H = \sqrt{\pi}\sigma$, $\sigma_H^2 = 4\sigma^2[1 - \pi\sigma^2]$, $C_H = (2/\sqrt{\pi})\sqrt{1 - \pi\sigma^2}$

Note that the variance of the Rayleigh variable is not equal to the square of the parameter, that is, $\sigma_H^2 \neq \sigma^2$. The reasons why σ is chosen to be the symbol of the parameter will be seen in Chapter 6.

2.2.5 Expected Values

Weighted averages occur so often in the study of random variables that the operation is given a special name, expectation, and a special notation, $E[\cdot]$. The expected value of a function of a random variable, $g(X)$, is given by

$$E[g(X)] = \int_{-\infty}^{+\infty} g(x) \cdot f_X(x)\, dx \tag{2.25}$$

The mean and variance of X are expressed as expected values as follows:

$$\mu_X = E[X] \tag{2.26}$$

$$\sigma_X^2 = E[(X - \mu_X)^2] \tag{2.27}$$

It is easily shown that expectation is a linear operation. That is, if α and β are constants and $g(X)$ and $h(X)$ are functions of X, the following is true:

$$E[\alpha \cdot g(X) + \beta \cdot h(X)] = \alpha \cdot E[g(X)] + \beta \cdot E[h(X)] \tag{2.28}$$

This fact may be used with Eq. (2.27) to show the following useful version of the calculation of the variance:

$$\sigma_X^2 = E[X^2] - \mu_X^2 \tag{2.29}$$

The derivation, as follows, involves expanding the square, $(X - \mu_X)^2$, applying Eq. (2.28), and simplifying the expression

$$\sigma_X^2 = E[(X - \mu_X)^2] = E[X^2 - 2\mu_X X + \mu_X^2] = E[X^2] - 2\mu_X E[X] + \mu_X^2$$

$$= E[X^2] - \mu_X^2 \tag{2.30}$$

The expected value of the square of X, $E[X^2]$, is the second moment of $f_X(x)$ taken about the origin of x. In general, the kth moment about the origin is $E[X^k]$.

The $E[X^2]$ is also known as the mean-square value. Its square root is known as the root-mean-square, or RMS, value. If the mean of X is zero, then the mean-square value is equal to the variance σ_X^2 and the RMS is equal to the standard deviation σ_X.

Linear transformations of random variables are important in our work, so it is useful to be able to calculate the mean and variance of a linear transformation quickly. Let Y be given by $Y = aX + b$. The mean of Y is

$$\mu_Y = E[aX + b] = aE[X] + b$$

$$= a\mu_X + b \tag{2.31}$$

The variance is

$$\sigma_Y^2 = E[(aX + b)^2] - (a\mu_X + b)^2$$

$$= (a^2 E[X^2] + 2abE[X] + b^2) - (a^2\mu_X^2 + 2ab\mu_X + b^2)$$

$$= a^2(E[X^2] - \mu_X^2)$$

$$= a^2\sigma_X^2 \tag{2.32}$$

Example 2.10: Standardized Variables. A special linear transformation is known as standardization. If X is a random variable with mean μ_X and variance σ_X^2, then the standardized variable Z is defined as

$$Z = \frac{X - \mu_X}{\sigma_X} \tag{2.33}$$

In words, X is standardized by subtracting its mean and then dividing by its standard deviation. We show below that Z has a mean equal to 0 and a variance equal to 1 regardless of the distribution of X:

$$\mu_Z = E\left[\frac{X - \mu_X}{\sigma_X}\right] = \frac{1}{\sigma_X} E[X - \mu_X] = 0 \tag{2.34}$$

$$\sigma_Z^2 = E\left[\left(\frac{X - \mu_X}{\sigma_X}\right)^2\right] = \frac{1}{\sigma_X^2} E[(X - \mu_X)^2] = 1 \tag{2.35}$$

2.2.6 Moment-Generating Functions

A special function of interest is $g(X) = e^{tX}$. The expected value of this function is called the moment-generating function $M_X(t)$. By writing the expectation out in integral form as

$$M_X(t) = E[e^{tX}]$$

$$= \int_{-\infty}^{+\infty} e^{tx} \cdot f_X(x)\, dx \tag{2.36}$$

it is apparent that $M_X(t)$ is the Laplace transform of $f_X(x)$. To see how this function generates moments, expand the exponential as a power series:

$$e^{tX} = 1 + tX + \frac{1}{2!} (tX)^2 + \frac{1}{3!} (tX)^3 + \cdots \tag{2.37}$$

and place the series into the expected-value operation. The result of the operation is

$$M_X(t) = 1 + tE[X] + \frac{1}{2!} t^2 E[X^2] + \frac{1}{3!} t^3 E[X^3] + \cdots \tag{2.38}$$

Thus, $M_X(t)$ is a function of the moments of $f_X(x)$ taken about the origin of X.

All the moments of $f_X(x)$ are conveniently recovered from $M_X(t)$ by the following formula:

$$E[X^k] = \frac{d^k}{dt^k} M_X(t)\Big|_{t=0} \tag{2.39}$$

In words, this formula says to take the kth derivative of $M_X(t)$ with respect to t and evaluate it at $t = 0$. The reader is encouraged to verify this technique.

Aside from calculating moments conveniently, the importance of the moment-generating function is this: If one knows all the moments of a PDF, one can specify $M_X(t)$. Since $M_X(t)$ is the Laplace transform of $f_X(x)$, there is a one-to-one relationship between the two functions and one could then recover the PDF from $M_X(t)$, at least in theory.[1] Thus, $f_X(x)$ is completely specified if all its moments are known.

Finally, if $g(X) = e^{itX}$, where the exponent itX is imaginary, the result is known as the characteristic function $\Phi_X(t)$ and is the Fourier transform of $f_X(x)$. The features of this function are similar to $M_X(t)$, namely that the moments can be recovered by differentiation and there is a one-to-one correspondence between $\Phi_X(t)$ and $f_X(x)$. The characteristic function has a mathematical advantage over the moment-generating function, namely, it can be shown to exist for all PDFs, whereas $M_X(t)$ may not exist. We will defer the discussion of Fourier transforms until Chapter 4.

Example 2.11: Moment-Generating Functions of Example Distributions. The moment-generating functions for the four example distributions are as follows:

Bernoulli variable K with parameter p: $M_K(t) = pe^t + (1 - p)$

Poisson variable N with parameter ν: $M_N(t) = \exp[\nu(e^t - 1)]$ (Note that the parameter is $\nu = \lambda\tau$, where τ is time; in this formula t is the dummy variable for the moment-generating function and is not time.)

Uniform variable Θ with range $(0, 2\pi)$:

$$M_\Theta(t) = \frac{1}{2\pi} \frac{e^{2\pi t} - 1}{t}$$

Rayleigh variable H with parameter σ:

$$M_H(t) = 1 + \sqrt{2\pi}\sigma t e^{1/2\sigma^2 t^2} \int_{-\infty}^{\sigma t} \frac{1}{\sqrt{2\pi}} e^{-1/2u^2} \, du$$

2.2.7 Gaussian (Normal) Random Variables

In our study of random vibrations, one particular distribution, namely the normal distribution, is so significant that it is worth special attention. This distribution is also known as the Gaussian distribution, after Carl Friedrich Gauss (1777–1855), who applied it to the study of the accuracy of measurements in

---

[1]However, the determination of a PDF from a truncated series of moments calculated from data is generally not a good approach.

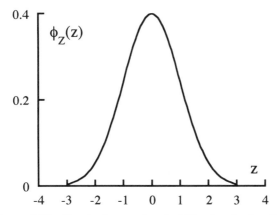

Figure 2.11 Standard normal probability density function.

the nineteenth century. However, as discussed in Section 2.5, Gauss was not the first to use this distribution, the general form of which was known over 100 years earlier.

The standard normal density function is shown in Figure 2.11 and is given by

$$f_Z(z) = \frac{1}{\sqrt{2\pi}} \exp\left(-\frac{1}{2} z^2\right) \qquad -\infty < z < +\infty \qquad (2.40)$$

The use of Z as the standard normal random variable is standard practice. As a standardized variable, its mean is 0 and its variance is 1, that is,

$$\mu_Z = 0 \qquad (2.41)$$

$$\sigma_Z^2 = 1 \qquad (2.42)$$

The moment-generating function is

$$M_Z(t) = \exp\left(\tfrac{1}{2} t^2\right) \qquad (2.43)$$

The CDF is shown in Figure 2.12 and is given by

$$F_Z(z) = \int_{-\infty}^{z} \frac{1}{\sqrt{2\pi}} \exp\left(-\frac{1}{2} \varsigma^2\right) d\varsigma \qquad (2.44)$$

The integral cannot be evaluated analytically. However, it has been widely tabulated; Section E.1 contains a short set of standard normal probability tables. Numerical approximations are also available in Appendix C.

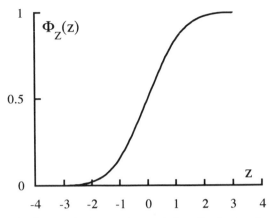

Figure 2.12 Standard normal cumulative distribution function.

The standard normal distribution occurs so often in probabilistic analysis that it is given its own symbols: $\phi(z)$ is the standard normal density function and $\Phi(z)$ is its cumulative.

The standard normal distribution can be generalized by a change of variable, substituting for Z:

$$Z = \frac{X - \mu_X}{\sigma_X} \tag{2.45}$$

The random variable X has a mean equal to μ_X and a standard deviation equal to σ_X. Its probability density function is

$$f_X(x) = \frac{1}{\sqrt{2\pi}\sigma_X} \exp\left[-\frac{1}{2}\left(\frac{x - \mu_X}{\sigma_X}\right)^2\right] \qquad -\infty < x < +\infty \tag{2.46}$$

Figures 2.13 and 2.14 show the effect of increasing the mean and standard deviation of X, respectively.

The moment-generating function is

$$M_X(t) = \exp(\tfrac{1}{2}\sigma_X^2 t^2 + \mu_X t) \tag{2.47}$$

Using this function it is easily shown that the central moments, that is, about μ_X, are

$$E[(X - \mu_X)^k] = \begin{cases} 0 & k \text{ odd} \\ 1 \cdot 3 \cdot 5 \cdots (k-1)\sigma_X^k & k \text{ even} \end{cases} \tag{2.48}$$

Note that all the moments are either zero or functions of the variance σ_x^2. Thus, a normal random variable is completely characterized by its mean and variance.

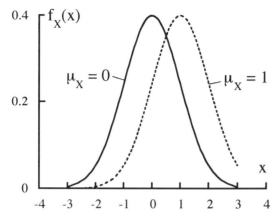

Figure 2.13 Normal density functions for two mean values.

Example 2.12: Using Standard Normal Probability Table. Find the following probabilities using the standard normal probability tables in Section E.1: $P(Z < 1)$, $P(Z < -1)$, and $P(-1 < Z < 1)$.

$$P(Z < 1) = 0.84134 \qquad\qquad \text{Directly from table}$$

$$\begin{aligned} P(Z < -1) &= P(Z > 1) && \text{By symmetry}\\ &= 1 - P(Z < 1) && \text{By complement}\\ &= 1 - 0.84134 && \text{From table}\\ &= 0.15866 \end{aligned}$$

$$\begin{aligned} P(-1 < Z < 1) &= P(Z < 1) - P(Z < -1) && \text{By definition}\\ &= 0.84134 - 0.15866\\ &= 0.68268 \end{aligned}$$

Thus, there is a 68% probability that the value of a normal random variable will be within plus or minus one standard deviation from the mean. The reader

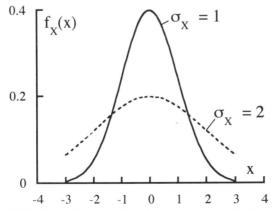

Figure 2.14 Normal density functions for two standard deviations.

may verify that the probabilities of being between $\pm 2\sigma$ and $\pm 3\sigma$ from the mean are 0.95450 and 0.99730, respectively.

Example 2.13: Probability Calculation for Arbitrary Normal Random Variable. Let X represent a random acceleration with a mean of $4.5g$ and a standard deviation of $3.0g$. What is the probability that X will be negative?

$$P(X < 0) = P\left(\frac{X - \mu_X}{\sigma_X} < \frac{0 - \mu_X}{\sigma_X}\right) = P\left(Z < \frac{0 - 4.5}{3.0}\right) = P(Z < -1.5)$$

$$= 1 - P(Z < 1.5) = 1 - 0.93319 = 0.06681$$

2.3 JOINTLY DISTRIBUTED RANDOM VARIABLES

If an experiment generates more than one random variable per observation, we say that the random variables occur together and are jointly distributed. This section presents the theory of jointly distributed random variables: their joint, marginal, and conditional distributions, a measure of their average variation together (their covariance), and a measure of their linear dependence on each other (their correlation coefficient).

2.3.1 Joint and Marginal Distributions

Referring back to the beginning of the chapter, we defined the probabilities of events A and B occurring separately, $P(A)$ and $P(B)$, and together, $P(A \cap B)$. When A and B are events involving random variables, such as $A = \{$a sample of random variable X has value $x\}$ and $B = \{$a sample of random variable Y takes on value $y\}$, we call $P(A)$ and $P(B)$ marginal probabilities and $P(A \cap B)$ their joint probability.

For our study of random vibrations, X might be the input to a dynamic system and Y a system response, both measured at the same time. Or X and Y might both be responses measured two different times. Random variables X and Y could be absolutely arbitrary and could even be completely unrelated (which turns out to be an important case to study). Our interest is in learning what some knowledge about one of the variables does to the conditional probability of the other variable, that is, their probabilistic dependence on one another.

The important concepts can be studied by considering the bivariate case, that is, two jointly distributed variables. The extensions to higher dimensions, that is, larger numbers of random variables, are straightforward.

The bivariate joint probability density function $f_{XY}(x, y)$ assigns probabilities to pairs of (x, y) that occur jointly. The probability that $\{X \approx x \cap Y \approx y\}$ is given by the heuristic

$$P(X \approx x \cap Y \approx y) = P(x < X \leq x + dx \cap y < Y \leq y + dy)$$

$$= f_{XY}(x, y) \cdot dx\, dy \tag{2.49}$$

That is, the joint probability is equal to the bivariate density function at point (x, y) times the area of an infinitesimal rectangular patch with sides dx and dy long.

The first and second axioms of probability are satisfied by the following two conditions:

$$1) \quad f_{XY}(x, y) \geq 0 \tag{2.50}$$

$$2) \quad \int \int f_{XY}(x, y) \, dx \, dy = 1 \tag{2.51}$$

In words, $f_{XY}(x, y)$ must be nonnegative and the total volume under the surface defined by $f_{XY}(x, y)$ must be equal to 1.

Figure 2.15 shows an example bivariate probability density function drawn in three dimensions. Figure 2.16 shows a convenient representation of the same PDF as a contour plot: the contours represent lines of equal probability density.

The bivariate CDF is defined as

$$F_{XY}(x, y) = P(X \leq x \cap Y \leq y) \tag{2.52}$$

The density function and the cumulative are related by

$$F_{XY}(x, y) = \int_{-\infty}^{y} \int_{-\infty}^{x} f_{XY}(u, v) \, du \, dv \tag{2.53}$$

$$f_{XY}(x, y) = \frac{\partial^2}{\partial x \, \partial y} F_{XY}(x, y) \tag{2.54}$$

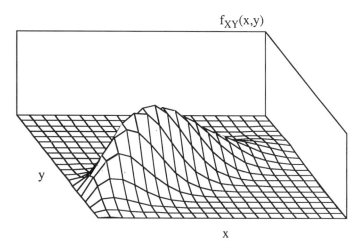

$f_{XY}(x,y)$

y

x

Figure 2.15   Example bivariate probability density function.

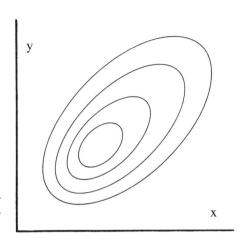

Figure 2.16 Contour plot representation of bivariate probability density function.

The marginal probability density functions, that is, for X and Y separately, can be recovered from the joint density function by integration. To find $f_X(x)$ from $f_{XY}(x, y)$ one fixes a particular value of X, x, and integrates the density for all values of Y at that x:

$$f_X(x) = \int_{-\infty}^{+\infty} f_{XY}(x, y)\, dy \tag{2.55}$$

We say that y is "integrated out" of the equation, since the remaining function has no dependence on y. The calculation is similar for the other marginal density function, $f_Y(y)$. The integration of the joint density function to find the marginal is illustrated in Figure 2.17.

Example 2.14: Joint Discrete Random Variables. The joint probability mass function of discrete random variables J and K, $p_{JK}(j, k)$, is given in the table below. The marginal distributions are shown in the margins of the table, to the right and below (hence, the name "marginal" distribution):

$p_{JK}(j, k)$		k			
		0	1	2	$p_J(j)$
	0	0.0	0.1	0.0	0.1
	1	0.1	0.2	0.1	0.4
j	2	0.1	0.2	0.1	0.4
	3	0.0	0.1	0.0	0.1
$p_K(k)$		0.2	0.6	0.2	

Example 2.15: Joint Continuous Random Variables. The joint PDF of continuous random variables R and Θ and the corresponding marginal density

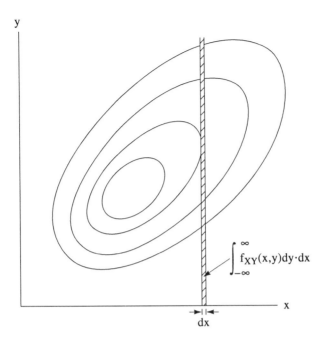

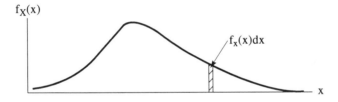

Figure 2.17 Marginal probability density function.

functions are given by

$$f_{R\Theta}(r, \theta) = \frac{1}{2\pi} r \exp\left(-\frac{1}{2} r^2\right) \qquad r > 0 \qquad 0 < \Theta < 2\pi$$

$$f_R(r) = r \exp\left(-\frac{1}{2} r^2\right) \qquad\qquad r > 0$$

$$f_\Theta(\theta) = \frac{1}{2\pi} \qquad\qquad\qquad 0 < \Theta < 2\pi$$

2.3.2 Conditional Distributions and Independence

Having defined $P(A \cap B)$, it follows that we should define the conditional probability $P(A|B)$ and discuss the meaning of independence for random variables. Recall that we defined the conditional probability of A given B as

$$P(A|B) = \frac{P(A \cap B)}{P(B)} \qquad (2.6)$$

The conditional PDF of X given the outcome of the random variable Y is y is denoted $f_{X|Y}(x|y)$. Heuristically, we have the following definition of the conditional density function:

$$P(X \approx x|Y \approx y) = \frac{P(X \approx x, Y \approx y)}{P(Y \approx y)} \qquad (2.56)$$

This is written with differentials as

$$f_{X|Y}(x|Y = y) \, dx = \frac{f_{XY}(x, y) \, dx \, dy}{f_Y(y) \, dy} \qquad (2.57)$$

The differentials cancel and leave the usual definition

$$f_{X|Y}(x|Y = y) = \frac{f_{XY}(x, y)}{f_Y(y)} \qquad (2.58)$$

In words, the conditional density function is a slice through the joint density function that is normalized so that the total area under the conditional density function is equal to 1. The normalization factor is equal to the marginal density function evaluated at the given value. This is illustrated in Figure 2.18.

Now, recall that if A and B are independent events, then

$$P(A|B) = P(A) \qquad (2.8)$$

$$P(A \cap B) = P(A) \cdot P(B) \qquad (2.9)$$

Similarly, random variables X and Y are independent if the analogous statements hold for their density functions, namely,

$$f_{X|Y}(x|Y = y) = f_X(x) \qquad (2.59)$$

$$f_{XY}(x, y) = f_X(x) \cdot f_Y(y) \qquad (2.60)$$

Equation (2.59) has the graphical interpretation that the joint density function $f_{XY}(x, y)$ can be sliced at any two values of y, and the slices are the same shape. This is shown in Figure 2.19.

Equation (2.60) has the analytical interpretation that X and Y are independent if the joint density function can be factored into two functions, one dependent only on x and the other dependent only on y. The converse is also true; that is, if they are independent, the joint density function can be factored.

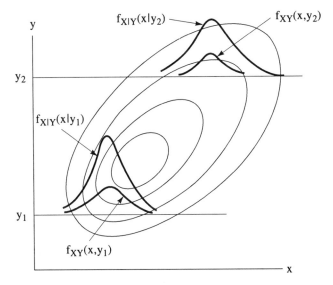

Figure 2.18 Conditional probability density functions for X given $Y = y_1$ and $Y = y_2$.

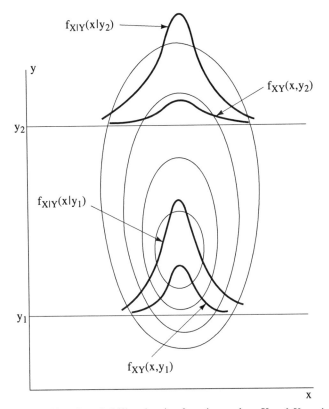

Figure 2.19 Conditional probability density functions when X and Y are independent.

Example 2.16: Conditional Probability for Joint Discrete Example. For the previous example with discrete J and K, the conditional probability mass functions for J given $k = 0$ and $k = 1$ are

$$j$$

	0	1	2	3
$p_{J|K}(j|k = 0)$	0	$\frac{1}{2}$	$\frac{1}{2}$	0
$p_{J|K}(j|k = 1)$	$\frac{1}{6}$	$\frac{1}{3}$	$\frac{1}{3}$	$\frac{1}{6}$

Example 2.17: Conditional Probability for Joint Continuous Example. For the previous example with continuous R and Θ, the joint density function is factorable into the two marginal densities. Hence, R and Θ are independent.

2.3.3 Expected Values, Covariance, and Correlation Coefficient

The expected value of a function of X and Y, $g(X, Y)$, is given by

$$E[g(X, Y)] = \int_{-\infty}^{+\infty} \int_{-\infty}^{+\infty} g(x, y)f_{XY}(x, y)\, dx\, dy \qquad (2.61)$$

The conditional expected value of $g(X, Y)$ is defined as

$$E[g(X, Y)|Y = y] = \int_{-\infty}^{+\infty} g(x, y)f_{X|Y}(x|Y = y)\, dx \qquad (2.62)$$

Note that the conditional expected value is essentially the expected value with just one random variable; the conditional density function is used instead of the marginal.

We are particularly interested in the joint moments of X and Y, $E[X^n Y^m]$. The sum of the exponents, $n + m$, is called the order of moment. Of greatest interest is the second order moment of X and Y about their centroid, that is, the expected value of $g(X, Y) = (X - \mu_X)(Y - \mu_Y)$. This moment is known as the covariance and denoted σ_{XY}:

$$\sigma_{XY} = E[(X - \mu_X)(Y - \mu_Y)]$$

$$= \int_{-\infty}^{+\infty} \int_{-\infty}^{+\infty} (x - \mu_X)(y - \mu_Y)f_{XY}(x, y)\, dx\, dy \qquad (2.63)$$

This calculation is illustrated in Figure 2.20. In the mechanical analogy, the covariance is the product of inertia of the joint density function.

By expanding the product and simplifying, it can be shown that

$$\sigma_{XY} = E[XY] - \mu_X\mu_Y \qquad (2.64)$$

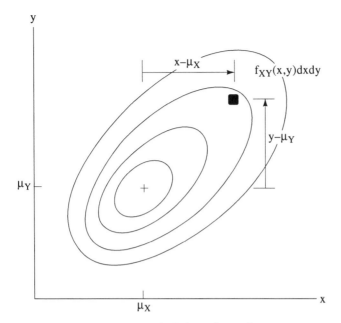

Figure 2.20 Calculation of covariance.

This equation is the most useful way to compute the covariance. The expected value of the product, $E[XY]$, is known as the correlation of X and Y and is of cardinal importance in our study of random processes in the next chapter.

Note the similarity between these equations and the equations for the variance of X, Eqs. (2.22) and (2.29). If Y were replaced by X in Eqs. (2.63) and (2.64), the variance of X would result. The physical interpretation of the covariance is also similar. Just as the variance measures deviations of X from its centroid, the covariance measures deviations of X and Y together from their centroid. Relatively large absolute values of σ_{XY} indicate that X and Y tend to vary from the centroid together; for example, if X is large, then Y will also be large. Values of σ_{XY} near zero indicate that they tend not to vary together.

To speak of "relatively" large or small values of σ_{XY}, we must normalize the covariance to eliminate units of measurement and to account for the marginal dispersions of X and Y. A nondimensional measure is given by the ratio known as the correlation coefficient ρ_{XY}:

$$\rho_{XY} = \frac{\sigma_{XY}}{\sigma_X \sigma_Y} \qquad (2.65)$$

where the ratio ranges between $+1$ and -1, as shown below.

Consider the quadratic function, $g(t) = (Ut - V)^2$, where $U = X - \mu_X$ and $V = Y - \mu_Y$. The expected value of $g(t)$ is given by

$$E[g(t)] = E[(Ut - V)^2] = E[U^2]t^2 - 2E[UV]t + E[V^2] \qquad (2.66)$$

A quadratic function such as this can have either two real roots, a repeated real root, or two imaginary roots, depending on the value of the characteristic function of the quadratic formula (i.e., $b^2 - 4ac$). Since $g(t)$ is a square, it is always positive and the roots of the above equation must be imaginary or repeated. Hence, the characteristic function must be less than or equal to zero. Substituting from Eq. (2.66) into the characteristic function, we get the inequality

$$(2E[UV])^2 - 4E[U^2]E[V^2] \leq 0 \tag{2.67}$$

$$E[UV]^2 \leq E[U^2]E[V^2] \tag{2.68}$$

Substitution for U and V gives

$$E[(X - \mu_X)(Y - \mu_Y)]^2 \leq E[(X - \mu_X)^2]E[(Y - \mu_Y)^2] \tag{2.69}$$

$$\sigma_{XY}^2 \leq \sigma_X^2 \sigma_Y^2 \tag{2.70}$$

This is known as Schwartz's inequality. It follows that the range of ρ_{XY} is

$$-1 \leq \rho_{XY} \leq 1 \tag{2.71}$$

which was to be shown.

Note that ρ_{XY} can also be written as

$$\rho_{XY} = E\left[\frac{X - \mu_X}{\sigma_X} \frac{Y - \mu_Y}{\sigma_Y}\right] \tag{2.72}$$

which is to say that the correlation coefficient is equal to the expected value of the product of the standardized variables for X and Y.

Example 2.18. For both previous examples, with discrete J and K and with continuous R and Θ, the covariances and correlation coefficients are equal to zero.

2.3.4 Linear Dependence

The correlation coefficient is a measure of the linear dependence of two random variables. If random variable Y is a linear function of X, $Y = aX + b$, the correlation coefficient is equal to ± 1. If X and Y are independent, the correlation coefficient is equal to zero. If Y is partially linearly related to X, the correlation coefficient is in the range -1 to $+1$ depending on the strength of the linear relation. If X and Y are nonlinearly related, the correlation coefficient can tell us nothing about that relationship. These points are shown below.

Suppose X and Y are linearly related by $Y = aX + b$, where a and b are constants. The correlation coefficient of X and Y is calculated as

$$\rho_{XY} = E\left[\frac{X - \mu_X}{\sigma_X} \frac{Y - \mu_Y}{\sigma_Y}\right]$$

$$= E\left[\frac{X - \mu_X}{\sigma_X} \frac{(aX + b) - (a\mu_X + b)}{|a|\sigma_X}\right] = \frac{a}{|a|} E\left[\left(\frac{X - \mu_X}{\sigma_X}\right)^2\right]$$

$$= \frac{a}{|a|} \tag{2.73}$$

Thus, if Y is perfectly linearly dependent on X, $\rho_{XY} = \pm 1$, depending on the sign of the constant a.

Next suppose that X and Y are independent. Then their correlation, $E[XY]$, is just the product of the marginal expected values because the joint density function may be factored into the product of the marginal density functions:

$$E[XY] = \int_{-\infty}^{+\infty} \int_{-\infty}^{+\infty} xy f_{XY}(x, y) \, dx \, dy = \int_{-\infty}^{+\infty} x f_X(x) \, dx \int_{-\infty}^{+\infty} y f_Y(y) \, dy$$

$$= \mu_X \mu_Y \tag{2.74}$$

It follows immediately that their covariance, $\sigma_{XY} = E[XY] - \mu_X \mu_Y$, and hence their correlation coefficient are equal to zero.

The converse of the above is not necessarily true. That is, if the covariance of X and Y is equal to zero, it is not necessarily true that X and Y are independent. We show this by a counterexample. Let the density function of X be symmetric about the origin, that is, $f_X(-x) = f_X(x)$. By symmetry, the expected value of X is equal to zero, $\mu_X = 0$. Let Y be related to X by a symmetric, nonlinear function such as $y = x^2$. The covariance is shown to be equal to zero as follows:

$$\sigma_{XY} = E[XY] - \mu_X \mu_Y = E[X \cdot X^2] = E[X^3] = \int_{-\infty}^{+\infty} x^3 f_X(x) \, dx$$

$$= 0 \tag{2.75}$$

The integral in the last step is equal to zero because $x^3 f_X(x)$ is antisymmetric; for each function evaluation at $+x$, there is a canceling evaluation at $-x$.

Hence, if the dependence between X and Y is nonlinear, their covariance might be equal to zero. The fact that $\sigma_{XY} = 0$, or $\rho_{XY} = 0$, does not mean that X and Y are independent. If $\sigma_{XY} = 0$, we say that X and Y are uncorrelated. If the correlation between X and Y is equal to zero, $E[XY] = 0$, we say that X and Y are orthogonal.

We now examine the significance of intermediate values of the correlation coefficient. Suppose Y is equal to a linear transformation of X plus another random variable, W:

$$Y = aX + b + W \tag{2.76}$$

Let W have a mean of zero and let it be uncorrelated with X, that is, $E[XW]$ = 0 and $\sigma_{XW} = 0$. Depending on the phenomenon being modeled, W is typically called *random error* or *random noise*. We would like to know the predictability of Y in the presence of this "noise." In other words, how much variability is left in the conditional distribution of Y given X?

The correlation coefficient of X and Y is developed as follows. The expected value and variance of Y are

$$\mu_Y = a\mu_X + b \tag{2.77}$$

$$\sigma_Y^2 = a^2\sigma_X^2 + \sigma_W^2 \tag{2.78}$$

The correlation of X and Y is given as

$$E[XY] = E[X(aX + b + W)] = aE[X^2] + b\mu_X \tag{2.79}$$

The covariance of X and Y is written as

$$\sigma_{XY} = E[XY] - \mu_X\mu_Y = (aE[X^2] + b\mu_X) - \mu_X(a\mu_X + b)$$

$$= a\sigma_X^2 \tag{2.80}$$

The correlation coefficient of X and Y is defined as

$$\rho_{XY} = \frac{\sigma_{XY}}{\sigma_X\sigma_Y} = \frac{a\sigma_X^2}{\sigma_X\sqrt{a^2\sigma_X^2 + \sigma_W^2}} = \text{sgn}(a)\sqrt{\frac{a^2\sigma_X^2}{a^2\sigma_X^2 + \sigma_W^2}} \tag{2.81}$$

where $\text{sgn}(a)$ is the sign, $+$ or $-$, attached to a. By dividing through by $a^2\sigma_X^2$ inside the radical, we may express the correlation coefficient in terms of a *noise-to-signal ratio*.

$$\rho_{XY} = \text{sgn}(a)\sqrt{\frac{1}{1 + r^2}} \tag{2.82}$$

where

$$r^2 = \frac{\sigma_W^2}{a^2\sigma_X^2} \tag{2.83}$$

Figure 2.21 shows a plot of ρ_{XY} versus r. It is seen that if there is no noise W, then $r = 0$, and there is perfect correlation, $\rho_{XY} = 1$. If the "signal" aX is twice as strong as the noise, then $r = \frac{1}{2}$ and $\rho_{XY} = 0.9$; the dependence is considered strong. If the noise is about the same "strength" as the signal aX, then $r = 1$ and $\rho_{XY} \approx 0.7$; the dependence of Y on X is considered moderate.

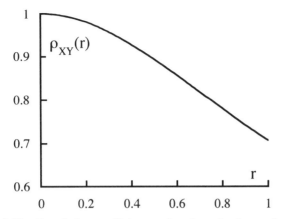

Figure 2.21 Correlation coefficient as function of noise-to-signal ratio.

Correlation coefficients less than 0.7 are indications of only weak dependence. The joint PDFs of X and Y for these values are illustrated in Figure 2.22.

2.3.5 Joint Normal Distribution

The most commonly encountered bivariate probability density function is the bivariate normal distribution, given by

$$f_{XY}(x, y) = \frac{1}{2\pi\sigma_X\sigma_Y\sqrt{1 - \rho^2}} \exp\left\{ -\frac{1}{2(1 - \rho^2)} \left[\left(\frac{x - \mu_X}{\sigma_X}\right)^2 \right.\right.$$

$$\left.\left. - 2\rho\left(\frac{x - \mu_X}{\sigma_X}\right)\left(\frac{y - \mu_Y}{\sigma_Y}\right) + \left(\frac{y - \mu_Y}{\sigma_Y}\right)^2 \right]\right\}$$

$$-\infty < x < +\infty \qquad -\infty < y < +\infty \qquad (2.84)$$

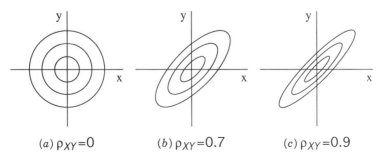

(a) $\rho_{XY}=0$ (b) $\rho_{XY}=0.7$ (c) $\rho_{XY}=0.9$

Figure 2.22 Bivariate normal density functions: (a) $\rho_{XY} = 0$; (b) $\rho_{XY} = 0.7$; (c) $\rho_{XY} = 0.9$.

The marginal distribution of X can be shown to be

$$f_X(x) = \frac{1}{\sqrt{2\pi}\sigma_X} \exp\left[-\frac{1}{2}\left(\frac{x - \mu_X}{\sigma_X}\right)^2\right] \qquad -\infty < x < +\infty \qquad (2.85)$$

This is recognized to be the ordinary univariate normal density function with mean μ_X and standard deviation σ_X. Similar results can be shown for Y.

The parameter ρ is the correlation coefficient between X and Y. If $\rho = 0$, the bivariate density can be factored into the product of the two marginal density functions. This means that if X and Y are jointly normally distributed, a correlation coefficient equal to zero implies that X and Y are independent. Recall that this is not generally true. Figure 2.22 illustrates the bivariate density function for several values of ρ.

The conditional density function for the bivariate normal distribution of Y given $X = x$ is

$$f_{Y|X}(y|x) = \frac{1}{\sqrt{2\pi}\sqrt{1 - \rho^2}\,\sigma_Y} \exp\left\{-\frac{1}{2}\left[\left(\frac{y - [\mu_Y + \rho(\sigma_Y/\sigma_X)\,(x - \mu_X)]}{\sqrt{1 - \rho^2}\,\sigma_Y}\right)^2\right]\right\}$$

$$(2.86)$$

The conditional mean and variance of Y given $X = x$ are defined as

$$\mu_{Y|X} = \mu_Y + \rho\,\frac{\sigma_Y}{\sigma_X}\,(x - \mu_X) \qquad (2.87)$$

$$\sigma_{Y|X}^2 = (1 - \rho^2)\sigma_Y^2 \qquad (2.88)$$

The bivariate normal density function can be easily generalized to more than two dimensions. Let there be n jointly distributed normal random variables X_i with mean values μ_i, variances σ_i^2, and covariances σ_{ij}. The general multivariate normal density function is

$$f_{X_1 X_2 \cdots X_n}(x_1, x_2, \ldots, x_n) = \frac{1}{(2\pi)^{n/2}|S|^{1/2}} \exp\left\{-\frac{1}{2}(\mathbf{x} - \mathbf{m})^{\mathrm{T}}\mathbf{S}^{-1}(\mathbf{x} - \mathbf{m})\right\}$$

$$(2.89)$$

where $\mathbf{S}$ is the symmetric covariance matrix:

$$\mathbf{S} = \begin{bmatrix} \sigma_1^2 & \sigma_{12} & \cdots & \sigma_{1n} \\ \sigma_{21} & \sigma_2^2 & \cdots & \sigma_{2n} \\ \vdots & \vdots & \ddots & \vdots \\ \sigma_{n1} & \sigma_{n2} & \cdots & \sigma_n^2 \end{bmatrix} \qquad (2.90)$$

and $\mathbf{x} - \mathbf{m}$ is the vector

$$
\mathbf{x} - \mathbf{m} = \begin{bmatrix} x_1 - \mu_1 \\ x_2 - \mu_2 \\ \vdots \\ x_n - \mu_n \end{bmatrix}
\tag{2.91}
$$

2.4 FUNCTIONS OF RANDOM VARIABLES

Suppose we have a function of a random variable, say $Y = g(X)$, such as $Y = X^2$ or $Y = \sin(X)$. The result of this function, Y, is also a random variable. One can directly determine the mean of Y, μ_Y, and other expected values by calculating the appropriate expected values using the PDF of X in the integration. However, determining the actual probability distribution of Y, $f_Y(y)$, is more complicated. This problem is known as transformation or change of variable.

This section presents two methods for transforming the distribution of a random variable: one through the CDF and one through the PDF. Then the density function method is extended to problems involving functions of more than one variable. Finally, since sums of random variables are important in our studies, special attention is paid to random sums.

2.4.1 Change of Variables by Cumulative Distribution Function

There are two main approaches that we might take: one through the CDF and one through the PDF. The most general approach is through the CDF. For the moment, let us restrict $y = g(x)$ to be a monotonically increasing function. The inverse function, which gives x as a function of y, is denoted as $x = g^{-1}(y)$. [This notation is in the same spirit as $y = \sin(x)$ and $x = \sin^{-1}(y)$.]

The CDF of Y can be related to the CDF of X through corresponding probability statements as follows:

$$
F_Y(y) = P(Y \le y) = P(g(X) \le y) = P(X \le g^{-1}(y))
$$

$$
= F_X(g^{-1}(y))
\tag{2.92}
$$

In words, to write the CDF of Y, first write the CDF of X and then substitute $g^{-1}(y)$ for x in the formula.

This procedure may be interpreted as the graphical construction of $F_Y(y)$ from $F_X(x)$, as shown in Figure 2.23. For a given value of X, say x_1, find the corresponding value of Y, $y_1 = g(x_1)$. Take the value of the cumulative distribution of X at x_1, $F_X(x_1)$, and plot this as the value of the cumulative distribution

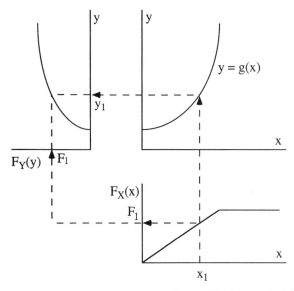

Figure 2.23   Graphical construction of $F_Y(y)$ from $F_X(x)$.

of Y at y_1, $F_Y(y_1) = F_X(x_1)$. From the illustration it is evident why this is called a mapping from the domain of X to the domain of Y.

The reader may verify that if $y = g(x)$ is monotonically decreasing, then

$$F_Y(y) = 1 - F_X(g^{-1}(y)) \tag{2.93}$$

If the function is not monotonic, the probability statement $P(Y \leq y)$ in terms of X will be more complicated and careful analysis is necessary.

Example 2.19: $y = \sin(x)$ by Cumulative Method. Let X be uniformly distributed between 0 and 2π, $F_X(x) = x/2\pi$, and let $y = \sin(x)$. As shown in Figure 2.24, the sine function is not monotonic in this interval; for each value of y there are two possible values of x. Note that the inverse function is $x = \sin^{-1}(y)$, where the reciprocal sine function returns angles in the range $(-\pi/2, +\pi/2)$. For y between 0 and 1, the two corresponding values of x are

$$x_1 = \sin^{-1}(y) \quad \text{and} \quad x_2 = \pi - \sin^{-1}(y)$$

For Y to be less than y, X must be either less than x_1 or greater than x_2. By examining the figure, the corresponding probability statements are

$$F_Y(y) = P(Y < y) = P(X < x_1 \cup X > x_2)$$
$$= P(X < x_1) + P(X > x_2)$$

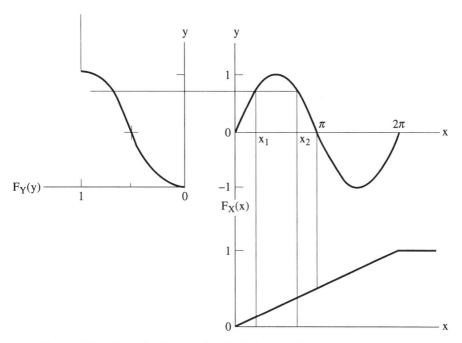

Figure 2.24 Example change of variable by cumulative method, $y = \sin(x)$.

Noting the symmetry about $x = \pi/2$, these statements can be written as

$$F_Y(y) = P(X < \sin^{-1}(y)) + P(X > \pi - \sin^{-1}(y))$$

$$= P(X < \sin^{-1}(y)) + P(\pi - \sin^{-1}(y) < X \leq \pi) + P(X > \pi)$$

$$= P(X < \sin^{-1}(y)) + P(X < \sin^{-1}(y)) + \tfrac{1}{2}$$

$$= \tfrac{1}{2} + 2P(X < \sin^{-1}(y))$$

Replace the probability statement by $F_X(x) = x/2\pi$ to get

$$F_Y(y) = \frac{1}{2} + \frac{1}{\pi} \sin^{-1}(y) \qquad -1 \leq y \leq +1$$

The reader may verify that, by similar analysis, the same equation holds for negative values of y, which justifies the stated limits on the equation.

2.4.2 Change of Variables by Probability Density Function

The second approach to transformation is through the PDF. The approach is easy to see conceptually for discrete random variables: For each value of X, we wish to find the corresponding value(s) of Y and assign the probability of that x to that y. The easiest way to perform this mapping is with a table for the joint probability mass function of X and Y:

$$p_{XY}(x, y = g(x)) = p_X(x) \qquad (2.94)$$

The probability mass function of Y is the marginal distribution calculated from the table. We call this an enumeration method, because it requires all possible values of the random variable to be enumerated, or named, one by one.

Example 2.20: Transformation of Discrete Random Variable by Density Method. Let $y = |x|$, where

$$p_X(x) = \begin{cases} \frac{1}{4} & x = -1, 0, 1, 2 \\ 0 & \text{otherwise} \end{cases}$$

The transformation is made using the following table:

		\(y = |x|\) 0	1	2	$p_X(x)$
	-1		$\frac{1}{4}$		$\frac{1}{4}$
	0	$\frac{1}{4}$			$\frac{1}{4}$
x	1		$\frac{1}{4}$		$\frac{1}{4}$
	2			$\frac{1}{4}$	$\frac{1}{4}$
	$p_Y(y)$	$\frac{1}{4}$	$\frac{1}{2}$	$\frac{1}{4}$	

Thus, the distribution of Y is

$$p_Y(y) = \begin{cases} \frac{1}{4} & y = 0 \\ \frac{1}{2} & y = 1 \\ \frac{1}{4} & y = 2 \\ 0 & \text{otherwise} \end{cases}$$

The approach for continuous random variables is similar. Heuristically, for each value of X, we wish to find the corresponding value of Y and map the probability from x to $y = g(x)$. Recalling Eq. (2.16), $P(X \approx x) = f_X(x)\, dx$, we write this using probability statements as

$$P(y < Y \le y + dy) = P(x < X \le x + dx) \qquad (2.95)$$

In terms of the density functions, this is

$$f_Y(y)\, dy = f_X(x)\, dx \qquad (2.96)$$

From this equation, it is apparent that there is a dependence on the relative size of the differential elements dy and dx. In general, the relative sizes are dependent on x and y.

The derivation of the actual formula requires the application of Leibniz's rule for differentiating an integral by the variable α when the limits of integration contain functions dependent on α, $a(\alpha)$ and $b(\alpha)$. Leibniz's rule states that if $R(\alpha)$ is defined as

$$R(\alpha) = \int_{a(\alpha)}^{b(\alpha)} r(\alpha, x) \, dx \tag{2.97}$$

then

$$\frac{d}{d\alpha} R(\alpha) = \int_{a(\alpha)}^{b(\alpha)} \frac{\partial}{\partial \alpha} r(\alpha, x) \, dx + r(\alpha, b(\alpha)) \frac{\partial}{\partial \alpha} b(\alpha)$$

$$- r(\alpha, a(\alpha)) \frac{\partial}{\partial \alpha} a(\alpha) \tag{2.98}$$

Let us again restrict our initial discussion to continuous monotonically increasing functions, $y = g(x)$. From Eq. (2.92) we know that $F_Y(y) = F_X(g^{-1}(y))$. Differentiation of this relationship yields

$$f_Y(y) = \frac{d}{dy} \int_{-\infty}^{g^{-1}(y)} f_X(x) \, dx \tag{2.99}$$

Application of Leibniz's rule gives

$$f_Y(y) = f_X(g^{-1}(y)) \cdot \frac{d}{dy} g^{-1}(y) \tag{2.100}$$

Hence, the required scaling relationship between the differential elements dx and dy is the derivative dx/dy. This is more easily remembered by the mnemonic

$$f_Y(y) = f_X(x) \cdot \frac{dx}{dy} \tag{2.101}$$

where it is understood that x is to be replaced by $g^{-1}(y)$ and the density function depends on y, not on x. Figure 2.25 illustrates the procedure.

If $y = g(x)$ is monotonically decreasing, it can be shown that the sign of the scale factor must be positive:

$$f_Y(y) = f_X(x) \cdot \left| \frac{dx}{dy} \right| \tag{2.102}$$

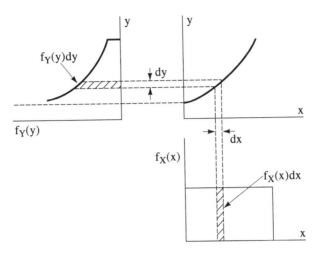

Figure 2.25 Graphical construction of $f_Y(y)$ from $f_X(x)$.

If the function is not monotonic, it must be broken into separate monotonically increasing and decreasing portions. Then this procedure can be executed separately on each piece and the results combined.

Example 2.21: $y = \sin(x)$ by Density Method. We repeat the previous example using the density function approach, illustrated in Figure 2.26. The derivative dx/dy is given by

$$\frac{dx}{dy} = \frac{d}{dy} \sin^{-1}(y) = \frac{1}{\sqrt{1 - y^2}}$$

Substitution into Eq. (2.102) gives, for each monotonic interval,

$$f_Y(y) = f_X(x)\left|\frac{dx}{dy}\right| = \frac{1}{2\pi} \frac{1}{\sqrt{1 - y^2}}$$

The total density function is equal to the sum of the densities across each monotonic interval. As seen in the previous example, for each value of y, there are two values of x. The density function has the same value at both values. Thus, the total density is twice the above:

$$f_Y(y) = \frac{1}{\pi\sqrt{1 - y^2}} \qquad -1 \le y \le +1$$

The reader may verify that this density agrees with the cumulative found in the previous example. Notice that this density function is singular at the extreme

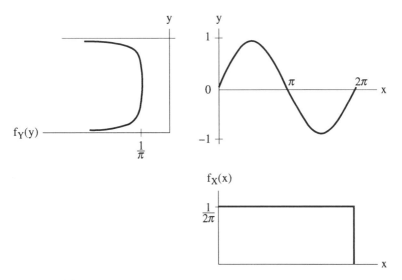

Figure 2.26 Example change of variable by density method, $y = \sin(x)$.

values $y = \pm 1$, implying that the density is infinite at the extremes. This is still a valid density function because its integral is equal to 1. This density function implies that, if one were to sample a sine wave at random times, it is more likely that the values observed would be near the extremes and less to be near zero. Physically, one might think of observing an oscillating mass at random times: It is far more likely to be seen near either of the extremes because it slows to zero velocity and is unlikely to be seen near the middle because it is moving at maximum velocity.

2.4.3 Multidimensional Change of Variables

Suppose we have two random variables X_1 and X_2 and two functions of those variables, $y_1 = g_1(x_1, x_2)$ and $y_2 = g_2(x_1, x_2)$. We wish to find the probability distribution of Y_1 and Y_2. The change-of-variable problem for multivariate distributions is only slightly more complicated than for single random variables. The density function approach is presented here because it handles all transformations of interest in this text.

The approach is basically the same as for functions of a single random variable. Heuristically, for each pair (x_1, x_2), we wish to find the corresponding pair (y_1, y_2) and map the probability density from x space to y space. We write the corresponding probability statement as

$$P(y_1 < Y_1 \leq y_1 + dy_1, y_2 < Y_2 \leq y_2 + dy_2)$$

$$= P(x_1 < X_1 \leq x_1 + dx_1, x_2 < X_2 \leq x_2 + dx_2) \qquad (2.103)$$

This leads to the following differential form:

$$f_{Y_1Y_2}(y_1, y_2)\, dy_1\, dy_2 = f_{X_1X_2}(x_1, x_2)\, dx_1\, dx_2 \qquad (2.104)$$

Again, a scale factor is required that relates the sizes of the differential rectangle in x space, dx_1 by dx_2, and the differential rectangle in y space, dy_1 by dy_2. The scale factor is known from advanced calculus to be the Jacobian determinant, or Jacobian, denoted $|J|$. Thus, the joint density at y_1 and y_2 is equal to the joint density at the corresponding x_1 and x_2 times the Jacobian:

$$f_{Y_1Y_2}(y_1, y_2) = f_{X_1X_2}(x_1, x_2) \cdot |J| \qquad (2.105)$$

where

$$|J| = \begin{vmatrix} \dfrac{\partial x_1}{\partial y_1} & \dfrac{\partial x_1}{\partial y_2} \\[2mm] \dfrac{\partial x_2}{\partial y_1} & \dfrac{\partial x_2}{\partial y_2} \end{vmatrix} = \frac{\partial x_1}{\partial y_1} \cdot \frac{\partial x_2}{\partial y_2} - \frac{\partial x_2}{\partial y_1} \cdot \frac{\partial x_1}{\partial y_2} \qquad (2.106)$$

In the equations above, x_1 and x_2 are to be replaced by the "inverse" functions in y_1 and y_2, say, $x_1 = h_1(y_1, y_2)$ and $x_2 = h_2(y_1, y_2)$.

For problems involving a reduction in the number of variables, one may first do the transformation in the same number of variables as the original space and then find the desired marginal distribution in the transformed space.

Example 2.22: Box–Muller Transformation. Let X and Y be independent normal random variables with means equal to zero and common variance σ^2. For example, X and Y may be the Cartesian coordinates of the random deviations from the center of a target. The joint density of X and Y is

$$f_{XY}(x, y) = \frac{1}{2\pi\sigma^2} \exp\left(-\frac{1}{2}\left(\frac{x^2 + y^2}{\sigma^2}\right)\right)$$

We wish to convert to polar coordinates R and Θ. The relationship between the two coordinate systems is

$$r = \sqrt{x^2 + y^2} \qquad \theta = \arctan\left(\frac{y}{x}\right)$$

The inverse relations are

$$x = r\cos\theta \qquad y = r\sin\theta$$

The Jacobian is

$$|J| = \begin{vmatrix} \dfrac{\partial}{\partial r} r \cos \theta & \dfrac{\partial}{\partial \theta} r \cos \theta \\ \dfrac{\partial}{\partial r} r \sin \theta & \dfrac{\partial}{\partial \theta} r \sin \theta \end{vmatrix} = \begin{vmatrix} \cos \theta & -r \sin \theta \\ \sin \theta & r \cos \theta \end{vmatrix} = r$$

Substitution into the transformation gives the joint density of R and Θ:

$$f_{R\Theta}(r, \theta) = \frac{1}{2\pi\sigma^2} \exp \left(-\frac{1}{2} \frac{r^2 \cos^2 \theta + r^2 \sin^2 \theta}{\sigma^2} \right) |r|$$

$$= \left(\frac{1}{2\pi} \right) \cdot \left[\frac{r}{\sigma^2} \exp \left(-\frac{1}{2} \frac{r^2}{\sigma^2} \right) \right]$$

This joint density is seen to be separable; hence, R and Θ are independent. Further inspection reveals that R has a Rayleigh distribution and Θ has a uniform distribution.

This transformation is named the Box–Muller transformation (Box et al., 1958). It is applied in numerical simulations going in the opposite direction of the above: A Rayleigh and a uniform random variable are generated by computer and two independent normal variables are found by transformation.

2.4.4 Sums of Random Variables

Sums of random variables are so ubiquitous in the study of random vibrations that they merit special attention. For example, let X_i be the ith of n random variables with marginal distributions $f_{X_i}(x_i)$, for example, the acceleration measured at different points on a body at the same time. The average, $\overline{X}$, is given by the sum

$$\overline{X} = \frac{1}{n} \sum_{i=1}^{n} X_i \tag{2.107}$$

It is apparent that $\overline{X}$ is a random variable; every different set of n observations will yield a different value of $\overline{X}$. The probability distribution of $\overline{X}$ depends on the joint distribution of the X_i's. Before discussing the change-of-variable problem, let us see what can be learned from the expected values.

First, let Y be an arbitrary linear combination of n random variables, X_i:

$$Y = a_0 + \sum_{i=1}^{n} a_i X_i \tag{2.108}$$

where a_0 and a_i are arbitrary constants. The reader may easily verify that, regardless of the distributions of X_i, the mean and variance are given by

$$\mu_Y = a_0 + \sum_{i=1}^{n} a_i \mu_{X_i} \tag{2.109}$$

$$\sigma_Y^2 = \sum_{i=1}^{n} a_i^2 \sigma_{X_i}^2 + \sum_{\substack{i=1 \\ i \ne j}}^{n} \sum_{j=1}^{n} a_i a_j \sigma_{X_i X_j} \tag{2.110}$$

If the X_i are independent, the covariances are zero and the variance simplifies to

$$\sigma_Y^2 = \sum_{i=1}^{n} a_i^2 \sigma_{X_i}^2 \tag{2.111}$$

Next consider the important special case of the sum of n independent and identically distributed random variables X_i with common distribution $f_X(x)$:

$$Y = \sum_{i=1}^{n} X_i \tag{2.112}$$

The mean and variance simplify to

$$\mu_Y = n\mu_X \tag{2.113}$$

$$\sigma_Y^2 = n\sigma_X^2 \tag{2.114}$$

Note that whereas the mean and variance increase linearly with n, the standard deviation increases with only the square root of n. Hence, the coefficient of variation of the sum, $C_Y = \sigma_Y/\mu_Y$, actually decreases by $1/\sqrt{n}$ as the number of variables in the sum increases:

$$C_Y = \frac{\sqrt{n}}{n} \frac{\sigma_X}{\mu_X} = \frac{1}{\sqrt{n}} C_X \tag{2.115}$$

This means that there is less relative variation in the sum of a large number of random variables than in the sum of a few random variables. Thus, if the goal is to estimate an average value, $\overline{X}$, the larger the sample size, the better. This fact is of great significance in the study of statistics.

The moment-generating function of the sum of independent identically distributed random variables is the nth power of the moment-generating function of X. This is easily seen as follows:

$$M_Y(t) = E\left[\exp(tY)\right] = E\left[\exp\left(t\sum_{i=1}^{n} X_i\right)\right] = E\left[\prod_{i=1}^{n} \exp(tX_i)\right]$$

$$= \prod_{i=1}^{n} E[\exp(tX_i)] = \prod_{i=1}^{n} M_{X_i}$$

$$= [M_{X_i}]^n \tag{2.116}$$

The switching of the order of the expected value and the product is justified by the fact that the joint density function factors into the product of the marginal density functions when the X_i are independent.

Now we consider the actual distribution of a sum of two random variables:

$$Z = X + Y \tag{2.117}$$

The first step is to transform the density from the space of X and Y to the space of Z and X. The relations between the two spaces are

$$z = x + y \quad \text{and} \quad x = x \tag{2.118}$$

$$y = z - x \quad \text{and} \quad x = x \tag{2.119}$$

The Jacobian scale factor is equal to 1. The joint density of X and Z is

$$f_{XZ}(x, z) = f_{XY}(x, z - x) \tag{2.120}$$

The marginal density function of Z is found by integrating over all values of X:

$$f_Z(z) = \int_{-\infty}^{+\infty} f_{XZ}(x, z) \, dx = \int_{-\infty}^{+\infty} f_{XY}(x, z - x) \, dx \tag{2.121}$$

When the limits on X are finite, some care will need to be exercised in performing the integration.

If X and Y are independent, the joint density function factors and we may write

$$f_Z(z) = \int_{-\infty}^{+\infty} f_X(x) f_Y(z - x) \, dx \tag{2.122}$$

This integral is a convolution integral. This result may also be expressed as the cumulative distribution as follows:

$$F_Z(z) = \int_{-\infty}^{z} f_Z(w) \, dw = \int_{-\infty}^{z} \left[\int_{-\infty}^{+\infty} f_X(x) f_Y(w - x) \, dx \right] dw$$

$$= \int_{-\infty}^{+\infty} \left[f_X(x) \int_{-\infty}^{z} f_Y(w - x) \, dw \right] dx$$

$$= \int_{-\infty}^{+\infty} \left[f_X(x) \int_{-\infty}^{z-x} f_Y(v) \, dv \right] dx$$

$$= \int_{-\infty}^{+\infty} f_X(x) F_Y(z - x) \, dx \tag{2.123}$$

In words, the probability that Z is less than (or equal to) z is equal to the integral over all values of x of the probability that X is any value, x, and Y is less than (or equal to) the difference between z and x.

The distribution of the sum of two random variables is generally different from the distributions of the addends and is generally difficult to determine in closed form. However, there are two special circumstances under which the distribution of a sum is easily determined. The first circumstance is if the two random variables are both normally distributed, then their sum is also normally distributed. (The proof of this is left to the reader.) The parameters of the resultant normal distribution can be calculated directly from the means, variances, and covariance of the addends. Certain other probability distributions also have this property, which is known as the reproductive or regenerative property.

The second circumstance is if the sum is of a large number of random variables. Then, under certain general conditions, the distribution of the sum may be shown to be asymptotically normal. This is stated mathematically as the central limit theorem. The theorem and the conditions under which it holds are discussed in the next section.

Example 2.23: Sum of Two Uniform Random Variables. Consider the sum $Z = X + Y$, where X and Y are independent and uniformly distributed on the interval $(0, 1)$:

$$f_{XY}(x, y) = 1 \qquad 0 < x < 1 \qquad 0 < y < 1$$

The region over which the joint density function is nonzero is a square. As shown in Figure 2.27, the line $z = x + y$ is a diagonal in the x–y plane and intersects the y axis at $(0, z)$ and the x axis at $(z, 0)$. For $z < 0$ and $z > 2$, the line does not cross the square and there is no probability that these values of Z occur. For $0 < z < 1$, the line is inside the square between the points $(0, z)$ and $(z, 0)$, that is, for $0 < x < z$. Hence, we have

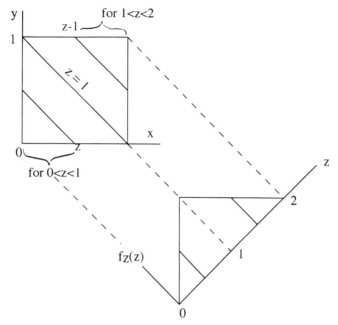

Figure 2.27 Probability density function of sum of two uniform random variables.

$$f_Z(z) = \int_{-\infty}^{+\infty} f_{XY}(x, z - x) \, dx = \int_0^z 1 \, dx = z \qquad 0 < z < 1$$

For $1 < z < 2$, the line is inside the square between the points $(z - 1, 1)$ and $(1, z - 1)$, that is, for $z - 1 < x < 1$. Thus, we have

$$f_Z(z) = \int_{-\infty}^{+\infty} f_{XY}(x, z - x) \, dx = \int_{z-1}^1 1 \, dx = 2 - z \qquad 1 < z < 2$$

Combining the above, the density function of Z is triangular, as shown in Figure 2.27:

$$f_Z(z) = \begin{cases} z & 0 < z < 1 \\ 2 - z & 1 < z < 2 \end{cases}$$

2.5 NORMAL, SAMPLING, AND EXTREME-VALUE DISTRIBUTIONS

For many practical cases in random vibrations, the random variable being observed has a normal, or Gaussian, probability distribution. This section shows that the normal distribution arises naturally as the limiting distribution of the sum of a large number of random variables and introduces the central limit theorem. Transformations of normal random variables are important in our subsequent discussion of signal analysis; this section presents the sampling distributions: chi-square, t, and F. Finally, since in design for random vibrations the distribution of the largest value in a sequence of observations is of practical importance, the limiting extreme-value distribution is presented. The treatments given in this section are not meant to be comprehensive, but rather are intended to dispel any mystery and provide intuitive justification for their later applications.

2.5.1 De Moivre–Laplace Approximation

The form of the normal distribution was first discovered by Abraham de Moivre (1667–1754) in about 1733 approximately as follows. Let K be an indicator variable, that is, $k = 1$ if the outcome is a success, with probability p and $k = 0$ with probability $1 - p$, and let K_i be the ith of n independent observations of K. Let X be the sum of these observations:

$$X = \sum_{i=1}^{n} K_i \tag{2.124}$$

In other words, X represents the number of successes in n independent and identically distributed Bernoulli trials. The distribution of X is known as the binomial distribution and is given by

$$p_X(x) = \binom{n}{x} p^x (1 - p)^{n-x} \qquad x = 0, 1, 2, \ldots, n \tag{2.125}$$

where $\binom{n}{x}$ is the binomial coefficient given by

$$\binom{n}{x} = \frac{n!}{x!(n - x)!} \tag{2.126}$$

The parameters of the distribution are n and p. The distribution takes its name from the presence of the binomial coefficient, which accounts for the number of different ways there are to order x successes in n trials, for example, $\{1, 0, 0, \ldots\}$, $\{0, 1, 0, \ldots\}$, and so on. The mean, variance, and moment-generating function of the binomial distribution are

$$\mu_X = np \tag{2.127}$$

$$\sigma_X^2 = np(1 - p) \tag{2.128}$$

$$M_X(t) = [pe^t + (1 - p)]^n \tag{2.129}$$

Note that the moment-generating function is the moment-generating function of the Bernoulli distribution raised to the nth power, as expected from Equation (2.116).

The binomial coefficient is difficult to calculate for large values of n. De Moivre replaced the factorials with Stirling's approximation:

$$n! \approx \sqrt{2\pi} e^{-n} n^{n+1/2} \qquad (2.130)$$

The result, after lengthy algebraic manipulation, is

$$p_X(x) \approx \frac{1}{\sqrt{2\pi} \sqrt{np(1-p)}} \exp\left[-\frac{1}{2} \left(\frac{x-np}{\sqrt{np(1-p)}} \right)^2 \right] \qquad (2.131)$$

Noting that $\mu_x = np$ and $\sigma_x^2 = np(1-p)$, we have

$$p_X(x) \approx \frac{1}{\sqrt{2\pi}\,\sigma_X} \exp\left[-\frac{1}{2} \left(\frac{x-\mu_X}{\sigma_X} \right)^2 \right] \qquad (2.132)$$

Thus, in the limit as n becomes large, the binomial probability approaches the normal density function. An interval width equal to 1, $\Delta x = 1$, is implicit so that $p_X(x) \approx \phi_X(x)\,\Delta x$. The approximation is known as the de Moivre–Laplace approximation.

2.5.2 Central Limit Theorem

It may be shown through similar limiting arguments that the same limiting distribution holds for other initial distributions. The generalization of this fact, now known as the central limit theorem, was first stated in 1812 by Pierre-Simon Laplace (1749–1827). Laplace stated that the limiting distribution of the sum of independent identically distributed random variables of arbitrary distribution is the normal distribution. The theorem has since been generalized, under certain general conditions, to not require identical distributions of the addends and to admit mild dependence between the addends.

For our purposes, we may state the central limit theorem as follows: Let $X_1, X_2, \ldots, X_n$ be a sequence of independent random variables with means $\mu_1, \mu_2, \ldots, \mu_n$ and variances $\sigma_1^2, \sigma_2^2, \ldots, \sigma_n^2$. Let S_n be the sum of this sequence:

$$S_n = \sum_{i=1}^{n} X_i \qquad (2.133)$$

Noting that the X_i are independent, the mean and variance are given by

$$\mu_{S_n} = \sum_{i=1}^{n} \mu_i \qquad (2.134)$$

$$\sigma_{S_n}^2 = \sum_{i=1}^{n} \sigma_i^2 \tag{2.135}$$

Then, in the limit as n goes to infinity, the standardized variable of S_n, $Z_n = (S_n - \mu_{S_n})/\sigma_{S_n}$, has the standard normal distribution:

$$\lim_{n \to \infty} f_{Z_n}(\zeta) = \phi(\zeta) \tag{2.136}$$

The conditions under which this occurs are known as the Lyapunov conditions, named for Aleksandr Lyapunov (1901). They may be informally summarized as follows:

1. The individual terms in the sum contribute a negligible amount to the variation in the sum.
2. It is very unlikely that any single term will contribute a disproportionately large amount to the sum.

In other words, no single variable (or a few variables) dominates either the total uncertainty or the total sum. All variables in the sum contribute comparably to the sum.

A proof of a simple version of this theorem is given in Appendix A. Practical experience gives the following guidelines for applying the central limit theorem:

1. If the density functions of X_i are bell shaped like the normal density function, the sum of four or more X_i appears normal.
2. If the density functions of X_i are flat like the uniform density, the sum of 12 or more appears normal.
3. If the density functions of X_i are concentrated in the tails like the distribution of the randomly sampled sine wave, the sum may not appear to be normal for fewer than 50 or 100 or more.

Figure 2.28 illustrates how the sum of uniformly distributed random variables approaches a normal density function. We note that before the Box–Muller and other algorithms were developed, the sum of 12 or more uniform variates was a standard algorithm for simulating normal random variables on computers.

Finally, we note that the central limit theorem is an example of one of four kinds of convergence that a sequence of random variables may have, namely, convergence in distribution. We discuss all four in Appendix A.

2.5.3 Distribution of Sample Mean: Normal

Suppose X is a random variable with mean μ_X, variance σ_X^2, and distribution $f_X(x)$, and suppose we conduct an experiment in which we sample this random

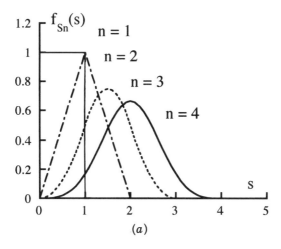

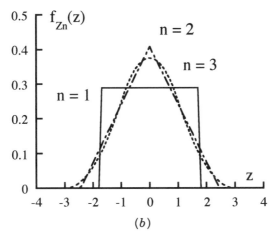

Figure 2.28 Convergence of the distribution of the sum of uniform random variables to the normal distribution; (a) densities of the sums, n = 1, 2, 3, 4; (b) densities of the standardized sums.

variable n times, letting X_i be the ith sample. If the experiment is carefully constructed, the X_i will be independent and identically distributed. Calculations made with these random variables are called sample statistics. In the following sections, we present the distributions associated with sample statistics known as the sampling distributions.

In a typical problem, we wish to estimate the unknown average μ_X. The sample mean or sample average $\overline{X}$ is a random variable given by the sum

$$\overline{X} = \frac{1}{n} \sum_{i=1}^{n} X_i \qquad (2.137)$$

Note that $\overline{X}$ is capitalized because it is a random variable. If the observed variable X is normal, then the sample mean $\overline{X}$ is automatically also normal because the normal distribution is reproductive. If X is not normal, then according to the central limit theorem, $\overline{X}$ will be approximately normal if the number of samples n is large enough. By inspection, the mean and variance of $\overline{X}$ are

$$\mu_{\overline{X}} = \frac{1}{n} n\mu_X = \mu_X \qquad (2.138)$$

$$\sigma_{\overline{X}}^2 = \frac{1}{n^2} n\sigma_X^2 = \frac{\sigma_X^2}{n} \qquad (2.139)$$

Thus, we conclude that the standardized variable

$$Z = \frac{\overline{X} - \mu_X}{\sigma_X/\sqrt{n}} \qquad (2.140)$$

has a standard normal distribution. This distribution may be used to determine how close our estimate of the mean μ_X is likely to be, assuming the variance σ_X^2 is known.

2.5.4   Distribution of Sample Variance: Chi-Square

The variance of a random variable σ_X^2 is estimated by the sample variance S_X^2 given by the sum

$$S_X^2 = \frac{1}{n-1} \sum_{i=1}^{n} (X_i - \overline{X})^2 = \frac{1}{n-1} \left(\sum_{i=1}^{n} X_i^2 - n\overline{X}^2 \right) \qquad (2.141)$$

Again, S_X^2 is capitalized because it is a random variable. Observe that in order to find the distribution of S_X^2, we need to know the distribution of the sum of the squares, X_i^2. If X has a standard normal distribution, this distribution can be developed exactly and is related to a chi-square distribution. (If X is not normal, then the following is an approximation.)

Let Z be a standard normal random variable and let the ith independent sample of this variable be Z_i. We define a random variable called chi-square with n degrees of freedom, χ_n^2, as the sum of n squares of Z_i:

$$\chi_n^2 = \sum_{i=1}^{n} Z_i^2 \qquad (2.142)$$

Consider the case $n = 1$, that is, chi-square with one degree of freedom, $\chi_1^2 = Z_1^2$. The cumulative distribution of χ_1^2 may be found as follows:

$$F_{\chi_1^2}(u) = P(\chi_1^2 \le u) = P(Z_1^2 \le u) = P(-\sqrt{u} < Z_1 \le \sqrt{u})$$

$$= \Phi(\sqrt{u}) - \Phi(-\sqrt{u}) \tag{2.143}$$

where $\Phi(\cdot)$ is the standard normal CDF. The density function is found by differentiation to be

$$f_{\chi_1^2}(u) = \frac{1}{\sqrt{2\pi u}} \exp\left(-\frac{1}{2}u\right) \quad u > 0 \tag{2.144}$$

The mean, variance, and moment-generating function are

$$\mu_{\chi_1^2} = 1 \tag{2.145}$$

$$\sigma_{\chi_1^2}^2 = 2 \tag{2.146}$$

$$M_{\chi_1^2}(t) = \frac{1}{\sqrt{1 - 2t}} \tag{2.147}$$

Noting that the samples are independent, the mean and variance of χ_n^2 for $n > 1$ are n times the mean and variance of χ_1^2, respectively:

$$\mu_{\chi_1^2} = n \tag{2.148}$$

$$\sigma_{\chi_n^2}^2 = 2n \tag{2.149}$$

Also, the moment-generating function is the nth power of the moment-generating function of χ_1^2:

$$M_{\chi_n^2}(t) = (1 - 2t)^{-n/2} \tag{2.150}$$

The density function may be shown by convolution to be

$$f_{\chi_n^2}(u) = \frac{1}{2^{n/2} \, \Gamma(n/2)} \, u^{(n/2 - 1)} \exp\left(-\frac{u}{2}\right) \quad u > 0 \tag{2.151}$$

where the operator $\Gamma(\cdot)$ is the gamma function, which is a continuous version of the factorial function $\Gamma(n) = (n - 1)!$ if n is an integer; otherwise,

$$\Gamma(x) = \int_0^\infty t^{x-1} \, e^{-t} \, dt \tag{2.152}$$

The gamma function cannot be integrated analytically, but Appendix C gives the equation of a numerical approximation. The chi-square cumulative also

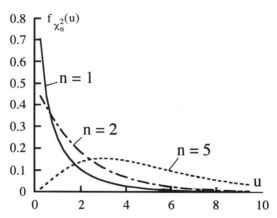

Figure 2.29 Probability density function for chi-squared distribution, $n = 1, 2, 5$.

cannot be established in closed form, but the distribution is extensively tabulated. A short table of the chi-square cumulative is contained in Section E.2. Several chi-square density functions are illustrated in Figure 2.29.

The chi-square distribution may be shown (by convolution) to be reproductive. This is also apparent when the definition of χ_n^2 is written as

$$\chi_n^2 = Z_1^2 + Z_2^2 + \cdots + Z_k^2 + Z_{k+1}^2$$
$$+ Z_{k+2}^2 + \cdots + Z_n^2 = \chi_k^2 + \chi_{n-k}^2 \tag{2.153}$$

Furthermore, since χ_n^2 is the sum of independent, identically distributed random variables, the chi-square distribution approaches the normal distribution for large values of n, say more than 25.

Returning to the distribution of the sample variance, S_X^2, Eq. (2.141) can be rewritten as

$$\frac{n-1}{\sigma_X^2} S_X^2 = \frac{1}{\sigma_X^2} \sum_{i=1}^{n} (X_i - \bar{X})^2$$

$$= \sum_{i=1}^{n} \left(\frac{X_i - \mu_X}{\sigma_X} - \frac{\bar{X} - \mu_X}{\sigma_X} \right)^2$$

$$= \sum_{i=1}^{n} \left[\left(\frac{X_i - \mu_X}{\sigma_X} \right)^2 - 2 \frac{X_i - \mu_X}{\sigma_X} \frac{\bar{X} - \mu_X}{\sigma_X} + \left(\frac{\bar{X} - \mu_X}{\sigma_X} \right)^2 \right]$$

$$= \sum_{i=1}^{n} \left(\frac{X_i - \mu_X}{\sigma_X} \right)^2 - n \left(\frac{\bar{X} - \mu_X}{\sigma_X} \right)^2 \tag{2.154}$$

Noting the mean and variance of $\bar{X}$ given in Eqs. (2.138) and (2.139), respectively, the last line can be written as

$$\frac{n-1}{\sigma_X^2} S_X^2 = \sum_{i=1}^{n} \left(\frac{X_i - \mu_X}{\sigma_X}\right)^2 - \left(\frac{\overline{X} - \mu_{\overline{X}}}{\sigma_{\overline{X}}}\right)^2 \qquad (2.155)$$

The summation is recognized to be the sum of the squares of n independent standard normal variables and, hence, is chi-square distributed with n degrees of freedom. The term to be subtracted is also the square of a standard normal variable and, hence, is chi-square distributed with one degree of freedom. Since the chi-square distribution is reproductive, we conclude that $(n-1)S_X^2/\sigma_X^2$ must be chi-square distributed with $n-1$ degrees of freedom:

$$\chi_{n-1}^2 = \frac{(n-1)S_X^2}{\sigma_X^2} \qquad (2.156)$$

We note that the positive square root of a chi-square variable is called chi, χ. Without going into details, the density function, mean, and variance of this variable are

$$f_{\chi_n^2}(v) = \frac{1}{\Gamma(n/2)} v \left(\frac{1}{2} v^2\right)^{(n/2-1)} \exp\left(-\frac{1}{2} v^2\right) \qquad v > 0 \quad (2.157)$$

$$\mu_{\chi_n^2} = \sqrt{2} \, \frac{\Gamma[(n+1)/2]}{\Gamma[n/2]} \qquad (2.158)$$

$$\sigma_{\chi_n^2}^2 = n - \mu_{\chi_n^2}^2 \qquad (2.159)$$

The Rayleigh distribution encountered earlier is chi distributed with $n = 2$.

Finally, the chi-square distribution is also known as a special form of another distribution known as the gamma distribution, given by

$$f_X(x) = \frac{\lambda}{\Gamma(r)} (\lambda x)^{r-1} \exp(-\lambda x) \qquad x > 0 \qquad (2.160)$$

The chi-square distribution results when $\lambda = \frac{1}{2}$ and $r = \frac{1}{2}n$. The mean, variance, and moment-generating function of the gamma distribution are

$$\mu_X = \frac{r}{\lambda} \qquad (2.161)$$

$$\sigma_X^2 = \frac{r}{\lambda^2} \qquad (2.162)$$

$$M_X(t) = \left(1 - \frac{t}{\lambda}\right)^{-r} \qquad (2.163)$$

2.5.5 Distributions of Related Ratios of Random Variables: t and F

It is shown above that $(\overline{X} - \mu_X)/(\sigma_X/\sqrt{n})$ has a standard normal distribution. However, the variance σ_X^2 is often unknown and must be estimated by the sample variance S_X^2. In this case, the variable given by $(\overline{X} - \mu_X)/(S_X/\sqrt{n})$ has a different distribution. This variable is shown as follows to be the ratio of a standard normal variable Z to the square root of a chi-squared variable with $n - 1$ degrees of freedom, χ_{n-1}^2, divided by its degrees of freedom:

$$\frac{\overline{X} - \mu_X}{S_X/\sqrt{n}} = \frac{(\overline{X} - \mu_X)/(\sigma_X/\sqrt{n})}{S_X/\sigma_X} = \frac{(\overline{X} - \mu_X)/(\sigma_X/\sqrt{n})}{\sqrt{(n-1)S_X^2/(n-1)\sigma_X^2}}$$

$$= \frac{Z}{\sqrt{\chi_{n-1}^2/(n-1)}} = T_{n-1} \qquad (2.164)$$

This variable is given the symbol T_{n-1} and the distribution is known as Student's t distribution.[2] In general, T_n is t distributed with n degrees of freedom and given by

$$T_n = \frac{Z}{\sqrt{\chi_n^2/n}} \qquad (2.165)$$

The density, mean, and variance are

$$f_T(t) = \frac{\Gamma[(n+1)/2]}{\sqrt{\pi n}\ \Gamma(n/2)\ [(t^2/n) + 1]^{(n+1)/2}} \qquad -\infty < t < +\infty \quad (2.166)$$

$$\mu_T = 0 \qquad (2.167)$$

$$\sigma_T^2 = \frac{k}{k-2} \qquad k > 2 \qquad (2.168)$$

As with the chi-squared distribution, the t distribution approaches the normal distribution as the number of degrees of freedom becomes large, say more than 25. Several t distributions are illustrated in Figure 2.30.

The last sampling distribution is for F, the ratio of two chi-squared variables. Suppose W is chi squared with u degrees of freedom and Y is chi squared with v degrees of freedom. The random variable $F_{u,v}$ is defined by

$$F_{u,v} = \frac{W/u}{Y/v} \qquad (2.169)$$

---

[2]Student is actually the pseudonym of an English statistician, William Sealy Gosset.

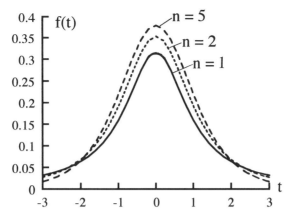

Figure 2.30 Probability density functions for t distribution, $n = 1, 2, 5$.

and has an F distribution with u degrees of freedom in the numerator and v degrees of freedom in the denominator. The distribution is designated F for Sir Ronald A. Fisher, the twentieth-century statistician who originated the analysis-of-variance (ANOVA) method. It is also referred to as Snedecor's F distribution. The density, mean, and variance for $F_{u,v}$ are

$$f_{F_{u,v}}(f) = \frac{\Gamma[(u+v)/2]}{\Gamma(u/2)\,\Gamma(v/2)} \frac{(u/v)^{u/2}\,f^{(u/2)-1}}{[(u/v)f+1]^{(u+v)/2}} \quad f > 0 \qquad (2.170)$$

$$\mu_{F_{u,v}} = \frac{v}{v-2} \qquad\qquad v > 2 \qquad (2.171)$$

$$\sigma^2_{F_{u,v}} = \frac{2v^2(u+v-2)}{u(v-2)^2\,(v-4)} \qquad v > 4 \qquad (2.172)$$

Several F distributions are illustrated in Figure 2.31.

Due to the gamma function, the cumulatives of the t and F distributions cannot be evaluated in closed form. However, they are used so often that they are extensively tabulated. Often these tables are in forms specially suited for their common application in interval estimation and hypothesis testing. Sections E.3 and E.4 contain short tables for these distributions covering values commonly used in random vibrations.

2.5.6 Distribution of Extreme Values

The distribution of the largest of n independent samples of a random variable is of interest in engineering design problems. For example, if the X_i represent peak accelerations experienced by a body, what is the probability distribution of the largest peak acceleration?

The distribution may be obtained exactly from first principles. Let Y_n rep-

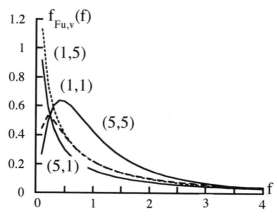

Figure 2.31 Probability density functions for F distribution (u, v): $(1, 1)$, $(1, 5)$, $(5, 1)$, $(5, 5)$.

resent the largest value in n independent samples of random variable X. The probability that Y_n is less than y is equal to the probability that all X_i are less than y. By independence, this is the product of the probabilities that each X_i is less than y. In equations:

$$F_{Y_n}(y) = P(Y_n \le y) = P(X_1 \le y, X_2 \le y, \ldots, X_n \le y)$$

$$= P(X_1 \le y)P(X_2 \le y) \cdots P(X_n \le y)$$

$$= F_{X_1}(y)F_{X_2}(y) \cdots F_{X_n}(y)$$

$$= [F_X(y)]^n \tag{2.173}$$

The product becomes the nth power in the final step because the X_i are independent and identically distributed. The corresponding probability density function is the derivative given by

$$f_{Y_n}(y) = n[F_X(y)]^{n-1} f_X(y) \tag{2.174}$$

For example, let X have a Rayleigh distribution. (It is shown in Chapter 6 that the Rayleigh distribution is appropriate for the peak values of certain random processes.) The cumulative and density functions for the largest of n independent observations of a Rayleigh variable are

$$F_{Y_n}(y) = \left[1 - \exp\left(-\frac{1}{2}\frac{y^2}{\sigma^2}\right)\right]^n \quad y \ge 0 \tag{2.175}$$

$$f_{Y_n}(y) = n\left[1 - \exp\left(-\frac{1}{2}\frac{y^2}{\sigma^2}\right)\right]^{n-1} \frac{y}{\sigma^2} \exp\left(-\frac{1}{2}\frac{y^2}{\sigma^2}\right) \tag{2.176}$$

It is common to approximate the exact distribution of an extreme value, such as above, with a distribution that is asymptotically equivalent for large values of n. It can be shown that many of these asymptotic distributions have the form of a double exponential, known as the extreme-value distribution:

$$F_{Y_n}(y) = \exp\{-\exp[-\alpha(y - \beta)]\} \tag{2.177}$$

$$f_{Y_n}(y) = \alpha \exp\{[-\alpha(y - \beta)] - e^{-\alpha(y-\beta)}\} \tag{2.178}$$

where β is the *characteristic largest value of* Y_n and α is inversely related to the standard deviation of Y_n. The parameters β and α are related to the number of samples n and the distribution of the individual X_i by

$$F_X(\beta) = 1 - \frac{1}{n} \quad \text{or} \quad \beta = F_X^{-1}\left(1 - \frac{1}{n}\right) \tag{2.179}$$

$$\alpha = nf_X(\beta) \tag{2.180}$$

The derivation of this distribution and its parameters is outside the scope of this text. The interested reader is referred to Gumbel (1958) or Ang and Tang (1984).

The corresponding mean and variance of the asymptotic distribution are

$$\mu_{Y_n} = \beta + \frac{\gamma}{\alpha} \tag{2.181}$$

$$\sigma_{Y_n}^2 = \frac{\pi^2}{6\alpha^2} \tag{2.182}$$

where γ is Euler's constant, given by

$$\gamma = -\int_0^\infty e^{-x} \ln x \, dx = 0.5772157 \ldots \tag{2.183}$$

For the example of a Rayleigh variable X, the parameters of the extreme-value distribution approximation are

$$\beta = \sigma\sqrt{2 \ln n} \tag{2.184}$$

$$\alpha = \frac{\sqrt{2 \ln n}}{\sigma} \tag{2.185}$$

The mean and variance of the corresponding extreme-value distribution are

$$\mu_{Y_n} = \sigma\left(\sqrt{2 \ln n} + \frac{\gamma}{\sqrt{2 \ln n}}\right) \tag{2.186}$$

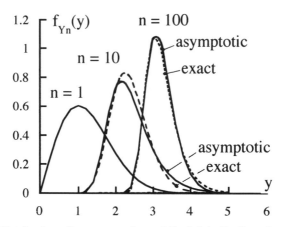

Figure 2.32 Distribution of extreme values of Rayleigh-distributed random variable, exact and asymptotic extreme-value distribution for n values of 1 (Rayleigh), 10, and 100.

$$\sigma_{Y_n}^2 = \frac{\pi^2}{6} \frac{\sigma^2}{2 \ln n} \tag{2.187}$$

Figure 2.32 illustrates the extreme-value distribution for this example.

PROBLEMS

2.1   Show that $P(A \cup B) = P(A) + P(B) - P(A \cap B)$ by expressing the set $\{A \cup B\}$ as the union of disjoint sets $\{A \cap \bar{B}\}$, $\{\bar{A} \cap B\}$, and $\{A \cap B\}$. Illustrate with a Venn diagram.

2.2   Find the probabilities (i) $P(X < 2)$, (ii) $P(X > 1)$, and (iii) $P(1 < X < 2)$ for X distributed according to the following distributions:

    a Poisson with parameter $\nu = 1$

    b Uniform between 0 and 2

    c Rayleigh with mean value equal to 1

2.3   Derive, by integration, the mean, variance, and coefficient of variation for the following distributions:

    a Bernoulli

    b Poisson

    c uniform

    d Rayleigh

    (*Hint:* The answers are given in the text.)

2.4 Derive the moment-generating functions for the following distributions:

a Bernoulli

b Poisson

c Uniform

d Rayleigh

(*Hint:* The answers are given in the text.)

2.5 Use the moment-generating functions to find the mean and variance of the following distributions:

a Bernoulli

b Poisson

c Uniform

d Rayleigh

2.6 Find the probabilities (i) $P(X < 2)$, (ii) $P(X > 1)$, and (iii) $P(1 < X < 2)$ for X distributed according to:

a Standard normal distribution

b Normal distribution with mean equal to 1 and variance equal to 4

2.7 Derive the moment-generating function for the normal distribution and show that the central moments of the normal distribution are dependent only on the variance.

2.8 Let X and Y be continuous positive random variables, i.e., $X > 0$ and $Y > 0$, and let $f_{XY}(x, y) = (y/A) \exp(-x + \frac{1}{2} y^2)$.

a Find the correct value of the normalizing constant A.

b Find the marginal probability density functions.

c Find the conditional probability density function for X given $y = 1$.

d Calculate the covariance and correlation coefficient.

2.9 Let $U = aX + b$ and $V = cY + d$. Show that the covariance of U and V is given by $\sigma_{UV} = ac\sigma_{XY}$.

2.10 Let $U = X + Y$ and $V = X - Y$, where X and Y are independent. Under what additional condition are U and V uncorrelated?

2.11 Verify the marginal and conditional distributions given in the text for the joint normal distribution.

2.12 Sketch the following in the xy plane:

a $|X| > 1$

b $|X| > 1$ and $|Y| > 1$

c $X + Y > 1$

d $X - Y > 1$

Calculate the corresponding probabilities if X and Y are independent standard normal random variables.

2.13 Let X have a uniform distribution on $(0, 1)$ and let $y = x^2$.

  a Construct the CDF of Y graphically.

  b Find the PDF for Y using the PDF method.

2.14 Let X have a uniform distribution on $(0, 1)$ and let $y = -\ln(x)$.

  a Construct the CDF of Y graphically.

  b Find the CDF of Y using the CDF method.

  c Find the PDF of Y using the PDF method.

2.15 Derive the general normal PDF by transformation from the standard normal PDF.

2.16 Find the moment-generating function of the sum of n independent and identically distributed Bernoulli indicator variables and show that this is the same as the moment-generating function of the binomial distribution.

2.17 Find the PDF of the sum of three independent and identically distributed random variables, uniformly distributed on $(0, 1)$.

2.18 Show that the sum of two independent normal random variables is normally distributed using the moment-generating functions.

2.19 Show that the sum of two correlated normal random variables is normally distributed using convolution. [*Hint:* You will need to complete the square: $a^2 \pm 2ab + b^2 = (a \pm b)^2$.]

2.20 Show that, similar to the de Moivre–Laplace approximation, the normal distribution is the limiting case for the Poisson distribution as the expected value goes to infinity.

Simulation

Many students today have access to a simple pseudo–random-number generator that generates numbers uniformly distributed on $(0, 1)$, e.g., via a calculator, a numerical library on a computer, or a spreadsheet on a personal computer. The following problems are designed to be done on a personal computer. A uniformly distributed random variable is presumed to be in the range $(0, 1)$. For histograms, let the number of bins be $\sqrt{n}$.

2.21 *Coin Tossing.* Generate a sample from a uniform distribution and subtract 0.5 from the value. If the outcome is greater than zero, say the outcome is heads, otherwise, tails. Repeat up to $n = 100$ times and let the running total number of heads be H. Plot H/n versus n. What do you observe about the "convergence" of H/n to the expected value?

2.22 *Signal to Noise.* Let X and N be independent uniform random variables and $Y = aX + N$, where $a = 10$. Plot xy pairs for 100 samples. Experiment using different values of a. Label you plots with the correlation coefficient between X and Y.

2.23 *Transformation to Sine.* Generate $n = 100$ samples of a random variable uniformly distributed on $(-\pi/2, +\pi/2)$. Take $V = \sin (U)$. Plot histograms of U and V and compare with the theoretical distributions.

2.24 *Rayleigh Variable.* Some nonuniform random variables are easily simulated by a simple CDF transformation from a uniform variable. Consider the Rayleigh distribution.

a Show that the CDF may be inverted as

$$x = \sigma\sqrt{-2 \ln [1 - F_X(x)]}$$

Now if we substitute $F = F_X(x)$ in the above and let F be a uniformly distributed random variable on $(0, 1)$, then X will have a Rayleigh distribution.

b Generate $n = 100$ samples of F and transform to X. Plot a histogram and compare with the theoretical distribution.

2.25 *Sums of Uniforms.* Let S_m be the sum of m independent uniform random variables. Generate $n = 100$ samples of S_m, standardize, and plot histograms for $m = 2, 3, 4$. Compare to the normal distribution.

2.26 *Sample Statistics.* Generate $m = 5, 10, 20, 40$ samples of a uniform random variable and calculate for each run:

a Sample mean

b Sample variance

Repeat for 100 runs for each m. Plot histograms and compare with the theoretical distributions.

2.27 *Extreme Values.* Generate $m = 5$ samples of a Rayleigh-distributed random variable and pick the largest sample. Repeat 100 times and plot the histogram of largest values. Compare versus the exact and approximate theoretical distributions. Try this for different values of m.

2.28 *Normal Distribution.* For a project, the student might wish to investigate three ways that one might generate samples of a normal random variable from samples of a uniform random variable U.

a Sum 12 samples of U and standardize the result (implicitly invoking the central limit theorem).

b Transform by the CDF method (problem 2.24) from uniform to normal using a polynomial approximation for the inverse of the standard normal CDF.

c Use the Box–Muller algorithm.

Write a "code" to do each of the three simulation methods and document the code, including theoretical background. Generate 100 or more samples with each method and compare the computational time required. Plot the histograms, and compare with the theoretical distribution.

3

RANDOM PROCESSES IN TIME DOMAIN

In this chapter, we extend our discussion of random variables to sequences of random variables, called random or stochastic processes. We present the theory of random processes and their characterization in the time domain. We define random processes in general; then we focus on the most important specialization for random vibrations: processes that are stationary, ergodic, and Gaussian. The role of correlation functions in characterizing random processes is discussed in detail. The chapter concludes with a brief introduction to three stochastic processes of practical importance, including a random process built up of an infinite sum of harmonics, thereby developing the rationale for the frequency-domain representation presented in Chapter 5.

There is a rich body of literature on random processes. For more general treatments, the reader is referred to such standard texts as Doob (1953); Parzen (1962); Papoulis (1965); Feller (1966); Hoel, Port, and Stone (1972); Box and Jenkins (1976); and Karlin and Taylor (1975, 1981).

3.1. RANDOM PROCESS DEFINITIONS

This section provides the basic definition of random processes and introduces the tools by which they are characterized. The special cases of stationary, ergodic, and Gaussian processes are defined.

3.1.1 State Spaces and Index Sets

A random process generates a sequence of random variables, say, $X_1, X_2, \ldots, X_n$. The term random process may be applied to either the process that generates the sequence or to the sequence itself. The set of possible values of the random variable $X(t)$ is called the state space Ω and the order of the sequence is tracked by the index t, which is an element of the index set T. The sequence is often a time sequence, for example, first, second, third and so on, but could also be along the length of a line, position in space, or some other parameter that indicates relative position. Thus, we formally state that $\{X(t) \in \Omega, t \in T\}$ is a random process defined on the state space Ω that evolves with respect to the index set T. For our purposes, we will be less formal and simply refer to $X(t)$ as the random process without repeating the details of the state space or index set once they have been defined. We will also assume that the index is time t unless an exception is specifically noted.

For example, in the preceding chapter a random variable X is sampled n times and X_i represents the ith sample in the sequence. Each sample is assumed to be independent of all other samples and all are identically distributed. Suppose that X takes on values 0 or 1 with probabilities $1 - p$ and p, respectively. The process is formally denoted $\{X(t) \in \{0, 1\}, t \in \{1, 2, \ldots, n\}\}$. Some sample realizations of this process for $n = 6$ are $\{0, 0, 1, 0, 0, 1\}$, $\{1, 0, 1, 0, 1, 1\}$, and $\{0, 0, 0, 0, 0, 1\}$.

Just as the values of the random variable $X(t)$ may be discrete or continuous, the index set T may also be either discrete or continuous. When T is discrete, it is common to note the index n as a subscript for example, X_n. The example above is a discrete-space, discrete-index process. Following are examples of the three other possible combinations of discrete- or continuous-state spaces and index sets. Sample realizations of each are illustrated in Figure 3.1.

- *Continuous State and Continuous Index.* In a typical random vibration situation, an accelerometer is mounted on a component in a missile. The acceleration measured at time t is a random process, $\{X(t) \in (-\infty, +\infty), t \in (0, +\infty)\}$.

- *Continuous State and Discrete Index.* The acceleration $X(t)$ is digitized, that is, converted to $\{X_n \in (-\infty, +\infty), n \in (0, 1, 2, \ldots)\}$, a series of values Δt apart: $X_0 = X(0)$, $X_1 = X(\Delta t)$, $X_2 = X(2 \Delta t)$, $\ldots$, $X_n = X(n \Delta t)$.

- *Discrete State and Continuous Index.* We are concerned with the number of times the acceleration crosses a critical value or threshold, for example, $x = 0$. Let t be the elapsed time from the start of our observation and let $\{N(t) \in (0, 1, 2, \ldots), t \in (0, \infty)\}$ be the random process that is the count of the random number of threshold crossings in the interval $(0, t]$.

In the material to follow, we adopt the notation $X(t)$ for the random process of interest and do not make a distinction between continuous- and discrete-state spaces or index sets unless it is necessary.

3.1.2 Ensembles and Ensemble Averages

Figure 3.2 shows a few sample realizations of an arbitrary continuous-state, continuous-index random process $X(t)$. The set of all possible sample realiza-

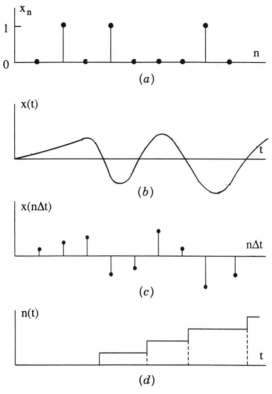

Figure 3.1 Sample realizations of random processes to illustrate four combinations of state spaces and index sets: (a) discrete-state space; discrete-index set; (b) continuous, continuous; (c) continuous, discrete; (d) discrete, continuous.

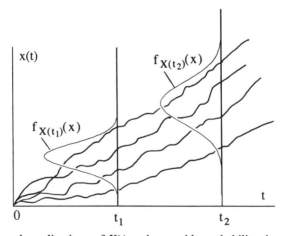

Figure 3.2 Sample realizations of $X(t)$ and ensemble probability density functions at t_1 and t_2.

tions of the random process is called the ensemble, literally, the process viewed as a whole. Simply put, the goal of the study of random processes is to understand the joint probability distributions of $X(t)$ at all times. The distribution of $X(t)$ at any given t, $f_{X(t)}(x)$, is called a marginal distribution of the process. It is evident from the figure that the complete probabilistic description of $X(t)$, including the mean, variance, and marginal probability distributions, changes or evolves with time.

If we know the distribution of $X(t)$, we may calculate the statistics of the ensemble, such as the mean and variance. The mean of the process at time t, $\mu_X(t)$, is calculated by taking the expected value across the ensemble, that is the ensemble average, using the marginal density function:

$$\mu_X(t) = E[X(t)] = \int_{-\infty}^{+\infty} x f_{X(t)}(x)\, dx \tag{3.1}$$

Similarly, the variance at time t, $\sigma_X^2(t)$, is calculated across the ensemble as:

$$\sigma_X^2(t) = E[\{X(t) - \mu_X(t)\}^2] = \int_{-\infty}^{+\infty} [x - \mu_X(t)]^2 f_{X(t)}(x)\, dx \tag{3.2}$$

Higher order moments can be calculated similarly. Note that these ensemble averages are taken at a particular point in time. The average over time, or temporal average, is a different kind of average that we discuss later in this section.

The joint distribution of $X(t)$ at all times t is the ultimate product of the study of random processes. Any number of times may be considered for a joint distribution, for example, $X(t_1)$, $X(t_2)$, $X(t_3)$, . . . , $X(t_n)$, but the most practical case is to consider $X(t)$ at two times, t_1 and t_2. As seen in Chapter 2, the correlation between two random variables is the most important tool for characterizing their joint distribution. Likewise, the most important joint measure in a random process is the correlation of the process with itself at two different times, $X(t_1)$ with $X(t_2)$. Denoted $R_X(t_1, t_2)$, this measure of correlation is called the autocorrelation function and is calculated by

$$R_X(t_1, t_2) = E[X(t_1)X(t_2)] = \int_{-\infty}^{+\infty} x_1 x_2 f_{X(t_1)X(t_2)}(x_1, x_2)\, dx \tag{3.3}$$

Figure 3.3 illustrates the calculation.

Related to the autocorrelation function, the autocovariance function $\sigma_{XX}(t_1, t_2)$ is defined as

$$\sigma_{XX}(t_1, t_2) = E[\{X(t_1) - \mu_X(t_1)\} \{X(t_2) - \mu_X(t_2)\}]$$

$$= E[X(t_1)X(t_2)] - \mu_X(t_1)\mu_X(t_2)$$

$$= R_X(t_1, t_2) - \mu_X(t_1)\mu_X(t_2) \tag{3.4}$$

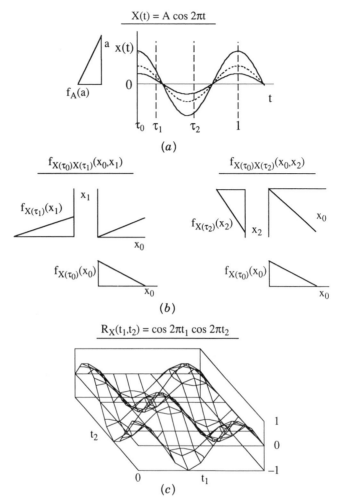

Figure 3.3 Example calculation of autocorrelation function $R_X(t_1, t_2)$: (a) sample realizations of $\{X(t) = A \cos 2\pi t: A > 0, t > 0, \sigma_A^2 + \mu_A^2 = 1\}$; (b) joint probability density functions of $X(t)$ at two pairs of time (τ_0, τ_1) and (τ_0, τ_2); (c) calculated $R_X(t_1, t_2) = \cos 2\pi t_1 \cos 2\pi t_2$.

and the autocorrelation coefficient function $\rho_{XX}(t_1, t_2)$ is defined as

$$\rho_{XX}(t_1, t_2) = \frac{\sigma_{XX}(t_1, t_2)}{\sigma_X(t_1)\sigma_X(t_2)} \tag{3.5}$$

Note that the variance function is contained within the autocovariance function because when $t_1 = t_2 = t$ we have

$$\sigma_{XX}(t, t) = E[\{X(t) - \mu_X(t)\}^2] = \sigma_X^2(t) \tag{3.6}$$

The autocorrelation function, autocovariance function, and autocorrelation coefficient function all provide measures of the linear dependence of $X(t_1)$ and $X(t_2)$. In general, this dependence varies with the particular times t_1 and t_2.

Finally, we note that the autocorrelation function is a positive definite function.[1] That is, if α_j, $j = 1, 2, \ldots, n$, are arbitrary real numbers, then

$$\sum_{j=1}^{n} \sum_{k=1}^{n} \alpha_j \alpha_k R_X(t_j, t_k) \geq 0 \qquad (3.7)$$

where the t_j are arbitrary times. This can be shown by letting Y be a random variable given by

$$Y = \alpha_1 X(t_1) + \alpha_2 X(t_2) + \cdots + \alpha_n X(t_n) = \sum_{j=1}^{n} \alpha_j X(t_j) \qquad (3.8)$$

The square of Y and its expected value must be nonnegative, that is, greater than or equal to zero. The expected value of Y^2 is

$$E[Y^2] = E\left[\left(\sum_{j=1}^{n} \alpha_j X(t_j) \right) \left(\sum_{k=1}^{n} \alpha_k X(t_k) \right) \right]$$

$$= E\left[\sum_{j=1}^{n} \sum_{k=1}^{n} \alpha_j \alpha_k X(t_j) X(t_k) \right]$$

$$= \sum_{j=1}^{n} \sum_{k=1}^{n} \alpha_j \alpha_k E[X(t_j) X(t_k)]$$

$$= \sum_{j=1}^{n} \sum_{k=1}^{n} \alpha_j \alpha_k R_X(t_j, t_k) \geq 0 \qquad (3.9)$$

This is the result we wished to show.

Example 3.1: Simple Example Random Processes—Definition. Sample functions of each of these examples are shown in Figures 3.4–3.7.

1. Let $X(t) = C$, where C is a random variable with distribution $f_C(c)$ (Figure 3.4). The process is actually a random constant, which does not change with time and is a degenerate process. The mean, variance, and autocorrelation functions of the process are constants:

$$\mu_X(t) = \mu_C$$

$$\sigma_X^2(t) = \sigma_C^2$$

$$R_X(t_1, t_2) = \sigma_C^2 + \mu_C^2$$

---

[1] Strictly speaking, it is nonnegative definite because the sum in Eq. (3.7) could be equal to zero.

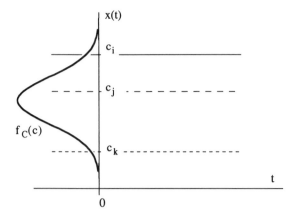

Figure 3.4 Sample realizations of $X(t) = C$, where C is a random variable.

2. Let $X(t) = Bt$, where B is a random variable with distribution $f_B(b)$ (Figure 3.5). The sample realizations of the process are lines of random slope through the origin of $x(t)$ and t. In this case, the parameters of the process depend on t:

$$\mu_X(t) = \mu_B t$$

$$\sigma_X^2(t) = \sigma_B^2 t^2$$

$$R_X(t_1, t_2) = (\sigma_B^2 + \mu_B^2)t_1 t_2$$

3. Let $X(t) = A \cos \omega t$, where A is a random amplitude with distribution $f_A(a)$ (Figure 3.6). Each sample realization of the process is a cosine

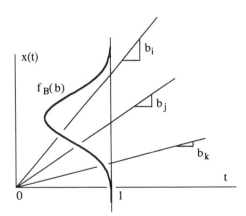

Figure 3.5 Sample realizations of $X(t)$ = Bt, where B is a random variable.

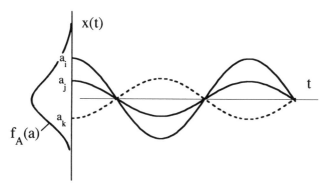

Figure 3.6 Sample realizations of $X(t) = A \cos \omega t$, where A is a random variable.

function having a random amplitude but the same phase as the others. The parameters depend on t:

$$\mu_X(t) = \mu_A \cos \omega t$$

$$\sigma_X^2(t) = \sigma_A^2 \cos^2 \omega t$$

$$R_X(t_1, t_2) = (\sigma_A^2 + \mu_A^2) \cos \omega t_1 \cos \omega t_2$$

Note that, for the purposes of the ensemble expectation, $\cos \omega t$ is treated as a constant.

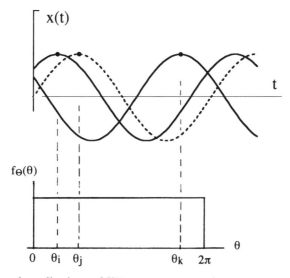

Figure 3.7 Sample realizations of $X(t) = \alpha \cos(\omega t - \Theta)$, where α is a constant and Θ is a random variable.

4. Let $X(t) = \alpha \cos(\omega t - \Theta)$, where α is a constant amplitude and Θ is a random phase with a uniform distribution between 0 and 2π (Figure 3.7). Each sample realization of the process is a cosine function having the same amplitude as the others but a random phase. The parameters are

$$\mu_X(t) = 0$$

$$\sigma_X^2(t) = \tfrac{1}{2}\alpha^2$$

$$R_X(t_1, t_2) = \tfrac{1}{2}\alpha^2[\cos \omega t_1 \cos \omega t_2 + \sin \omega t_1 \sin \omega t_2]$$

3.1.3 Stationary Processes

The probability distributions of all physically realizable random processes evolve with respect to the index parameter. Practically, this means the mean and variance of the process depend on t and the autocorrelation function depends on both t_1 and t_2. However, for many situations in random vibrations, the probability distributions do not appear to evolve over the time intervals of interest to the engineer. In these cases, the process is called stationary, and important simplifications in the mathematical analysis may be made.

Formally, a process is strictly stationary if the joint probability distribution of $\{X(t_1), X(t_2), X(t_3), \ldots, X(t_n)\}$ is identical to the joint distribution of the same variables displaced an arbitrary amount of time, h, that is, to $\{X(t_1 + h), X(t_2 + h), X(t_3 + h), \ldots, X(t_n + h)\}$. This is a difficult condition to demonstrate in practice.

A more relaxed form of stationarity occurs when the mean and variance of the process are constants, independent of t, and the autocorrelation function depends only upon the difference between t_1 and t_2. $R_X(t_1, t_2) = R_X(t_2 - t_1)$. Then the process is called weakly stationary (or covariance stationary or stationary in the wide sense). The mean and variance are expressed as constants:

$$\mu_X(t) = \mu_X \tag{3.10}$$

$$\sigma_X^2(t) = \sigma_X^2 \tag{3.11}$$

The stationary autocorrelation function is written in terms of the difference between t_1 and t_2, known as the lag, $\tau = t_2 - t_1$:

$$R_X(\tau) = R_X(t, t + \tau) = E[X(t)X(t + \tau)] \tag{3.12}$$

The stationary autocovariance and autocorrelation coefficient functions are given by

$$\sigma_{XX}(\tau) = R_X(\tau) - \mu_X^2 \tag{3.13}$$

$$\rho_{XX}(\tau) = \frac{\sigma_{XX}(\tau)}{\sigma_X^2} \tag{3.14}$$

The positive definiteness of a stationary autocorrelation function is stated as

$$\sum_{j=1}^{n} \sum_{k=1}^{n} \alpha_j \alpha_k R_X(t_k - t_j) \geq 0 \qquad (3.15)$$

The significance of this property is that the Fourier transform of the autocorrelation function can be shown to be positive. This is shown to be an important property in Chapter 5.

Stationarity is a useful simplification in modeling. However, we never encounter true stationary processes, because stationarity implies that a process continues from the infinite past to the infinite future. All physically realizable random processes have beginnings and endings and are nonstationary. So when we apply stationarity in our analyses, we really mean that the results we obtain are adequately accurate for our purposes.

Practically, the application of stationarity depends on the problem. One often wishes to know the response of a structure or components in an environment. If the environment persists long enough for the response to reach a "steady state," stationarity is commonly applied. For example, an offshore platform may have a first natural period several seconds long; for it, a stationary environment might last for as short as a few minutes and could be hours or days long. On the other hand, an aerospace component may have a first natural period measured in hundredths or thousands of a second; for it, a stationary environment might be less than a second long. Judgment must be applied.

Random vibrations that must be treated as nonstationary generally fall into three kinds: (1) transient random processs, such as due to shock or impact, which start and end in quiescent states; (2) random processes driven by deterministically varying phenomena, such as due to the spin up of rotating machinery; and (3) random processes more loosely coupled to external phenomena that evolve over time, such as due to an aircraft performing a maneuver. Special analytical procedures have been developed for each kind of nonstationary vibration, but these are outside the scope of our presentation. The interested reader is referred to the growing body of specialist literature and research found in conferences and journals.

Example 3.2: Simple Example Random Processes—Stationary

1. The trivial process $X(t) = C$ is stationary and independent of τ.
2. The process $X(t) = Bt$ is nonstationary.
3. The process $X(t) = A \cos \omega t$ is nonstationary. However, since the ensemble parameters are periodic with period $2\pi/\omega$, the process is periodically stationary.
4. The process $X(t) = \alpha \cos (\omega t - \Theta)$ is stationary and its autocorrelation function may be written as

$$R_X(\tau) = \tfrac{1}{2}\alpha^2 \cos \omega\tau$$

3.1.4 Ergodic Processes

Another important simplification in the analysis of a random process may be made if the process is ergodic. An ergodic process is a process for which the averages taken across the ensemble are the same as the averages taken in time from any single sample realization.

Let $X(t, k)$ be the kth sample realization of a random process $X(t)$. The temporal average of $X(t, k)$ is defined as

$$\langle X(t, k) \rangle = \lim_{T \to \infty} \frac{1}{2T} \int_{-T}^{+T} X(t, k) \, dt \tag{3.16}$$

The symbol $\langle \cdot \rangle$ indicates a temporal average. A process is ergodic in the mean if

$$\langle X(t, k) \rangle = E[X(t)] = \mu_X \tag{3.17}$$

Similarly, a process is ergodic in the variance if

$$\langle \{X(t, k) - \mu_X\}^2 \rangle = E[\{X(t) - \mu_X\}^2] = \sigma_X^2 \tag{3.18}$$

And a process is ergodic in the autocorrelation function if

$$\langle X(t, k)X(t + \tau, k) \rangle = E[X(t)X(t + \tau)] = R_X(\tau) \tag{3.19}$$

A process that satisfies at last these three ergodic conditions is called weakly ergodic. A process for which all possible ensemble averages are equal to the corresponding temporal averages is called strongly ergodic. A process that is not ergodic is called nonergodic.

Note that, for a process to be ergodic, it must be stationary. However, the converse is not true; a stationary process is not necessarily ergodic. The conditions under which a stationary process is also ergodic are the subject of the ergodic theorem. The interested reader will find this discussed briefly in Appendix A. For our purposes here, it is sufficient to say that many practical processes encountered in random vibrations are at least weakly ergodic.

Many other practical problems in random vibrations involve random processes that are nonergodic but have both ergodic and nonergodic aspects. For example, the random vibration of a representative component may be ergodic, but from one component to another there might also be significant variation of parameters, such as support stiffness and damping, which makes the response of each component different. Nonergodic variation could easily dominate a problem if it is especially sensitive to small parameter variations or if the variations are large and not well controlled.

The study of such problems is outside our scope. The interested reader is referred to the growing body of literature on probabilistic or reliability-based design (e.g., Madsen, Krenk, and Lind, 1986; Thoft-Christiansen and Baker, 1982; or Ang and Tang, 1984).

Example 3.3: Simple Example Random Processes—Ergodic

1. The trivial process $X(t) = C$ is stationary but nonergodic. It is apparent that the time average of every outcome merely gives the constant to which that sample is equal and sheds no light on the probability distribution of C.
2. The processes $X(t) = Bt$ and $X(t) = A \cos \omega t$ are nonstationary and therefore cannot be ergodic.
3. The process $X(t) = \alpha \cos (\omega t - \Theta)$ is stationary and ergodic. Each sample realization of this process is identical except for phase.

3.1.5 Gaussian Processes

A random process for which the joint distribution of $\{X(t_1), X(t_2), X(t_3), \ldots, X(t_n)\}$ is a joint normal distribution is called a Gaussian, or normal, process. Powerful simplifications are possible if the process is Gaussian:

- A normal random variable is completely characterized by its mean μ_X and variance σ_X^2; as was shown in Chapter 2, all higher moments of the density function are dependent on the variance. By Eq. (2.89), all joint probability distributions of a Gaussian random process are completely determined once the mean and autocorrelation functions are known.
- It follows that, if a Gaussian process is weakly stationary, it is also strictly stationary. Furthermore, if a Gaussian process is weakly ergodic, it is also strongly ergodic.
- The derivatives and integrals of a Gaussian process are also Gaussian processes. The rationale for this is presented subsequently.

As with the concept of stationarity and ergodicity, judgment must be applied in assuming a process to be Gaussian. A quick check of the histogram of sampled data is usually sufficient to confirm a Gaussian marginal distribution. In fact, it is often difficult to statistically prove that a process is not Gaussian.

There are, however, many physical circumstances for which a Gaussian process may not be a good model. One example is the height of ocean waves in shallow water; these appear to be trochoidal, with flat troughs and peaked crests. Another example is the response of a nonlinear system, for example, one with strain hardening or softening stiffness.

If the random process is known to be non-Gaussian and its distributions are known, the equations developed in this text may be adapted to the correct distributions. Otherwise, more advanced methods, such as discussed in Lin (1967), Nigam (1983), or Soong and Grigoriu (1993), are required.

Example 3.4: Simple Example Random Processes—Gaussian

1. The trivial process $X(t) = C$ is Gaussian if C has a Gaussian distribution.
2. The process $X(t) = Bt$ is Gaussian if B has a Gaussian distribution.

3. The process $X(t) = A \cos \omega t$ is Gaussian if A has a Gaussian distribution.
4. The process $X(t) = \alpha \cos (\omega t - \Theta)$ is non-Gaussian, regardless of the distribution of θ.

3.2   CORRELATION

This section discusses the autocorrelation function in greater detail and provides a physical understanding of the information it contains. The cross-correlation function is defined and applied to the linear transformation of a random process. The relationship between a stationary random process and its derivative is explored.

3.2.1   Characteristics of Autocorrelation Function

The correlation function was introduced by physicist Geoffrey I. Taylor (1920) to describe the statistical nature of the turbulent flow of a fluid. Correlation functions continue to be fundamental tools in the mathematical theory of turbulence and other nonlinear phenomena, as well as in the study of random processes. Let us consider the general shape of the autocorrelation function $R_X(\tau)$ of a stationary random process.

Bounds. For $X(t)$, a stationary random process, we may rewrite the autocorrelation function as

$$R_X(\tau) = E[X(t)X(t + \tau)] = \sigma_{XX}(\tau) + \mu_X^2 = \rho_{XX}(\tau)\sigma_X^2 + \mu_X^2 \quad (3.20)$$

Since $\rho_{XX}(\tau)$ is a correlation coefficient, its value is always between $+1$ and -1, that is, $|\rho_{XX}(\tau)| \leq 1$. It immediately follows that the autocorrelation function is bounded by

$$-\sigma_X^2 + \mu_X^2 \leq R_X(\tau) \leq \sigma_X^2 + \mu_X^2 \quad (3.21)$$

The maximum value is the mean-square value and is found at $\tau = 0$. This may be seen by observing that, at $\tau = 0$,

$$R_X(0) = E[X(t)^2] = \sigma_X^2 + \mu_X^2 \quad (3.22)$$

Thus

$$|R_X(\tau)| \leq R_X(0) \quad (3.23)$$

Symmetry. The autocorrelation function is symmetric about $\tau = 0$. This may be seen by substituting $s = t + \tau$:

$$R_X(\tau) = E[X(t)X(t + \tau)] = E[X(s - \tau)X(s)] = R_X(-\tau) \quad (3.24)$$

Limiting Value. For most random processes of interest in random vibrations, the correlation coefficient goes to zero for large values of τ. That is, the values of $X(t)$ and $X(t + \tau)$ become uncorrelated when widely separated in time. So, for these processes, the limit for large lags is equal to the square of the mean value:

$$\lim_{\tau \to \infty} R_X(\tau) = \mu_X^2 \tag{3.25}$$

Scale of Fluctuation. The autocorrelation function also measures how long the correlation persists in the process: The faster an autocorrelation function decays to zero, the less the sample realizations of the process remain correlated. A numerical measure of the persistence of correlation in a random process is given by the area under the autocorrelation coefficient function, called the scale of fluctuation θ:

$$\theta = \int_{-\infty}^{\infty} \rho_X(\tau) \, d\tau = \frac{1}{\sigma_X^2} \int_{-\infty}^{\infty} R_X(\tau) \, d\tau \tag{3.26}$$

The scale of fluctuation has units of time lag τ. For time lags smaller than θ, we may expect significant correlation, and for time lags much larger than θ, we expect very little correlation.

The reader should beware that the scale of fluctuation may not be interpreted as the width of an equivalent correlation coefficient function of unit height:

$$\rho_X(\tau) = \begin{cases} 1 & |\tau| < \frac{1}{2}\theta \\ \frac{1}{2} & |\tau| = \frac{1}{2}\theta \\ 0 & \text{otherwise} \end{cases} \tag{3.27}$$

This would-be autocorrelation function is invalid because it is not positive definite.

A typical autocorrelation function with the characteristics shown above is shown in Figure 3.8. These conditions, plus positive definiteness, form the constraints on a function to be a valid autocorrelation function. These constraints are sometimes difficult to satisfy in measured data due to numerical errors.

For reasons that are discussed in Chapter 5, it is customary when analyzing measured data to shift the mean value of the process to zero. This has no effect on the analysis of the variations about the mean value and the mean value can always be added back later when needed. If the mean value of $X(t)$ is zero ($\mu_X = 0$), the autocorrelation function and the autocovariance function are the same.

It is also often customary to further normalize the data by dividing by its standard deviation to improve the numerical analysis. Thus, the process may

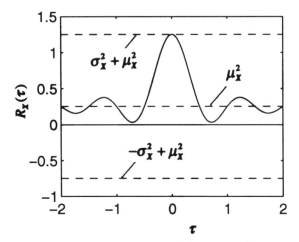

Figure 3.8 General features of autocorrelation function $R_X(\tau)$ for stationary random process.

be standardized, just as a random variable is standardized, by subtracting its mean and dividing by its standard deviation. For a standardized process, the autocorrelation function, the autocovariance function, and the autocorrelation coefficient function are all the same. The reader will find all three referred to as the autocorrelation function in different texts, which could result in confusion if not careful.

Example 3.5: Autocorrelation Function of Low-Pass Random Process. An important autocorrelation function has the form given by the formula

$$R_X(\tau) = \sigma_X^2 \frac{\sin (\omega_c \tau)}{\omega_c \tau}$$

where ω_c is called the cutoff frequency. This autocorrelation function may be considered the primary building block of all stationary random processes. (The function sin $(x)/x$ is sometimes called the sinc function.) Note that it is equal to σ_X^2 at $\tau = 0$, has its first zero crossing at $\omega_c \tau = \pi$ (or $\tau = \pi/\omega_c$), and has subsequent zeros at every $\omega_c \tau = n\pi$, $n = 1, 2, \ldots$. Figure 3.9 shows this autocorrelation function and Figure 3.10 shows a sample realization.

3.2.2 Cross-Correlation Function and Linear Transformations

Let us define a function that measures the correlation between two different random processes, say $X(t)$ and $Y(t)$. This function is called the cross-correlation function $R_{XY}(t_1, t_2)$ and is defined as

$$R_{XY}(t_1, t_2) = E[X(t_1)Y(t_2)] \tag{3.28}$$

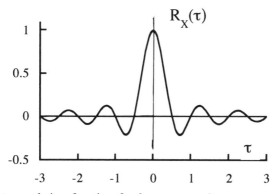

Figure 3.9 Autocorrelation function for low-pass random process with cutoff frequency $f_0 = 1$ Hz.

If both processes are stationary, the cross-correlation function depends only upon the time lag between t_1 and t_2, $\tau = t_2 - t_1$:

$$R_{XY}(\tau) = E[X(t)Y(t + \tau)] \tag{3.29}$$

The order of the subscripts in $R_{XY}(\tau)$ is important. Reversing the order of subscripts defines a second cross-correlation function, $R_{YX}(\tau)$:

$$R_{YX}(\tau) = E[Y(t)X(t + \tau)] \tag{3.30}$$

The two cross-correlation functions are not equal. For stationary processes, they are reflections of one another about the origin, which may be seen as

$$R_{XY}(\tau) = E[X(t)Y(t + \tau)] = E[X(s - \tau)Y(s)] = E[Y(s)X(s - \tau)]$$
$$= R_{YX}(-\tau) \tag{3.31}$$

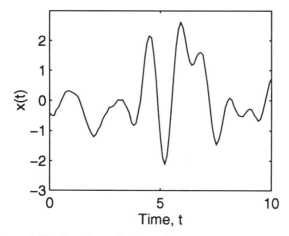

Figure 3.10 Sample realization of low-pass random process.

The cross-correlation function $R_{XY}(\tau)$ is bounded by

$$-\sigma_X\sigma_Y + \mu_X\mu_Y \leq R_{XY}(\tau) \leq \sigma_X\sigma_Y + \mu_X\mu_Y \qquad (3.32)$$

The maximum value of this function does not usually occur at $\tau = 0$. The time lag at which the maximum occurs often has some physical significance.

We consider below the effect of a simple time delay and of the sum of two random processes on the auto- and cross-correlation functions.

Delay and Attenuation. Consider a stationary random process $X(t)$ and let it be subject to some delay δ and attenuation by a factor a. This might be the case, for example, if $X(t)$ is an electrical signal transmitted along a long wire. Then $Y(t)$ is given by

$$Y(t) = aX(t - \delta) \qquad (3.33)$$

The cross-correlation function is

$$R_{XY}(\tau) = E[X(t)Y(t + \tau)] = E[X(t)\{aX(t - \delta + \tau)\}]$$

$$= aR_X(\tau - \delta) \qquad (3.34)$$

In this case, the cross-correlation function is simply the autocorrelation function shifted by δ and scaled by a. The autocorrelation of $Y(t)$ is given by:

$$R_Y(\tau) = E[Y(t)Y(t + \tau)] = E[\{aX(t - \delta)\}\{aX(t - \delta + \tau)\}]$$

$$= a^2R_X(\tau) \qquad (3.35)$$

As should be expected, $Y(t)$ has the same autocorrelation function as $X(t)$, except the total mean-square is reduced by a^2.

Sum of Two Processes. Let $Z(t)$ be the sum of two stationary random processes $X(t)$ and $Y(t)$:

$$Z(t) = X(t) + Y(t) \qquad (3.36)$$

The cross-correlation function of $Z(t)$ and one of its constituents, say $X(t)$, is given by

$$R_{ZX}(\tau) = E[Z(t)X(t + \tau)] = E[\{X(t) + Y(t)\}X(t + \tau)]$$

$$= R_X(\tau) + R_{YX}(\tau) \qquad (3.37)$$

This result simplifies if $X(t)$ and $Y(t)$ are uncorrelated, that is, $R_{XY}(\tau) = R_{YX}(\tau) = 0$. Then the cross-correlation is simply equal to the autocorrelation function of $X(t)$:

$$R_{ZX}(\tau) = R_X(\tau) \qquad (3.38)$$

The autocorrelation of $Z(t)$ is given by

$$R_Z(\tau) = E[Z(t)Z(t + \tau)] = E[\{X(t) + Y(t)\}\{X(t + \tau) + Y(t + \tau)\}]$$

$$= R_X(\tau) + R_{XY}(\tau) + R_{YX}(\tau) + R_Y(\tau) \qquad (3.39)$$

If $X(t)$ and $Y(t)$ are uncorrelated, the autocorrelation of $Z(t)$ is the sum of the autocorrelation functions of $X(t)$ and $Y(t)$:

$$R_Z(\tau) = R_X(\tau) + R_Y(\tau) \qquad (3.40)$$

Example 3.6: Nondispersive Propagation. Suppose $X(t)$ is a stationary random process that is transmitted as a signal through a transmission channel and the propagation is nondispersive, that is, the propagation speed is not dependent upon frequency. (Examples of nondispersive propagation include radio waves and compression waves in air, water, or other material. Examples of dispersive propagation include surface waves on the ocean and bending waves in a structure.) The signal propagates a distance d through the channel with a speed r. At the other end of the channel the signal is received as $Y(t)$ after a delay d/r and attenuation by a factor a. Also suppose a zero-mean uncorrelated source of noise, $N(t)$, is added to the signal before it is received.

The received signal is defined by

$$Y(t) = aX\left(t - \frac{d}{r}\right) + N(t)$$

The autocorrelation function of $Y(t)$ is given by

$$R_Y(\tau) = a^2 R_X(\tau) + R_N(\tau)$$

The cross-correlation function between $X(t)$ and $Y(t)$ is given by

$$R_{XY}(\tau) = aR_X\left(\tau - \frac{d}{r}\right)$$

In practice, it is often the case that one of the characteristics of the transmission channel, either d or r, is unknown. Then the situation described above is an experiment to determine the other parameter. The cross-correlation function is measured and its maximum value found at, say, τ_m; then the unknown constant is found by solving $\tau_m = d/r$. This, in essence, is how radar works. The interested reader will find more on this subject in Bendat and Piersol (1986).

3.2.3 Derivatives of Stationary Processes

We wish to establish the cross-correlation function between a random process $X(t)$ and its derivative process $\dot{X}(t)$. For our purposes, we assume that each sample function of $X(t)$ is differentiable and that the process $\dot{X}(t)$ is simply the family of derivatives of the sample functions of $X(t)$. This assumption is reasonable for the elementary study of random vibrations but is unnecessarily restrictive for advanced work. In Appendix A, we take a closer look at the existence of the derivative process.

Consider the derivative of the autocorrelation function $R_X(\tau)$ with respect to τ. The expected-value operation is an integral of the product $X(t)X(t + \tau)$ times the density function and is independent of τ. So the order of the differentiation and expectation operations may be exchanged as follows:

$$\frac{d}{d\tau} R_X(\tau) = \frac{d}{d\tau} E[X(t)X(t + \tau)] = E\left[X(t) \frac{d}{d\tau} X(t + \tau)\right]$$

$$= E[X(t)\dot{X}(t + \tau)]$$

$$= R_{X\dot{X}}(\tau) \tag{3.41}$$

The resulting expected value is the definition of the cross-correlation function of $X(t)$ and $\dot{X}(t)$, $R_{X\dot{X}}(\tau)$. Thus, the first derivative of the autocorrelation function of $X(t)$ is the cross-correlation function for $X(t)$ and its derivative $\dot{X}(t)$.

By inspection, we see that the second derivative with respect to τ in Eq. (3.41) gives the cross-correlation function of $X(t)$ and $\ddot{X}(t)$, $R_{X\ddot{X}}(\tau)$:

$$\frac{d^2}{d\tau^2} R_X(\tau) = R_{X\ddot{X}}(\tau) \tag{3.42}$$

However, note that we may change the location of τ in Eq. (3.41) as follows:

$$E[X(t)\dot{X}(t + \tau)] = E[X(t - \tau)\dot{X}(t)] \tag{3.43}$$

and now the second derivative with respect to τ gives

$$\frac{d^2}{d\tau^2} R_X(\tau) = E\left[\frac{d}{d\tau} X(t - \tau)\dot{X}(t)\right] = E[-\dot{X}(t - \tau)\dot{X}(t)]$$

$$= -R_{\dot{X}}(\tau) \tag{3.44}$$

Hence, the second derivative of the autocorrelation function is also equal to minus the autocorrelation function of the derivative process.

The reader may verify that the following are also true:

$$\frac{d^3}{d\tau^3} R_X(\tau) = -R_{\dot{X}\ddot{X}}(\tau) \tag{3.45}$$

$$\frac{d^4}{d\tau^4} R_X(\tau) = R_{\ddot{X}}(\tau) \tag{3.46}$$

An important fact about the derivative of a stationary process may be deduced from Eq. (3.41). Note that $R_X(\tau)$ is an even function, that is, it is symmetric about the origin, and it has a maximum value at the origin. Therefore, the derivative of $R_X(\tau)$ is zero at the origin, positive immediately to the left of the origin, and negative immediately to the right. This means that $R_{X\dot{X}}(\tau)$ is an odd function (antisymmetric) and is equal to zero at the origin. Thus, we have

$$R_{X\dot{X}}(0) = 0 \tag{3.47}$$

which means that any stationary process $X(t)$ and its derivative process $\dot{X}(t)$ are uncorrelated at any given instant of time t.

There is another important point to be made if the process is Gaussian. Let $X(t, k)$ be the kth sample realization of $X(t)$. Its derivative is defined as the limit:

$$\dot{X}(t, k) = \frac{d}{dt} X(t, k) = \lim_{h \to 0} \frac{X(t + h, k) - X(t, k)}{h} \tag{3.48}$$

We see that this is the sum of two normally distributed random variables (one variable is negative). Such a sum is also normally distributed, because the normal distribution is reproductive. Thus, if $X(t)$ is Gaussian, its derivative $\dot{X}(t)$ is also Gaussian. If $X(t)$ is not Gaussian, the distribution of $\dot{X}(t)$ must be determined.

3.3 SOME ESSENTIAL RANDOM PROCESSES

This section presents two random processes that relate directly to our study of random vibrations. The harmonic process forms the basis for the spectral decomposition presented in Chapter 5. The Poisson process is used as a model of the time to failure of a system in Chapter 10.

3.3.1 Harmonic Processes

Let $X(t)$ be a random process defined by

$$X(t) = A \cos \omega t + B \sin \omega t \tag{3.49}$$

where A and B are independent and identically distributed random variables with means equal to zero and variances equal to σ^2. The reader may verify that the mean of $X(t)$ is equal to zero. For our purposes, the probability distribution for A and B is arbitrary, but if A and B were normally distributed, $X(t)$ would be a Gaussian process.

The autocorrelation function of $X(t)$ is equal to

$$R_X(\tau) = \sigma^2 \cos \omega\tau \qquad (3.50)$$

This is shown as follows:

$$R_X(\tau) = E[X(t)X(t + \tau)]$$

$$= E[(A \cos \omega t + B \sin \omega t)\{A \cos \omega(t + \tau) + B \sin \omega(t + \tau)\}]$$

$$= E[A^2]\cos \omega t \cos \omega(t + \tau) + E[B^2]\sin \omega t \sin \omega(t + \tau) \qquad (3.51)$$

The above uses the fact that A and B are independent, so $E[AB] = 0$. Next note that $E[A^2] = E[B^2] = \sigma^2$ and use the trigonometric identity for the cosine of the sum of two angles: $\cos(\alpha \pm \beta) = \cos \alpha \cos \beta \mp \sin \alpha \sin \beta$. This yields

$$R_X(\tau) = \sigma^2 \cos [\omega t - \omega(t + \tau)]$$

$$= \sigma^2 \cos(-\omega\tau)$$

$$= \sigma^2 \cos \omega\tau \qquad (3.52)$$

The last step comes from the fact that the cosine is an even function (symmetric about the origin).

Each sample realization of this process is a sinusoid with a random amplitude and a random phase. The process is stationary but not ergodic. Notice that the autocorrelation function captures the most important information defining the process, namely its mean and variance and the frequency of the harmonic; but all information about the phase of individual realizations is lost.

To expand on the above, let us create a process that is the sum of m of the harmonic processes above, each with its own characteristic frequency and variance. Let $\omega_1, \omega_2, \ldots, \omega_m$ be m distinct frequencies, let $A_1, A_2, \ldots, A_m$ and $B_1, B_2, \ldots, B_m$ be independent random variables all having zero mean, and let each pair of A_k and B_k have common variance equal to σ_k^2. For each k we may define a harmonic process as above:

$$X_k(t) = A_k\cos \omega_k t + B_k\sin \omega_k t \qquad (3.53)$$

Now we form a new random process that is the sum of all the $X_k(t)$:

$$X(t) = \sum_{k=1}^{m} X_k(t) = \sum_{k=1}^{m} (A_k\cos \omega_k t + B_k\sin \omega_k t) \qquad (3.54)$$

Since the coefficients A_k and B_k are independent, the random processes $X_k(t)$ are independent. Thus, the autocorrelation function of $X(t)$ is the sum of the individual autocorrelation functions:

$$R_X(\tau) = \sum_{k=1}^{m} R_{X_k}(\tau) = \sum_{k=1}^{m} \sigma_k^2 \cos \omega_k \tau \qquad (3.55)$$

And the total variance of the process, σ^2, is the sum of the individual variances:

$$\sigma^2 = \sum_{k=1}^{m} \sigma_k^2 \qquad (3.56)$$

Let $p(\omega_k)$ represent the portion of the total variance contributed by the process with frequency ω_k:

$$p(\omega_k) = \frac{\sigma_k^2}{\sigma^2} \qquad \sum_{k=1}^{m} p(\omega_k) = 1 \qquad (3.57)$$

Then we may write the autocorrelation function as

$$R_X(\tau) = \sigma^2 \sum_{k=1}^{m} p(\omega_k) \cos \omega_k \tau \qquad (3.58)$$

Figure 3.11a depicts the coefficients as a function of discrete frequencies.

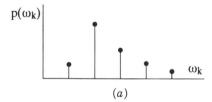

(a)

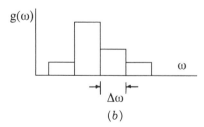

(b)

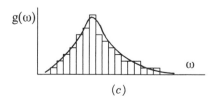

(c)

Figure 3.11 Representation of coefficients of autocorrelation function of sum of harmonic processes: (a) as fractions of total variance of discrete frequencies; (b) as density function over continuous frequencies with $p(\omega_k) = g(\omega_k)\Delta\omega$; ($c$) as limit, $\Delta\omega \to d\omega$.

Now suppose the number of frequencies in the sum becomes infinite, $m \to \infty$, in such a way that their spacing becomes infinitesimally close together, $\omega_{k+1} - \omega_k \to d\omega$. Analogous to the probability density function, the discrete weighting function $p(\omega_k)$ must be replaced by densities $p(\omega_k) \to g(\omega)d\omega$. In Eq. (3.58), the sum representing the autocorrelation function of $X(t)$ must be replaced by an integral:

$$R_X(\tau) = \sigma^2 \int_0^\infty g(\omega)\cos \omega\tau \, d\omega \tag{3.59}$$

Figures 3.11*b* and 3.11*c* depict the coefficients as density functions over continuous frequencies.

The reader may recognize Eq. (3.59) as the Fourier cosine transform of $g(\omega)$. Since Fourier transform pairs are unique, $g(\omega)$ is a unique alternative representation of the autocorrelation function $R_X(\tau)$, carrying exactly the same information about the random process. In Chapter 5, $g(\omega)$ will be formally defined and named the spectral density function because it distributes the variance of $X(t)$ as a density across the frequency spectrum.

3.3.2 Poisson Process

The Poisson process is commonly used to model the "arrival" of events such as cars on a highway, telephone calls to a switch, and earthquakes on a geologic fault. We use it in random vibrations to model the occurrence of events such as a random acceleration crossing a damaging threshold. Figure 3.1*d* shows a sample realization.

Let $N(t)$ be a random process that counts the number of events of interest or "arrivals" occurring in time interval $(0, t]$. It has a discrete sample space and continuous index set, namely, $\{N(t) \in (0, 1, 2, . . .), t \in (0, \infty)\}$. Under the following conditions these arrivals form a Poisson process and the marginal distribution of $N(t)$ is governed by a Poisson distribution:

1. The numbers of arrivals in nonoverlapping intervals are independent random variables. [If $t_1 < t_2 < t_3 < t_4$, then (t_1, t_2) and (t_3, t_4) are nonoverlapping intervals, while (t_1, t_3) and (t_2, t_4) are overlapping.]
2. There exists a positive quantity, say $\lambda > 0$, such that for a small interval of time Δt the following are true:

   a. The probability of exactly one arrival in Δt is proportional to Δt:

   $$P(N(t + \Delta t) = n + 1 | N(t) = n) = \lambda \cdot \Delta t \tag{3.60}$$

   b. the probability of no arrivals in Δt is 1 minus the above:

   $$P(N(t + \Delta t) = n | N(t) = n) = 1 - \lambda \cdot \Delta t \tag{3.61}$$

    c. The probability of more than one arrival in Δt is negligible and vanishes as Δt goes to zero.

3. The process starts at time zero with a count of zero, that is, $N(0) = 0$.

We find the probability distribution of $N(t)$ at time t, $p_N(n, t)$, by setting up and solving a differential equation. Consider the time interval from t to $t + \Delta t$. At the end of the interval, there are two ways in which the count can be n arrivals: Either the interval started with n arrivals and there were no new arrivals in Δt or the interval started with $n - 1$ arrivals and there was one new arrival in Δt. This gives us the following probability equation:

$$p_N(n, t + \Delta t) = P(N(t + \Delta t) = n)$$

$$= P(\{N(t) = n \cap \text{ no new arrivals in } \Delta t\}$$

$$\cup \{N(t) = n - 1 \cap \text{ one new arrival in } \Delta t\})$$

$$= p_N(n, t) \cdot [1 - \lambda \cdot \Delta t] + p_N(n - 1, t) \cdot [\lambda \cdot \Delta t] \quad (3.62)$$

We may rearrange this equation, divide by Δt, and take the limit as Δt goes to zero to get a differential equation:

$$p_N(n, t + \Delta t) - p_N(n, t) = -p_N(n, t) \cdot [\lambda \cdot \Delta t] + p_N(n - 1, t)$$

$$\cdot [\lambda \cdot \Delta t] \quad (3.63)$$

$$\lim_{\Delta t \to 0} \frac{p_N(n, t + \Delta t) - p_N(n, t)}{\Delta t} = -\lambda \cdot p_N(n, t) + \lambda \cdot p_N(n - 1, t) \quad (3.64)$$

$$\frac{d}{dt} p_N(n, t) = -\lambda \cdot p_N(n, t) + \lambda \cdot p_N(n - 1, t) \quad (3.65)$$

Note that, since n is by definition nonnegative, the equation for $n = 0$ is

$$\frac{d}{dt} p_N(0, t) = -\lambda \cdot p_N(0, t) \quad (3.66)$$

The reader may verify that the solution for $n = 0$ is

$$p_N(0, t) = e^{-\lambda t} \qquad t \geq 0 \quad (3.67)$$

The remainder of the solution may be shown by induction to be

$$p_N(n, t) = \frac{(\lambda t)^n e^{-\lambda t}}{n!} \qquad n \geq 0, t \geq 0 \quad (3.68)$$

As stated above, this is a Poisson distribution, where the expected number of arrivals at time t is $\mu_N(t) = \lambda t$.

The autocorrelation function may be shown to be

$$R_N(t_1, t_2) = \begin{cases} \lambda t_1 + \lambda^2 t_1 t_2 & 0 \le t_1 \le t_2 \\ \lambda t_2 + \lambda^2 t_1 t_2 & 0 \le t_2 \le t_1 \end{cases} \tag{3.69}$$

The time between arrivals in a Poisson process can be shown to have an exponential distribution. Let Y be the random time of the first arrival. The event that the first arrival is after time y is the same as the event that there have been no arrivals up to time y. We may write

$$P(Y > y) = P(N(y) = 0) = \frac{(\lambda y)^0 e^{-\lambda y}}{0!} = e^{-\lambda y} \tag{3.70}$$

The cumulative distribution of Y is

$$F_Y(y) = P(Y \le y) = 1 - e^{-\lambda y} \tag{3.71}$$

which is recognized as the exponential distribution.

Thus, the expected time one would have to wait for the first arrival is $1/\lambda$. Since the arrivals are independent, this distribution is also the distribution of the waiting time between arrivals. The exponential distribution is so intimately tied to the Poisson process that an alternate definition of a Poisson process is a process with stationary, independent arrivals with exponential waiting times between arrivals.

PROBLEMS

3.1 Identify and classify the state space and index set for the following random processes:

a Temperature measured hourly at the airport.

b Elevation of sea surface measured continuously at a wave staff

c Daily closing price of a stock

d Number of significant defects in an optical fiber starting at one end

3.2 Explain the "interesting features" of the autocorrelation function in Figure 3.3. What is the autocovariance function of this process and what does it imply?

3.3 Verify the means, variances, and autocorrelation functions given in the four simple example processes in the text.

3.4   Consider a deterministic square wave of period T, $X_{square}(t)$, given by

$$X_{square}(t) = \begin{cases} 1 & 0 < t < T/2 \\ -1 & T/2 < t < T \\ X_{square}(t - T) & t > T \\ X_{square}(t + T) & t < 0 \end{cases}$$

Determine the means, variances, and autocorrelation functions for the following random process variations of this function:

a Square-wave process of random amplitude
b Square-wave process of random phase
c Square-wave process of random amplitude and phase

3.5   For the example process $X(t) = \alpha \cos(\omega t - \Theta)$, derive the stationary autocorrelation function $R_X(\tau)$ from the general autocorrelation function $R_X(t_1, t_2)$.

3.6   Find the scale of fluctuation for the low-pass random process.

3.7   A local average process $Y_T(t)$ is defined by Vanmarcke (1984) as

$$Y_T(t) = \frac{1}{T} \int_{t-T/2}^{t+T/2} X(u)\, du$$

that is, each sample outcome of $Y_T(t)$ is the average of a sample of $X(t)$ over a time segment of length T. Show that the mean and variance of $Y_T(t)$ are

$$\mu_{Y_T}(t) = \mu_X(t)$$

$$\sigma_{Y_T}^2(t) = \frac{2}{T} \int_0^T \left(1 - \frac{\tau}{T}\right) R_X(\tau)\, d\tau$$

3.8   Find the autocorrelation and cross-correlation functions for $W(t) = X(t)Y(t)$, where $X(t)$ and $Y(t)$ are uncorrelated.

3.9   Define the random process $\dot{X}(t)$ for which the sample realizations are the derivatives of the sample realizations of $X(t) = \alpha \cos(\omega t - \Theta)$. Find the cross-correlation function and verify that $X(t)$ and $\dot{X}(t)$ are independent at $\tau = 0$.

3.10 Verify the mean, variance, and autocorrelation functions given for the Poisson process.

Simulation

Many students today have access to a simple pseudo-random-number generator that generates numbers uniformly distributed on (0, 1), e.g., via a calculator, a numerical library on a computer, or a spreadsheet on a personal computer. The following problems are designed to be done on a personal computer. A uniformly distributed random variable is presumed to be in the range (0, 1). For histograms, let the number of bins be $\sqrt{n}$.

3.11 *White noise.* A discrete index version of white noise is easily simulated by computer. Simply generate independent, identically distributed samples of a random variable X_k. For example, generate and save 100 samples of a uniform random variable; repeat 20 times. Calculate and plot the sample mean and variance functions for the ensemble ($n = 20$). Calculate the temporal mean and variance of each sample realization of the process. Calculate the value of the sample autocorrelation function for $t_1 = 0$ and $t_2 = 1$.

3.12 *Autoregressive and Moving-Average Processes.* Let X_k be a discrete-index white noise and define the process Z_k to be

$$Z_k = X_k + a_1 Z_{k-1} + a_2 Z_{k-2}$$

This is known as an autoregressive process of order 2, AR(2), because it is determined by the current value of the input process X_k and its own two most recent values. Another process, W_k, is defined by

$$W_k = X_k + b_1 X_{k-1} + b_2 X_{k-2}$$

This is known as a moving-average process of order 2, MA(2), because it is a weighted average of the current and the two most recent values of X_k. A process that has both autoregressive and moving-average terms is known as an ARMA process. Let $a_1 = b_1 = \frac{1}{2}$ and $a_2 = b_2 = \frac{1}{4}$. Generate sample realizations of Z_k and W_k and qualitatively compare the time histories.

3.13 *Serial Correlation.* Generate one sample realization with length $n = 24$, $k = 0, 1, \ldots, n - 1$, of any stationary discrete-index random process Y_k, for instance $Y_k = A \sin(2\pi k/T) + X_k$, where $A = 1$, $T = 6$, and X_k is a discrete-index white noise. Calculate the sample autocorrelation function, $j = 0, 1, \ldots, n - 1$, for this realization as

$$\hat{R}_Y(j) = \frac{1}{n - j} \sum_{k=1}^{n-j} Y_k Y_{k+j}$$

(This is easily implemented on a spreadsheet by copying the samples from one column to another, shifting the new column down j cells, forming a third column containing the indicated products, and summing down the product column.) Note that for each j the sum is divided by $n - j$, which is the actual number of products that are averaged due to the shifting. Since the number of products averaged decreases as the lag increases, we can expect that the quality of our estimates for large values of j will be relatively poor. In fact, in practice, only the first quarter of the autocorrelation function is considered usable.

3.14 *Sum of Harmonics.* For a project, the advanced student might wish to construct a simulation of the sum of harmonics random process, i.e., generate and add together sine functions with unit amplitude and random phase. Plot the ensemble and temporal histograms of the sample realizations. Experiment to see how many sines must be added together before the process appears to be Gaussian. Experiment with sines of unequal amplitudes. Document your code and work.

4

FOURIER TRANSFORMS

Many readers are already familiar with the decomposition of a periodic function into a Fourier series. This chapter introduces the extension of this approach to the decomposition of nonperiodic functions by the Fourier transform. The basic theorems for doing Fourier transforms are presented and the equivalence of the time and frequency domains is discussed. The further extension of Fourier transforms to include nontraditional generalized functions is presented, the most important being the Dirac delta function. The analysis of measured data using the discrete Fourier transform is discussed in Chapter 13.

There are two types of references on Fourier transforms, tables and texts. In addition to the tables in standard mathematical handbooks, such as Abramowitz and Stegun (1964) and Beyer (1978), there are specialized books of tables, such as Campbell and Foster (1948). Textbooks devoted to the theory and application of Fourier transforms include Bracewell (1978), Wiener (1933), Bochner (1959), Papoulis (1962), and Champeney (1973).

4.1 FOURIER TRANSFORM

The Fourier transform and the Fourier series were invented by Baron Jean-Baptiste-Joseph Fourier in about 1807. Fourier applied these transforms to solve problems in heat conduction. One of his original example problems is an iron ring that has been heated by a fire at one spot. In this problem, Fourier decomposes the temperature distribution around the ring as a series of cosine functions and describes the heat flow around the ring with changes in the amplitudes of the cosines, the rate of change being dependent on the period of the cosine. His work was met with such great skepticism by his contemporaries that he delayed its publication until 1822. In particular, the question of the convergence of a Fourier series for discontinuous functions, for example, a square wave, was in doubt. The convergence issue was finally resolved theoretically by Josiah Willard Gibbs in 1899. Today Fourier series analysis is a common part of the undergraduate engineering curriculum.

This section begins with a review of the Fourier series, then presents the Fourier transform as the limit of a Fourier series of a function with an infinitely long period. The role of symmetry is discussed. Some theorems useful for performing transforms are given.

4.1.1 Fourier Series

A function $g(t)$ is periodic if there is a number T, called the period, for which the function repeats itself identically:

$$g(t) = g(t + T) \tag{4.1}$$

The square wave shown in Figure 4.1a is an example. The sine and cosine are also examples.

Under certain general conditions, a periodic function $g(t)$ can be represented by the sum of an infinite number of sine and cosine functions:

$$g(t) = \frac{a_0}{2} + \sum_{n=1}^{\infty} \left[a_n \cos\left(\frac{2\pi nt}{T}\right) + b_n \sin\left(\frac{2\pi nt}{T}\right) \right] \tag{4.2}$$

where the coefficients a_n and b_n are given by

$$a_n = \frac{2}{T} \int_{-T/2}^{T/2} g(t)\cos\left(\frac{2\pi nt}{T}\right) dt \quad n = 0, 1, \ldots$$

$$b_n = \frac{2}{T} \int_{-T/2}^{T/2} g(t)\sin\left(\frac{2\pi nt}{T}\right) dt \quad n = 1, 2, \ldots \tag{4.3}$$

We say that for every periodic function in the time domain t (that meets the required conditions) there is an equivalent representation as a Fourier series in the frequency domain $2\pi n/T$.

The Fourier series is more compactly expressed using complex coefficients as

$$g(t) = \sum_{n=-\infty}^{\infty} C_n e^{i2\pi nt/T} \tag{4.4}$$

where the complex coefficients C_n are given by

$$C_n = \frac{1}{T} \int_{-T/2}^{T/2} g(t) e^{-i2\pi nt/T} dt \tag{4.5}$$

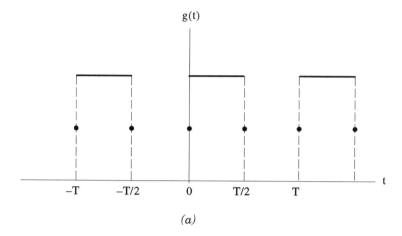

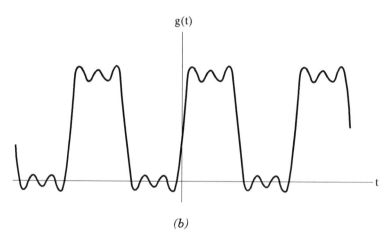

Figure 4.1 Periodic square wave: (a) analytical function; (b) Fourier series approximation with $n = 5$.

To show that these equations are valid, take Eq. (4.4) representing $g(t)$ as a sum, multiply both sides by $\exp(-i2\pi mt/T)$, and integrate over one period of the function:

$$\int_{-T/2}^{T/2} g(t)e^{-i2\pi mt/T}\,dt = \int_{-T/2}^{T/2} \left(\sum_{n=-\infty}^{\infty} C_n e^{i2\pi nt/T} \right) e^{-i2\pi mt/T}\,dt$$

$$= \sum_{n=-\infty}^{\infty} C_n \left(\int_{-T/2}^{T/2} e^{i2\pi(n-m)t/T}\,dt \right) \qquad (4.6)$$

Recall Euler's formula, $e^{i\theta} = \cos\theta + i\sin\theta$, and substitute for the exponential:

$$\int_{-T/2}^{T/2} g(t)e^{-i2\pi mt/T}\,dt = \sum_{n=-\infty}^{\infty} C_n \left[\int_{-T/2}^{T/2} \cos\left(\frac{2\pi(n-m)t}{T}\right) \right.$$
$$\left. + i\sin\left(\frac{2\pi(n-m)t}{T}\right) dt \right] \qquad (4.7)$$

Note that when n and m are different, the integrals of $\cos\theta$ and $\sin\theta$ are over a full period and are equal to zero. The only nonzero term in the sum arises from the cosine term when n and m are equal, for which $\cos 0 = 1$ and the integral is equal to T. Thus, we have

$$\int_{-T/2}^{T/2} g(t)e^{-i2\pi mt/T}\,dt = C_m T \qquad (4.8)$$

Solving for C_m gives

$$C_m = \frac{1}{T}\int_{-T/2}^{T/2} g(t)e^{-i2\pi mt/T}\,dt \qquad (4.9)$$

which was to be shown.

The conditions under which it is possible to find a Fourier series are known as the Dirichlet conditions:

- The function $g(t)$ is continuous over its period, except possibly at a finite number of points at which it may have finite jumps.
- The function $g(t)$ has a finite number of maxima and minima over its period.

The sum of the Fourier series converges to $g(t)$ everywhere except where $g(t)$ has discontinuities. At these discontinuities, the sum converges to the average value of the left-hand and right-hand limits, that is, $\frac{1}{2}[g(t^+) + g(t^-)]$.

Example 4.1: Square Wave. The square wave with period T shown in Figure 4.1a may be written as an equation as

$$g(t) = \begin{cases} 0 & -\frac{1}{2}T < t < 0 \\ 1 & 0 < t < \frac{1}{2}T \\ \frac{1}{2} & t = -\frac{1}{2}T,\, 0,\, \frac{1}{2}T \\ g(t+T) & \text{otherwise} \end{cases}$$

Note that at the discontinuities, $g(t)$ is equal to the average value of the left and right sides of the discontinuities. The first Fourier coefficient a_0 is equal to twice the average value of the function: $a_0 = 1$. The coefficients a_n for the cosine terms are all equal to zero, since the positive and negative contributions over the half period from 0 to $\frac{1}{2}T$ cancel. The coefficients b_n for the sine terms may be shown to be

$$b_n = \begin{cases} \dfrac{2}{n\pi} & n \text{ odd} \\ 0 & n \text{ even} \end{cases}$$

The Fourier series is thus

$$g(t) = \frac{1}{2} + \frac{2}{\pi} \sin\left(\frac{2\pi t}{T}\right) + \frac{2}{3\pi} \sin\left(\frac{6\pi t}{T}\right) + \frac{2}{5\pi} \sin\left(\frac{10\pi t}{T}\right) + \cdots$$

The sum of first four terms of the series (for up to $n = 5$) are shown in Figure 4.1b.

4.1.2 Fourier Transforms

The Fourier transform may be approached as the limit of the Fourier series of a function with an infinitely long period. Below we write $g(t)$ as a Fourier series [Eq. (4.4)], substitute Eq. 4.5 for the coefficients C_n, and let T go to infinity:

$$g(t) = \lim_{T \to \infty} \sum_{n=-\infty}^{\infty} \left(\frac{1}{T} \int_{-T/2}^{T/2} g(t) e^{-i2\pi nt/T} \, dt \right) e^{i2\pi nt/T} \qquad (4.10)$$

Let $\Delta\omega = 2\pi/T$ be the frequency in radians of the periodic function. Substituting $\Delta\omega$, the above equation becomes

$$g(t) = \lim_{T \to \infty} \sum_{n=-\infty}^{\infty} \left(\frac{\Delta\omega}{2\pi} \int_{-T/2}^{T/2} g(t) e^{-in\Delta\omega t} \, dt \right) e^{in\Delta\omega t} \qquad (4.11)$$

Now, we let T go to infinity and let the number of harmonics in the sum, n, go to infinity as well so that $n\,\Delta\omega = \omega$ remains constant. The frequency spacing $\Delta\omega$ becomes infinitiesimally small, $d\omega$. The sum becomes an integral:

$$g(t) = \int_{-\infty}^{\infty} \left(\frac{1}{2\pi} \int_{-\infty}^{\infty} g(t) e^{-i\omega t} \, dt \right) e^{i\omega t} \, d\omega \qquad (4.12)$$

We formally define the Fourier transform pair $g(t)$ and $G(\omega)$ from these nested integrals to be

$$G(\omega) = \frac{1}{2\pi} \int_{-\infty}^{\infty} g(t)e^{-i\omega t}\, dt \qquad (4.13)$$

$$g(t) = \int_{-\infty}^{\infty} G(\omega)e^{i\omega t}\, d\omega \qquad (4.14)$$

The function $G(\omega)$ is called the Fourier transform of $g(t)$, and Eq. (4.13), involving $e^{-i\omega t}$, is known as the forward Fourier transform. Equation (4.14), involving $e^{+i\omega t}$, is known as the inverse Fourier transform. The transform pair is unique, that is, there is a unique correspondence between a function in the time domain t and its transform in the frequency domain ω.

Applying Euler's formula, the Fourier transform is sometimes written as

$$G(\omega) = \frac{1}{2\pi} \int_{-\infty}^{\infty} g(t)\cos(\omega t)\, dt - i\frac{1}{2\pi} \int_{-\infty}^{\infty} g(t)\sin(\omega t)\, dt \qquad (4.15)$$

The first integral alone is known as the Fourier cosine transform and the second integral alone is the Fourier sine transform. Other definitions of the Fourier transform pair are possible. In particular, it is popular to place the factor $1/2\pi$ in the inverse transform rather than the forward transform. The reader should always be careful to note the location of this factor.

There are two general conditions that a function $g(t)$ must meet in order to have a Fourier transform:

- The function must be absolutely integrable, that is, $\int_{-T/2}^{T/2}|g(t)|\, dt < \infty$.
- Any discontinuities must be finite.

At a discontinuity, the same definition of $g(t)$ as for Fourier series holds, that is, $g(t) = \frac{1}{2}[g(t^{+}) + g(t^{-})]$. These conditions are generally met by most analytic functions of interest to us. Exceptions are functions such as $g(t) = \cos(t)$, for which the absolute integral is not finite. The transforms of such functions are handled as limiting cases by using the Dirac delta function discussed later in this chapter.

The conditions for the existence of a Fourier transform are also not met by stationary random processes, because they are not absolutely integrable (they continue forever). We resolve this problem in Chapter 5.

Regarding our notation, if the function is denoted $g(t)$, with a lowercase letter, its Fourier transform will usually be denoted $G(\omega)$, using a capital letter. At times, we will find it convenient to introduce $\mathcal{F}\{\cdot\}$ to mean the operation of taking the Fourier transform, that is, $\mathcal{F}\{g(t)\} = G(\omega)$. In this notation, the inverse Fourier transform is denoted $\mathcal{F}^{-1}\{\cdot\}$, that is, $\mathcal{F}^{-1}\{G(\omega)\} = g(t)$. [This is in the same spirit that $\sin^{-1}(x)$ is the inverse function of $\sin(\theta)$.]

Example 4.2: Rectangle Function (Boxcar). Let $g(t)$ be the rectangular function shown in Figure 4.2 and given by

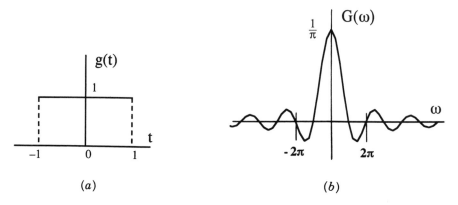

Figure 4.2 Rectangle boxcar function: (*a*) in time domain; (*b*) Fourier transform.

$$g(t) = \begin{cases} 1 & |t| < 1 \\ \frac{1}{2} & |t| = 1 \\ 0 & |t| > 1 \end{cases}$$

This function is sometimes called a boxcar (as in a train) and may be thought of as a rectangular window in time t through which any other function $h(t)$ may be viewed such that the product $g(t) \cdot h(t)$ is

$$g(t) \cdot h(t) = \begin{cases} h(t) & |t| < 1 \\ \frac{1}{2}h(t) & |t| = 1 \\ 0 & |t| > 1 \end{cases}$$

We will find this concept useful later when we discuss data sampling and analysis.

The Fourier transform $G(\omega)$ is found by direct substitution of $g(t)$ into Eq. (4.13):

$$G(\omega) = \frac{1}{2\pi} \int_{-1}^{+1} 1 \cdot e^{-i\omega t} \, dt$$

$$= \frac{\sin \omega}{\pi \omega}$$

This transform is shown in Figure 4.2. Note that the values of $g(t)$ at the points of discontinuity make no contribution to the integral and so are quite arbitrary in defining the forward transform. The reader should verify that the inverse transform [Eq. (4.14)] recovers the original function. It will then be seen that the above definition of $g(t)$ at the points of discontinuity is correct.

4.1.3 Symmetry

An arbitrary function can be uniquely split into two other functions, one symmetric and one antisymmetric about the origin. Certain economies result in Fourier transform theory by taking advantage of these symmetries.

A function is said to be symmetric, or even, if $g(t) = g(-t)$, for example, the cosine function. A function is said to be antisymmetric, or odd, if $g(t) = -g(-t)$, for example, the sine function.

An arbitrary function can be uniquely expressed as the sum of an even function $e(t)$ and an odd function $o(t)$:

$$g(t) = e(t) + o(t) \tag{4.16}$$

where the even and odd parts are defined by

$$e(t) = \tfrac{1}{2}[g(t) + g(-t)] \qquad o(t) = \tfrac{1}{2}[g(t) - g(-t)] \tag{4.17}$$

Direct substitution of Eq. (4.17) into Eq. (4.16) will show that the sum is equal to $g(t)$. That the decomposition is unique can be shown as follows. Suppose that there are two decompositions possible, say,

$$g(t) = e_1(t) + o_1(t) = e_2(t) + o_2(t) \tag{4.18}$$

Rearranging terms, we may write

$$e_1(t) - e_2(t) = o_2(t) - o_1(t) \tag{4.19}$$

But the reader may verify that the sum or difference of two even functions must be even and the sum or difference of two odd functions must be odd. The only way for the above equation to be true is for both sides to be equal to zero. Therefore, $e_1(t) = e_2(t)$, $o_1(t) = o_2(t)$ and the decomposition must be unique.

Now consider the Fourier transform of the decomposition of $g(t)$ into even and odd parts [Eq. (4.16)]:

$$G(\omega) = \frac{1}{2\pi} \int_{-\infty}^{\infty} [e(t) + o(t)]e^{-i\omega t} \, dt$$

$$= \frac{1}{2\pi} \int_{-\infty}^{\infty} e(t)\cos(\omega t) \, dt + \frac{1}{2\pi} \int_{-\infty}^{\infty} o(t)\cos(\omega t) \, dt$$

$$- i\frac{1}{2\pi} \int_{-\infty}^{\infty} e(t)\sin(\omega t) \, dt - i\frac{1}{2\pi} \int_{-\infty}^{\infty} o(t)\sin(\omega t) \, dt \quad (4.20)$$

The integrals of $o(t)\cos(\omega t)$ and $e(t)\sin(\omega t)$ are equal to zero, since the product of an even function and an odd function is odd and the integral of an odd function from $-\infty$ to $+\infty$ is equal to zero. The transform is reduced to

$$G(\omega) = \frac{1}{2\pi} \int_{-\infty}^{\infty} e(t)\cos(\omega t) \, dt - i\frac{1}{2\pi} \int_{-\infty}^{\infty} o(t)\sin(\omega t) \, dt \quad (4.21)$$

Observe that, in the first integral, $\cos(\omega t)$ is an even function of ω, that is, $\cos(\omega t) = \cos(-\omega t)$. So integrating with respect to t for $+\omega$ and $-\omega$ gives the same result and the first integral produces and even function of ω. Conversely, for the second integral, $\sin(\omega t)$ is an odd function of ω, and the integral produces an odd function of ω. Thus, the cosine transform produces the even part of the transform, and the sine transform produces the odd part. We write

$$G(\omega) = E(\omega) - iO(\omega) \quad (4.22)$$

where

$$E(\omega) = \frac{1}{2\pi} \int_{-\infty}^{\infty} e(t)\cos(\omega t) \, dt \quad (4.23)$$

$$O(\omega) = \frac{1}{2\pi} \int_{-\infty}^{\infty} o(t)\sin(\omega t) \, dt \quad (4.24)$$

If $g(t)$ is an even function, its Fourier transform is even, and if $g(t)$ is an odd function, its Fourier transform is odd. Furthermore, if $g(t)$ is real valued, Eq. (4.21) indicates that the real part of the transform is even and the imaginary part is odd; that is, if $G(\omega) = a + ib$, then $G(-\omega) = a - ib$. Thus, $G(-\omega)$ is the complex conjugate of $G(\omega)$, $G^*(\omega)$. This kind of symmetry is called Hermitian.

4.1.4 Basic Theorems

Let functions $g(t)$ and $h(t)$ have Fourier transforms $G(\omega)$ and $H(\omega)$, respectively. The following theorems are useful for deriving Fourier transforms. They may be easily proved by substitution into the definitions [Eqs. (4.13) and (4.14)].

Linearity. Fourier transformation is a linear operation. That is, the transform of a linear combination of functions is the linear combination of their transforms:

$$\mathcal{F}\{a\cdot g(t) + b\cdot h(t)\} = a\cdot G(\omega) + b\cdot H(\omega) \tag{4.25}$$

Shifting. Recall that the tangent of the phase angle of a complex number is the ratio of the imaginary part to the real part. Shifting the origin of $g(t)$ by a constant τ produces a phase shift in the transform. The amount of the phase shift is equal to $-\omega\tau$, which increases in proportion to frequency, that is, the higher frequencies of the transform are shifted more than the lower frequencies:

$$\mathcal{F}\{g(t - \tau)\} = G(\omega)\cdot e^{-i\omega\tau} = G(\omega)\cdot[\cos(\omega\tau) - i\sin(\omega\tau)] \tag{4.26}$$

Scaling (Similarity). Scaling the t axis by α amounts to a compression of the function in t. The resulting transform is stretched to higher frequencies and is decreased in amplitude, but the area under the transform remains the same:

$$\mathcal{F}\{g(\alpha t)\} = \frac{1}{|\alpha|} \cdot G\left(\frac{\omega}{\alpha}\right) \tag{4.27}$$

Shifting and Scaling. When shifting and scaling $g(t)$, first shift the origin by τ, then scale by α. The transform will be the straightforward combination of the two immediately previous results:

$$\mathcal{F}\{g(\alpha(t - \tau))\} = \frac{1}{|\alpha|} \cdot G\left(\frac{\omega}{\alpha}\right) \cdot e^{-i\omega t} \tag{4.28}$$

Transform of a Transform. Suppose we substitute t for ω in $G(\omega)$ so that we have a function $G(t)$. The forward transform of $G(t)$ is not $g(\omega)$, but rather

$$\mathcal{F}\{G(t)\} = \frac{1}{2\pi} g(-\omega) \tag{4.29}$$

Differentiation. The transform of the derivative of $g(t)$ is equal to the product of the transform of $g(t)$ times $i\omega$. From the definition of a derivative as a limit, we have

$$\mathcal{F}\left\{\frac{d}{dt} g(t)\right\} = \frac{1}{2\pi} \int_{-\infty}^{\infty} \frac{d}{dt} g(t)e^{-i\omega t}\, dt$$

$$= \frac{1}{2\pi} \int_{-\infty}^{\infty} \lim_{h\to 0} \frac{g(t + h) - g(t)}{h} e^{-i\omega t}\, dt$$

$$= \lim_{h\to 0} \frac{1}{h}\left\{\frac{1}{2\pi} \int_{-\infty}^{\infty} g(t + h)e^{-i\omega t}\, dt - \frac{1}{2\pi} \int_{-\infty}^{\infty} g(t)e^{-i\omega t}\, dt\right\}$$

$$= \lim_{h\to 0} \frac{1}{h} \{G(\omega)e^{i\omega h} - G(\omega)\}$$

$$= i\omega G(\omega) \tag{4.30}$$

(L'Hopital's rule is used in the last step.) Thus, in the frequency domain, the differentiation of $g(t)$ magnifies the contribution of higher frequencies. Differentiation also shifts the phase of the transform by 90°.

Example 4.3: Application of Fourier Transform Theorems. Suppose we have the function

$$g_0(t) = \frac{\sin t}{\pi t}$$

Notice that this is the Fourier transform of the rectangular time window given in the first example of this chapter, but with t substituted for ω. The Fourier transform may be written directly from Eq. (4.29) as

$$G_0(\omega) = \begin{cases} \dfrac{1}{2\pi} & |\omega| < 1 \\[2mm] \dfrac{1}{4\pi} & |\omega| = 1 \\[2mm] 0 & |\omega| > 1 \end{cases}$$

This may be interpreted as a rectangular window in the frequency domain. Suppose the origin of $g_0(t)$ is shifted to the right by τ, so that

$$g_1(t) = g_0(t - \tau) = \frac{\sin(t - \tau)}{\pi(t - \tau)}$$

The transform is now given by

$$G_1(\omega) = \begin{cases} \dfrac{1}{2\pi} e^{-i\omega\tau} = \dfrac{1}{2\pi} [\cos(\omega\tau) - i\,\sin(\omega\tau)] & |\omega| < 1 \\[2mm] \dfrac{1}{4\pi} e^{-i\omega\tau} = \dfrac{1}{4\pi} [\cos(\tau) - i\,\sin(\tau)] & |\omega| = 1 \\[2mm] 0 & |\omega| > 1 \end{cases}$$

Finally, suppose that $g(t)$ is compressed by α, so that

$$g_2(t) = g_0(\alpha t) = \frac{\sin(\alpha t)}{\pi(\alpha t)}$$

The transform is now given by

$$G_2(\omega) = \begin{cases} \dfrac{1}{2\pi|\alpha|} & |\omega/\alpha| < 1 \\[2mm] \dfrac{1}{4\pi|\alpha|} & |\omega/\alpha| = 1 \\[2mm] 0 & |\omega/\alpha| > 1 \end{cases}$$

Notice that the area under the Fourier transform remains constant: as α increases (implying compression in the time domain), the amplitude of the frequency window decreases, but the width of the window increases.

4.1.5 Correlation, Convolution, and Windowing

Let functions $g(t)$ and $h(t)$ have Fourier transforms $G(\omega) = H(\omega)$, respectively. Three special transformations of the product of $g(t)$ times $h(t)$ are of special importance to our study: the correlation integral, the convolution integral, and the windowing operation. The two integrals differ only slightly in their expression, but the differences in their physical interpretations and transforms are vast.

Correlation. The correlation integral is defined by

$$u(\tau) = \int_{-\infty}^{\infty} g(t)h(t + \tau)\, dt \tag{4.31}$$

This integral can be likened to a correlation coefficient. It describes the linear dependence of one function on the other as a function of time shift τ. (Indeed, the Fourier transform integral itself is actually the correlation between a function and the sine and cosine functions.) Its transform is derived by changing the order of integration as follows:

$$U(\omega) = \frac{1}{2\pi} \int_{-\infty}^{\infty} \left(\int_{-\infty}^{\infty} g(t)h(t + \tau)\, dt \right) e^{-i\omega\tau}\, d\tau$$

$$= \int_{-\infty}^{\infty} g(t) \left(\frac{1}{2\pi} \int_{-\infty}^{\infty} h(t + \tau)e^{-i\omega\tau}\, d\tau \right) dt \tag{4.32}$$

We recognize the inner integral to be the Fourier transform of $h(t + \tau)$. Applying the shift theorem [Eq. (4.26)], we have

$$U(\omega) = \int_{-\infty}^{\infty} g(t)H(\omega)e^{+i\omega\tau}\, dt$$

$$= 2\pi G(-\omega)H(\omega) \tag{4.33}$$

If $g(t)$ and $h(t)$ are real functions, $G(\omega)$ has Hermitian symmetry and $G(\omega) = G^*(\omega)$. The final result is

$$U(\omega) = 2\pi G^*(\omega)H(\omega) \tag{4.34}$$

In words, the Fourier transform of the correlation integral of two functions is equal to the product of the complex conjugate of the transform of the first function times the transform of the second. The factor 2π is required to maintain the proper units.

When $h(t) = g(t)$, the correlation is called the autocorrelation of $g(t)$. Then we have the following Fourier transform pair:

$$u(\tau) = \int_{-\infty}^{\infty} g(t)g(t + \tau)\, dt \tag{4.35}$$

$$U(\omega) = 2\pi|G(\omega)|^2 \tag{4.36}$$

The Fourier transform of the autocorrelation integral of $g(t)$ is proportional to the squared modulus of $G(\omega)$. Be sure to note that this autocorrelation is an integral over time t of a function that meets the criteria to have a Fourier transform, namely, it is absolutely integrable. This result is not true for a stationary random process because a stationary process is not absolutely integrable and has no Fourier transform. Also, the autocorrelation function of a random process is an ensemble average and is not a time integral unless the process is ergodic.

Convolution. The convolution integral is defined by

$$v(\tau) = \int_{-\infty}^{\infty} g(t)h(\tau - t)dt \tag{4.37}$$

This integral appears frequently in engineering because it represents the response of a linear system with impulse response $h(t)$ to an input function $g(t)$. The Fourier transform of the convolution integral, $V(\omega)$, is found by the same steps as above to be

$$V(\omega) = 2\pi G(\omega)H(\omega) \tag{4.38}$$

In words, the Fourier transform of the convolution of two functions in the time domain is equal to the product of the Fourier transforms of the functions in the frequency domain. (The factor 2π is required to maintain the proper units.) The transform of the impulse response, $H(\omega)$, is called the filter function. This result is of great practical importance to us. For example, the acceleration measured by a transducer passes through amplifiers, transmission lines, and so on. The analysis of such a series of filters is greatly simplified by working in the frequency domain. Furthermore, in Chapter 7, we discuss the dynamic response of mechanical systems in the context of linear systems, see that the convolution integral appears, and make use of this result.

Windowing. The windowing operation is defined by

$$w(t) = g(t)h(t) \tag{4.39}$$

Figure 4.2 illustrates a simple rectangular window $g(t)$ through which $h(t)$ may be captured. Other useful window functions are described in Chapter 13.

The Fourier transform of the windowed function $w(t)$ may be derived by applying the transform of a transform theorem [Eq. (4.29)] to the transform of a convolution [Eq. (4.38)]. The result is

$$W(\omega) = \int_{-\infty}^{\infty} G(\theta)H(\omega - \theta) \, d\theta \tag{4.40}$$

In words, the Fourier transform of the windowed function is equal to the convolution of the transforms of the window function and the original function.

4.2 DIRAC DELTA FUNCTION

Some functions that are useful to the study of random vibrations do not meet the conditions required for them to have proper Fourier transforms. The most important of these for random vibrations is the impulse or Dirac delta function. This section defines the delta function and presents its properties, including its Fourier transform. The generalized transforms of the cosine and sine functions are also given.

4.2.1 Definition and Properties of Delta Function

We are interested in describing the behavior of a function $g(t)$ that, in one way or another, violates the criteria to have a Fourier transform. Typically, the function exhibits some sort of transfinite behavior, either acting so suddenly or in such a small space that the details of the function cannot be seen. Or the function may act for so long or over such a great space that for practical purposes the function continues infinitely. In these cases, the requirement for absolute integrability is not met.

We approach such functions by defining a sequence of functions $g_n(t, \theta)$ dependent on some parameter, say θ. This sequence of functions is defined so that, as we let θ go to a limit, typically zero or infinity, the sequence behaves in the limit like the "function" of interest. We refer to the limit of the sequence of functions $g_n(t, \theta)$ as a generalized function.

The most important generalized function for us is the impulse function or Dirac delta function, $\delta(t)$. This may be thought of as the limit of a sequence of rectangular windows with constant area (Figure 4.3):

$$\delta(t) = \lim_{\theta \to 0} g_n(t, \theta) = \begin{cases} \dfrac{1}{\theta} & |t| < \tfrac{1}{2}\theta \\[2mm] \dfrac{1}{2\theta} & |t| = \tfrac{1}{2}\theta \\[2mm] 0 & |t| > \tfrac{1}{2}\theta \end{cases} \tag{4.41}$$

As θ tends to zero, the function's amplitude goes to infinity, but the area, equal to $\theta \cdot 1/\theta$, remains constant. In the limit, we may think of this as a function that is zero everywhere except at the origin, where it is singular. Yet the area under the function is equal to 1. We write

$$\delta(t) = 0 \qquad t \neq 0 \tag{4.42}$$

$$\int_{-\infty}^{\infty} \delta(t)\, dt = 1 \tag{4.43}$$

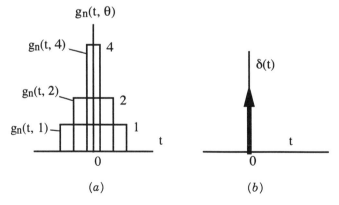

Figure 4.3 Dirac delta function: (a) limit of sequence of unit area rectangles of decreasing width; (b) symbolic representation.

Many other sequences of functions $g_n(t, \theta)$ besides the rectangle may be used to define the delta function. For instance, the Gaussian function is often used, especially since its deriviatives of all orders exist. The precise definition of $g_n(t, \theta)$ is, for our purposes, arbitrary. Once the function has shrunk beyond the limits of our resolving power, the local details of the function do not matter.

The net effect of this function is to contribute a unit "impulse," $\delta(t)dt$, to the system of equations in which it appears. For example, the delta function may be used in describing the probability density function of a mixed random variable, that is, one that is continuous, but also takes on discrete values with finite probability. In Chapter 5, the delta function is used to define the auto-correlation function of white noise and the spectral density function of a "narrow-band" process. In Chapter 7, we use the delta function to determine the response of a dynamic system to an impulsive force, that is, one that suddenly imparts momentum to the system. In the following section, we use the delta function to help express the Fourier transform of functions that ordinarily do not meet the conditions normally required for existence.

The usefulness of the delta function ultimately comes from its behavior inside integrals. When integrated directly, the delta function gives us the Heaviside step function:

$$H(t) = \begin{cases} 0 & t < 0 \\ 1 & t > 0 \end{cases} \tag{4.44}$$

This is suggested by the integral

$$H(t) = \int_{-\infty}^{t} \delta(x) \, dx \tag{4.45}$$

Again, the behavior of the delta function at $t = 0$ is undefined.

The most important use of the delta function is to integrate the product of a function $g(t)$ and $\delta(t)$. This integral is defined by the behavior of the limiting sequence near origin. We write the integal as the limit of the integral of the defining sequence $g_n(t, \theta)$ given in Eq. (4.41):

$$\int_{-\infty}^{\infty} g(t)\delta(t) \, dt = \lim_{\theta \to 0} \int_{-\infty}^{\infty} g(t)g_n(t, \theta) \, dt = \lim_{\theta \to 0} \int_{-\theta/2}^{\theta/2} g(t) \frac{1}{\theta} \, dt \tag{4.46}$$

As θ approaches zero, the area under the integrand is given by the mean-value theorem of calculus as the height of the integrand, $g(0) \cdot 1/\theta$, times the width of the interval, θ. This gives

$$\int_{-\infty}^{\infty} g(t) \, \delta(t) \, dt = \lim_{\theta \to 0} g(0) \frac{1}{\theta} \theta = g(0) \tag{4.47}$$

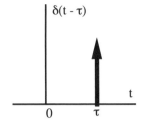

Figure 4.4   Dirac delta function at $t = \tau$.

Thus, the integral of a function times a delta function is equal to the value of the function at $t = 0$:

$$\int_{-\infty}^{\infty} g(t)\delta(t) \, dt = g(0) \tag{4.48}$$

This is known as the sifting property of the delta function. Heuristically, the delta function "sifts" out the value of $g(t)$ at $t = 0$.

The sifting property can be extended by shifting the location of the singularity. Let $\delta(t - \tau)$ be a delta functon that "fires" at $t = \tau$, that is, it is zero everywhere except when $t - \tau = 0$ (Figure 4.4). The sifting property is then

$$\int_{-\infty}^{\infty} g(t)\delta(t - \tau) \, dt = g(\tau) \tag{4.49}$$

A scaling theorem for $\delta(\alpha t)$ can be developed from the limiting sequence or by considering a change of variable as follows. Let $u = \alpha t$ and the equation becomes

$$\int_{-\infty}^{\infty} g(t)\delta(\alpha t) \, dt = \int_{-\infty}^{\infty} g(u/\alpha)\delta(u) \, \frac{du}{|\alpha|}$$

$$= g(0) \cdot \frac{1}{|\alpha|} \tag{4.50}$$

We may think of the following as the definition for the scale and shift of a delta function:

$$\delta(\alpha(t - \tau)) = \frac{1}{|\alpha|} \delta(t - \tau) \tag{4.51}$$

4.2.2   Fourier Transform of Delta Function

Recall that the two conditions for a Fourier transform to exist are that the function must be absolutely integrable, that is, $\int_{-\infty}^{\infty} |g(t)| \, dt < \infty$, and that any discontinuities must be finite. These conditions are violated by certain functions that are of interest to us in random vibrations. Some functions that do not have Fourier transforms in the traditional sense include the following:

- $g(t)$ equal to a constant
- $g(t) = \cos(\omega t)$ or any other harmonic function
- $g(t)$ equal to any periodic function, since a periodic function may be written as a Fourier series of sines and cosines

It is possible for us to handle these functions as generalized functions. If $g_n(t, \theta)$ is a generalized function having a limit of $g(t)$, we define the Fourier transform of $g(t)$, $G(\omega)$, to be the corresponding limit of the sequence of Fourier transforms, $G_n(\omega, \theta)$, of the generalized function.

Consider the rectangular window of width 2θ:

$$g_n(t, \theta) = \begin{cases} 1 & |t| < \theta \\ \dfrac{1}{2} & |t| = \theta \\ 0 & |t| > \theta \end{cases} \tag{4.52}$$

As θ tends to infinity, $g_n(t, \theta)$ is a limiting sequence that becomes $g(t) = 1$. The Fourier transform of $g_n(t, \theta)$ is given by

$$\begin{aligned} G_n(\omega, \theta) &= \frac{1}{2\pi} \int_{-\theta}^{+\theta} 1 \cdot e^{-i\omega t}\, dt \\ &= \frac{\sin(\omega\theta)}{\pi\omega} \end{aligned} \tag{4.53}$$

This transform is shown in Figure 4.5, plotted for several different values of θ. Note that at $\omega = 0$, $G_n(\omega, \theta)$ has a maximum value of θ/π. As θ increases, the oscillations become more rapid, the oscillations die out in less time, and the amplitude at the origin grows larger. The area under the function, however, remains constant, equal to 1. In the limit, this sequence is another defining sequence for the delta function.

Thus, in the limit, we have the following Fourier transform pair:

$$g(t) = 1 \tag{4.54}$$

$$G(\omega) = \delta(\omega) \tag{4.55}$$

In words, the Fourier transform of a constant is a delta function. By the linearity theorem, it is apparent that if $g(t)$ is a constant other than 1, the transform is a delta function multiplied by that constant.

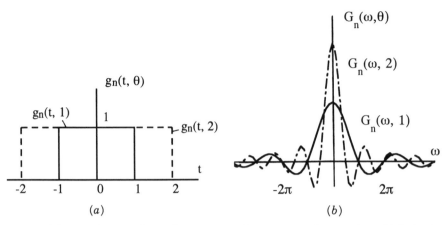

Figure 4.5 Fourier transform of a constant; (a) sequence of rectangles which, in the limit, define a constant; (b) corresponding sequence of Fourier transforms define a Dirac delta function.

By the transform of a transform theorem, we have the following Fourier transform pair:

$$g(t) = \delta(t) \tag{4.56}$$

$$G(\omega) = 1/2\pi \tag{4.57}$$

In words, the Fourier transform of a delta function is a constant, $1/2\pi$.

4.2.3 Transforms of Cosine and Sine Functions

We next show that the cosine and sine functions have Fourier transforms that are pairs of delta functions. Let $g(t)$ be the even impulse pair given by

$$g(t) = \tfrac{1}{2}\delta(t + 1) + \tfrac{1}{2}\delta(t - 1) \tag{4.58}$$

There are two impulses of amplitude $1/2$, one firing at $t = -1$ and one firing at $t = +1$. The Fourier transform is calculated using the sifting property:

$$G(\omega) = \frac{1}{2\pi} \int_{-\infty}^{\infty} \left[\frac{1}{2}\delta(t + 1) + \frac{1}{2}\delta(t - 1) \right] e^{-i\omega t}\, dt = \frac{1}{2\pi} \left[\frac{1}{2}(e^{+i\omega} + e^{-i\omega}) \right]$$

$$= \frac{1}{2\pi} \cos(\omega) \tag{4.59}$$

In words, the Fourier transform of an even impulse pair is a cosine.

By the transform of a transform theorem, the transform pair for a cosine

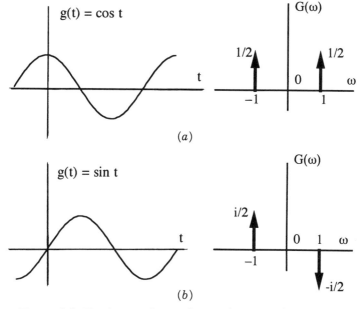

Figure 4.6 Fourier transform pairs: (a) for cos t; (b) for sin t.

function is

$$g(t) = \cos(t) \tag{4.60}$$

$$G(\omega) = \tfrac{1}{2}\delta(\omega + 1) + \tfrac{1}{2}\delta(\omega - 1) \tag{4.61}$$

By the shift theorem, the transform pair for a sine function is

$$g(t) = \sin(t) = \cos(t - \tfrac{1}{2}\pi) \tag{4.62}$$

$$G(\omega) = i\tfrac{1}{2}\delta(\omega + 1) - i\tfrac{1}{2}\delta(\omega - 1) \tag{4.63}$$

The transforms of the cosine and sine functions are shown in Figure 4.6. The reader should verify these formulas. Remember that a delta function times a constant is a weighted delta function.

PROBLEMS

For the problems below, the Fourier transform of a function, for example, $g(t)$, is denoted by the corresponding capital letter, in this case $G(\omega)$.

4.1 Derive the Fourier series for the square wave given in the text.

4.2   Find the Fourier series for the Haversine function $h(t)$, defined by

$$h(t) = \tfrac{1}{2}[1 - \cos(t)] = \sin^2(\tfrac{1}{2}t)$$

4.3   Derive the Fourier transform for the rectangle (boxcar) function given in the text.

4.4   Derive the Fourier transform for the function that is a single cycle of a Haversine:

$$g(t) = \begin{cases} \tfrac{1}{2}[1 - \cos(t)] & 0 < t < 2\pi \\ 0 & \text{otherwise} \end{cases}$$

4.5   Let $g(t)$ be the Gaussian function, $g(t) = e^{-t^2/2}$. Show that its Fourier transform is $G(\omega) = (1/\sqrt{2\pi})\, e^{-\omega^2/2}$. *Hint:* Complete the square using $t^2 + 2i\omega t + (i\omega t)^2 = (t + i\omega)^2$.

4.6   Show that:
   a The sum of two even functions is even.
   b The sum of two odd functions is odd.
   c The product of an even and an odd function is odd.
   d The product of two even functions is even.
   e The product of two odd functions is even.
   f The definite integral from $-\infty$ to $+\infty$ of an odd function is zero.

4.7   Prove the theorems in the text for:
   a Linearity
   b Shifting
   c Scaling
   d Shifting and scaling
   e Transform of a transform

4.8   *Modulation theorem.* Show that $g(t)\cos(\omega_0 t)$ has the Fourier transform $\tfrac{1}{2}G(\omega - \omega_0) + \tfrac{1}{2}G(\omega + \omega_0)$. [This is of practical significance for radio and television where $g(t)$ represents the signal of interest and $\cos(\omega_0 t)$ is the carrier wave.]

4.9   *Rayleigh's theorem.*
   a Show that the Fourier transform of $g^*(t)$ is $G^*(-\omega)$. *Hint:* Write the function in terms of real and imaginary parts of the even–odd decomposition.
   b Show that $u(0) = \int_{-\infty}^{\infty} U(\omega)d\omega$.

c Prove the following well-known theorem:

$$\int_{-\infty}^{\infty} g(t)h^*(t)\, dt = 2\pi \int_{-\infty}^{\infty} G(\omega)H^*(\omega)\, d\omega$$

This is known as the power theorem. If $h^*(t)$ is replaced by $g^*(t)$ and $H^*(t)$ by $G^*(t)$, Rayleigh's theorem (also known as Plancherel's theorem) results:

$$\int_{-\infty}^{\infty} |g(t)|^2\, dt = 2\pi \int_{-\infty}^{\infty} |G(\omega)|^2\, d\omega$$

Hint: Write $u(\tau) = \int_{-\infty}^{\infty} g(t)h^*(t + \tau)\, dt$ and consider $u(0)$.

4.10 *Parseval's theorem.* Let $g(t)$ be a real and periodic function and a_n and b_n be the coefficients of its Fourier series. Prove Parseval's theorem:

$$\frac{1}{T}\int_{-T/2}^{T/2} [g(t)]^2\, dt = a_0^2 + \frac{1}{2}\sum_{n=1}^{\infty} (a_n^2 + b_n^2)$$

This corresponds to Rayleigh's theorem for Fourier series.

4.11 A continuous signal that has been "clipped" is one that has an upper or lower limit or both. There is a finite probability that the value of the signal is equal to the limit. This probability may be represented by a delta function in the probability density function. Suppose $X(t)$ is a zero-mean Gaussian random process that has been clipped at $\pm 3\sigma$. Sketch its probability density function and cumulative distribution function. Label the probabilities concentrated in the delta functions.

Discrete Fourier Transform

Many students have access to a computer program, e.g., a sophisticated spreadsheet, that has a function call to a discrete Fourier transform, probably using a fast Fourier transform (FFT) algorithm. A DFT is a Fourier transform of a discrete indexed series x_j, and an FFT is an algorithm that implements a DFT in a computationally efficient manner. The student may wish to defer these problems until after studying Chapter 13, which presents the theory of DFTs and FFTs in detail. Appendix D contains example FFT code in FORTRAN and in C.

Following are problems that exercise a DFT program. As with continuous Fourier transforms, a normalizing factor must appear in either the forward (minus-i) transform or the inverse (plus-i) transform; these problems assume the following discrete Fourier transform pair:

$$X_k = \sum_{j=0}^{n-1} x_j e^{-i2\pi kj/n} \qquad k = 0, \ldots, n-1$$

$$x_j = \frac{1}{n} \sum_{k=0}^{n-1} X_k e^{+i2\pi jk/n} \qquad j = 0, \ldots, n-1$$

We assume that n is a power of 2, i.e., $n = 2, 4, 8, 16, 32, 64, \ldots$. Note that the first $\frac{1}{2}n + 1$ values of the transform X_k correspond to frequency $f = 0$ and the positive frequencies, whereas the rest correspond to negative frequencies, as shown in the example below for $n = 8$. The reasons for this are explained in Chapter 13. The length of the time series is $T = n \Delta t$, and the frequency interval is given by $\Delta f = 1/T$. Here, we assume $\Delta t = 1$, so $T = n$ and $\Delta f = 1/n$:

k	0	1	2	3	4	5	6	7
f_k	0	Δf	$2 \Delta f$	$3 \Delta f$	$4 \Delta f$	$-3 \Delta f$	$-2 \Delta f$	$-\Delta f$

Let $n = 16$ and find the DFT for

4.12 *Simple harmonics.* Let $M = 8$ and
  a $x_j = \cos(2\pi j/M)$
  b $x_j = \sin(2\pi j/M)$.

  The DFT should be (a) a real and even impulse pair and (b) an imaginary and odd impulse pair at $f_k = \pm n \Delta f/M$. The magnitude of each impulse should be $\frac{1}{2}M$. Experiment with $M = 2, 4, 6$.

4.13 *Unit impulse and constant.*
  a Let $x_j = 0$ for all j except $j = 0$, where $x_0 = 1$.
  b Let $x_j = 1$ for all j.
  Experiment with $n = 16, 64$.

4.14 *Square wave.* Let $x_j = 1$ for $j = 0, \ldots, \frac{1}{2}n$, and $x_j = 0$ for $j = \frac{1}{2}n + 1, \ldots, n - 1$. Experiment with $n = 16, 64$.

5

RANDOM PROCESSES IN FREQUENCY DOMAIN

Stationary random vibrations are more usefully studied in the frequency domain than in the time domain. This chapter discusses the characterization of a stationary random process in the frequency domain by the spectral density function and pairs of random processes by the cross-spectral density function. The spectral density function is defined, its mathematical properties are discussed, and aspects of its practical application are presented. We introduce the white-noise and bandpass processes. The cross-spectral density function is defined and applied to linear transformations and derivatives of random processes.

References on the frequency-domain approach to stationary random processes fall mainly into two categories. In the first category are those dealing with random processes and communications theory, such as Doob (1953), Papoulis (1965), and Jenkins and Watts (1968). In the second category are those that are primarily random vibrations texts, which are cited in Chapter 1.

5.1 SPECTRAL DENSITY FUNCTION

Physicists and engineers have been studying spectral decompositions of physical phenomena ever since Sir Isaac Newton (1643–1727) split sunlight into colors in about 1700. The earliest function most closely resembling the spectral density function we use today was developed by Sir Arthur Schuster in a series of papers (1898, 1900, 1906) that investigated the presence of periodicities in meteorological, magnetic, and optical phenomena. Schuster's "periodogram" in its final form is proportional to our spectral density function but is plotted in terms of period rather than frequency.

This section defines the spectral density function and describes its mathematical properties. The relationship between the spectral density function and the finite Fourier transforms of sample realizations of a random process is explained. Some aspects of the practical application of the spectral density function are presented.

5.1.1 Definition

In Chapter 3, we construct a random process that is the sum of individual harmonic processes:

$$X(t) = \sum_{k=1}^{m} X_k(t) = \sum_{k=1}^{m} (A_k\cos \omega_k t + B_k\sin \omega_k t) \qquad (3.54)$$

As the number of harmonic components, m, goes to infinity, the autocorrelation function of this process may be represented as a Fourier cosine transform:

$$R_X(\tau) = \sigma^2 \int_0^{\infty} g(\omega)\cos(\omega\tau) \, d\omega \qquad (3.59)$$

We now formally define the Fourier transform pair:

$$S_X(\omega) = \frac{1}{2\pi} \int_{-\infty}^{\infty} R_X(\tau)e^{-i\omega\tau} \, d\tau \qquad (5.1)$$

$$R_X(\tau) = \int_{-\infty}^{\infty} S_X(\omega)e^{+i\omega\tau} \, d\omega \qquad (5.2)$$

The Fourier transform of the autocorrelation function is called the spectral density function $S_X(\omega)$. Note that we continue to subscript the function with the random process of interest. Equations (5.1) and (5.2) are often called the Wiener–Khinchine relations, named after mathematicians Norbert Wiener (1894–1964) and A. I. Khinchine who independently proved the relationship (Wiener 1930; Khinchine 1934).[1]

In order for this transform pair to exist in the traditional sense, the autocorrelation function must be absolutely integrable $[\int_{-\infty}^{\infty} |R_X(\tau) \, d\tau| < \infty]$. Recall that the limit of $R_X(\tau)$ as $\tau \to \infty$ is μ_X^2, so if the mean is not equal to zero, the integral does not converge. Hence, before the spectral analysis is performed, the mean value of the process should be made equal to zero by subtracting the

---

[1]Weiner's paper is motivated by the extension of Fourier transform theory to stationary random processes for which the ordinary Fourier transform does not exist, as will be shown in the next section; he is directly concerned with the spectral density function as a characterization of a random process. Khinchine is mostly concerned with the mathematical conditions for the existence of the autocorrelation function.

mean from sample realizations of $X(t)$. The effect of a nonzero mean value can always be added back when necessary.

One sometimes encounters the integral of the spectral density function, known as the spectral distribution function, which can be shown to be

$$\Psi(\omega) = \int_{-\infty}^{\omega} S_X(w) \, dw = \int_{-\infty}^{\infty} R_X(\tau) \frac{e^{-i\omega\tau} - 1}{-i\tau} \, d\tau \qquad (5.3)$$

If a random process has any finite harmonic components or a nonzero mean, they will appear as Dirac delta functions in the spectral density function and as jumps in the spectral distribution function. The delta function for a nonzero mean value would have a weight equal to μ_X^2.

The spectral density function is real, symmetric and positive, and its area is equal to the process's mean-square value. We show these properties below.

Symmetry. The autocorrelation function of real-valued stationary random process $X(t)$ is real and symmetric. From our study of Fourier transforms we know that its transform, the spectral density function, must also be real valued and symmetric:

$$S_X(-\omega) = S_X(\omega) \qquad (5.4)$$

(A negative frequency may be understood in terms of rotating vectors. If positive frequency is rotation in the counterclockwise direction, negative frequency is rotation in the opposite, clockwise direction.)

This symmetry may be used to simplify the computation of $S_X(\omega)$. The full Fourier transform may be replaced by the cosine transform pair:

$$S_X(\omega) = \frac{1}{\pi} \int_0^{\infty} R_X(\tau)\cos(\omega\tau) \, d\tau \qquad (5.5)$$

$$R_X(\tau) = 2 \int_0^{\infty} S_X(\omega)\cos(\omega\tau) \, d\omega \qquad (5.6)$$

Positive Valued. The spectral density function is positive for all ω. One may infer this from our heuristic development of Eq. (3.59): $g(\omega)$ is defined to be the fraction of the total mean square contributed by the harmonic process with frequency ω and is always greater than or equal to zero.

We may also show that this is true because the autocorrelation function is positive definite. Recall the definition of the positive definite property:

$$\sum_{j=1}^{n} \sum_{k=1}^{n} \alpha_j \alpha_k R_X(t_k - t_j) \geq 0 \qquad (3.15)$$

We may generalize the equation by replacing the coefficients α with a function $g(t)$ (which may contain delta functions) and by replacing the sums with integrals:

$$\int_{-\infty}^{\infty} \int_{-\infty}^{\infty} g(s)g(t)R_X(t - s)\ ds\ dt \geq 0 \qquad (5.7)$$

The integration produces a constant, but we wish to take a Fourier transform that requires the result to be a function. So we introduce a dummy variable, u, to make the left side of the inequality a function of u, $q(u)$:

$$q(u) = \int_{-\infty}^{\infty} \int_{-\infty}^{\infty} g(s)g(u + t)R_X(u + t - s)\ ds\ dt \geq 0 \qquad (5.8)$$

Applying the shift theorem, the reader may show that taking the Fourier transform yields

$$Q(\omega) = \frac{1}{2\pi} \int_{-\infty}^{\infty} \int_{-\infty}^{\infty} \int_{-\infty}^{\infty} g(s)g(u + t)R_X(u + t - s)e^{-i\omega u}\ ds\ dt\ du$$

$$= 4\pi^2 |G(\omega)|^2 S_X(\omega) \geq 0 \qquad (5.9)$$

Since $|G(\omega)|^2$ is positive, $S_X(\omega)$ must also be positive, which was to be shown. This result was first shown by Salomon Bochner (1933).

We note that when working with actual data, the direct calculation of the autocorrelation function will sometimes lead to an invalid, non–positive definite result due to numerical errors. This situation is recognized when the spectral density function has negative values. This is one reason why the recommended analytical procedure is to calculate the spectral density function first, then to get the autocorrelation function by inverse transformation.

Mean-Square Value. Since $g(\omega)$ in Eq. (3.59) is defined to be the limit of the fractions p_k that add up to 1, the integral of $\sigma_{X}^2 g(\omega)$ over all its values must be equal to σ_X^2. Likewise, the integral of $S_X(\omega)$ can be shown to be equal to σ_X^2 by evaluating Eq. (5.2) at $\omega = 0$:

$$R_X(0) = E[X^2(t)] = \sigma_X^2 = \int_{-\infty}^{\infty} S_X(\omega)e^0 d\omega = \int_{-\infty}^{\infty} S_X(\omega)\ d\omega \qquad (5.10)$$

Thus, assuming the mean is equal to zero, the area under the spectral density function is equal to the mean-square or variance:

$$\sigma_X^2 = \int_{-\infty}^{\infty} S_X(\omega)\ d\omega \qquad (5.11)$$

As mentioned earlier, the square root of the mean-square is known as the root-mean-square, or RMS.

Example 5.1: Spectral Density Function of Low-Pass Random Process. The autocorrelation function of a low-pass random process is given in Chapter 3 as

$$R_X(\tau) = \sigma_X^2 \frac{\sin(\omega_c \tau)}{\omega_c \tau}$$

The corresponding spectral density is

$$S_X(\omega) = \begin{cases} \sigma_X^2/2\omega_c & |\omega| < \omega_c \\ \sigma_X^2/4\omega_c & |\omega| = \omega_c \\ 0 & |\omega| > \omega_c \end{cases}$$

This process is known as a low-pass random process because it is the result of passing purely random noise through a low-pass linear filter. All frequency content above the cutoff frequency ω_c is eliminated.

By inspection, the area of the spectral density function is equal to σ_X^2. If the process were Gaussian with a mean equal to zero, the probability that $X < x$ at a given instant would be given by $\Phi(x/\sigma_X)$, for example, $P(X < 3\sigma_X) \approx 0.9987$.

5.1.2 Relationship with Fourier Transform of $X(t)$

Our definition of the spectral density function as the Fourier transform of the autocorrelation function may appear to be somewhat awkward and indirect. One might wish to try a more direct approach such as taking the Fourier transform of the random process $X(t)$ directly; call it $X(\omega)$. We show in this section that there is such a relationship, which is in accord with our intuition, and that it leads us back to the same definition we have given.

We would like to define the Fourier transform of the random process $X(t)$ to be $X(\omega)$:

$$X(\omega) = \int_{-\infty}^{\infty} X(t)e^{-i\omega t} \, dt \qquad (5.12)$$

However, if the process is stationary, $X(t)$ continues forever and $\int_{-\infty}^{\infty}|X(t)| \, dt$ is not finite; so this Fourier transform does not exist.

We may approach this problem by first defining the finite Fourier transform[2]

---

[2]Note the absence of the factor $1/2\pi$ in the definition of the finite Fourier transform.

of $X(t)$ over the interval $(0, T)$, $X(\omega, T)$, which does exist:

$$X(\omega, T) = \int_0^T X(t)e^{-i\omega t} \, dt \qquad (5.13)$$

Later, we will take the limit at T goes to infinity to obtain the result equivalent to starting with the transform of $X(t)$.

Let $Y(\omega) = X(\omega, T)$ be a complex-valued random process evolving in ω. Its expected value is easily shown to be equal to zero. Its variance is

$$\sigma_Y^2(\omega) = E[|X(\omega, T)|^2] = E[X(\omega, T)X^*(\omega, T)]$$

$$= E\left[\int_0^T X(t)e^{-i\omega t} \, dt \int_0^T X(s)e^{+i\omega s} \, ds \right]$$

$$= E\left[\int_0^T \int_0^T X(t)X(s)e^{-i\omega(t-s)} \, ds \, dt \right] \qquad (5.14)$$

The region of integration over s and t is the square shown in Figure 5.1a.

We change the variables of integration: Let $\tau = t - s$ and integrate on τ and t; the new region of integration is the parallelogram shown in Figure 5.1b. We have

$$\sigma_Y^2(\omega) = E\left[\int_0^T \int_{t-T}^t X(t)X(t - \tau)e^{-i\omega\tau} \, d\tau \, dt \right] \qquad (5.15)$$

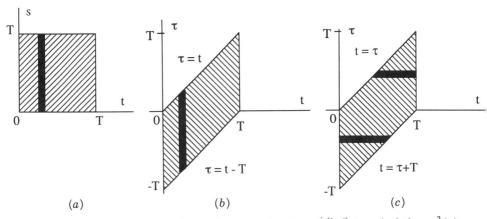

Figure 5.1 Region over which to integrate $X(t)X(s)e^{-i\omega(t-s)}$ in calculating $\sigma_Y^2(\omega)$: (a) in original coordinates (t, s); (b) in transformed coordinates (t, τ) where $\tau = t - s$; (c) in (t, τ) integrating first on t.

The autocorrelation function appears when we interchange the order of integration and take the expected value inside:

$$
\sigma_Y^2(\omega) = \int_0^T \int_{t-T}^t E[X(t)X(t-\tau)]e^{-i\omega\tau}\, d\tau\, dt
$$

$$
= \int_0^T \int_{t-T}^t R_X(\tau)e^{-i\omega\tau}\, d\tau\, dt \tag{5.16}
$$

[Note that $R_X(-\tau) = R_X(\tau)$.]

Since the integrand is not dependent on t, we would like to switch the order of integration and integrate first on t and then on τ. There are two regions of integration, as shown in Figure 5.1c. For $\tau \in (-T, 0)$ we integrate t from 0 to $\tau + T$; for $\tau \in (0, T)$ we integrate t from τ to T. Performing the integration, we have

$$
\sigma_Y^2(\omega) = \int_{-T}^0 R_X(\tau)e^{-i\omega\tau}\left(\int_0^{\tau+T} dt\right) d\tau + \int_0^T R_X(\tau)e^{-i\omega\tau}\left(\int_\tau^T dt\right) d\tau
$$

$$
= \int_{-T}^0 R_X(\tau)e^{-i\omega\tau}(\tau + T)\, d\tau + \int_0^T R_X(\tau)e^{-i\omega\tau}(T - \tau)\, d\tau
$$

$$
= \int_{-T}^T R_X(\tau)e^{-i\omega\tau}(T - |\tau|)\, d\tau \tag{5.17}
$$

Note that this integral is not finite as T goes to infinity.

We may force the integral to have a finite limit by dividing the equation by $2\pi T$:

$$
\frac{1}{2\pi T}\sigma_Y^2(\omega) = \frac{1}{2\pi}\int_{-T}^T R_X(\tau)e^{-i\omega\tau}\left(1 - \frac{|\tau|}{T}\right) d\tau \tag{5.18}
$$

It is evident from Figure 5.2 that as T goes to infinity, $1 - |\tau|/T$ goes to 1.

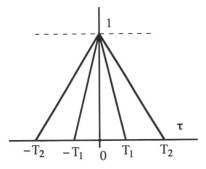

Figure 5.2 Function $1 - |\tau|/T$ for T_1 and T_2, $T_2 > T_1$.

The result is

$$\lim_{T \to \infty} \frac{1}{2\pi T} \sigma_Y^2(\omega) = \lim_{T \to \infty} \frac{1}{2\pi} \int_{-T}^{T} R_X(\tau) e^{-i\omega\tau} \left(1 - \frac{|\tau|}{T}\right) d\tau$$

$$= \frac{1}{2\pi} \int_{-\infty}^{\infty} R_X(\tau) e^{-i\omega\tau} \, d\tau$$

$$= S_X(\omega) \tag{5.19}$$

Going back to our starting point [Eq. (5.14)], we have the final result:

$$S_X(\omega) = \lim_{T \to \infty} \frac{1}{2\pi T} E[|X(\omega, T)|^2] \tag{5.20}$$

Thus, the spectral density function, as we have previously defined it, corresponds to the ensemble average of the squared moduli of the Fourier transforms of $X(t)$. This result suggests that we may estimate the spectral density function from a "large number" of "long" sample realizations of $X(t)$ roughly as follows:

- Obtain n sample realizations of $X(t)$ with length T.
- For each realization, take its finite Fourier transform and compute its squared modulus.
- Compute the average squared modulus at each frequency ω.
- Divide by $2\pi T$ to get the proper normalization and units.

This gives us the equation

$$\hat{S}_X(\omega, T, n) = \frac{1}{2\pi T} \frac{1}{n} \sum_{k=1}^{n} |X_k(\omega, T)|^2 \tag{5.21}$$

Note that taking longer records does not necessarily improve the estimate. The record length needs to only be long enough for the limiting operation above to be valid. This can be tested practically by applying Eq. (5.18) to the autocorrelation function and lengthening T until satisfactory convergence to a limit is achieved. To improve the estimate, one must increase the number of sample realizations that are averaged, n. This is discussed in Chapter 14.

5.1.3 Practical Issues

There are some issues regarding the practical application of the spectral density function. These mainly relate to the fact that spectral decomposition of electrical waveforms was established long before Wiener made the link to the autocorrelation function.

Early spectral analyzers were analog devices that operated on one sample waveform, $x(t)$, at a time (Figure 5.3). The waveform, represented by a current,

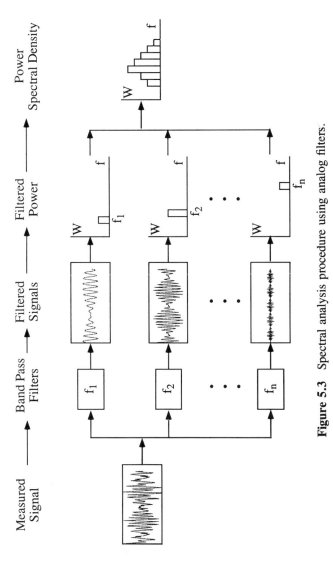

Figure 5.3 Spectral analysis procedure using analog filters.

$i(t)$, or a voltage, $v(t)$, is iteratively filtered to a signal with only a "single" frequency. The filtered signal's mean-square value is then measured and plotted. Since i^2 and v^2 are proportional to the average power dissipated in a resistive load, this is called the "power" at that frequency. In practice, it is impossible to develop a physical filter that outputs only a single frequency; the output is actually a signal that contains many frequencies but is dominated by frequencies within a narrow bandwidth, Δf. Thus, the measured power is divided by the width of the band, Δf, making the units i^2 per unit frequency or v^2 per unit frequency. Hence, $S_X(\omega)$ is called the *power spectral density* (PSD) or the *power spectrum*.

Repeating this for the whole spectrum of frequencies in the waveform produces a picture of the frequency content. The peaks in the plot correspond to frequencies that tend to dominate the waveform. Note that the resolution of the peaks is limited to the quality of the filters. There will always be leakage of power into the measurement from frequencies outside the intended frequency band.

Due to the tradition of analog spectral analysis, we have two differences between the mathematical definition of the spectral density function and its practical implementation: (1) in practice, the units of frequency are changed from radians per second to cycles per second, or hertz; and (2) negative frequencies are discarded. One will find one change or the other or both made in the literature, meaning that there are four versions of the spectral density function mentioned, as illustrated in Figure 5.4. However, one thing will always be true: the area under the spectral density function will always be equal to the mean-square of $X(t)$.

To change the units from radians per second to hertz, the frequency is divided by 2π, that is, $f = \omega/2\pi$. To compensate, the corresponding density is multiplied by 2π. The designation $S_X(f)$ is used to indicate the two-sided spectral density in hertz:

$$S_X(f) = 2\pi S_X(\omega) \tag{5.22}$$

If the negative frequencies are discarded, the density must be multiplied by 2 to compensate. The designation $G_X(\omega)$ is used to indicate the one-sided spectral density in radians per second:

$$G_X(\omega) = \begin{cases} 2S_X(\omega) & \omega \geq 0 \\ 0 & \omega < 0 \end{cases} \tag{5.23}$$

If both changes are made, the density must be multiplied by 4π to compensate. The designation $W_X(f)$ [and sometimes $G_X(f)$] is used to indicate the one-sided spectral density in hertz:

$$W_X(f) = \begin{cases} 4\pi S_X(\omega) & \omega, f \geq 0 \\ 0 & \omega, f < 0 \end{cases} \tag{5.24}$$

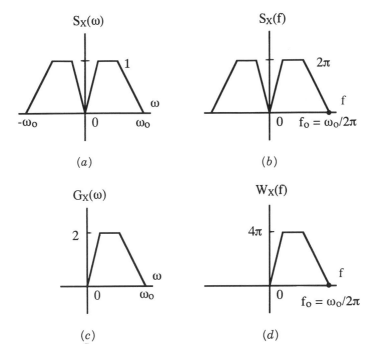

Figure 5.4 Four representations of spectral density function: (*a*) $S_X(\omega)$, two sided with frequency in radians per second; (*b*) $S_X(f)$, two sided with frequency in hertz; (*c*) $G_X(\omega)$, one sided with frequency in radians per second; (*d*) $W_X(f)$, one sided with frequency in hertz. All have area equal to σ_X^2.

The spectral density function is also commonly described in terms of octaves (oct) and decibles (dB). As in music, an octave is a doubling of frequency. The increase in octaves from f_1 to f_2 is

$$\text{oct} = \log_2 \frac{f_2}{f_1} = \frac{1}{\log_{10} 2} \log_{10} \frac{f_2}{f_1} \qquad (5.25)$$

A bel is the common logarithm of the ratio of two measurements of power and is named after Alexander Graham Bell (1847–1922). A decibel is one-tenth of a bel and is given by

$$\text{dB} = 10 \log_{10} \frac{W_X(f_2)}{W_X(f_1)} \qquad (5.26)$$

Thus, a doubling of spectral density corresponds to an increase of approximately 3 dB, since $10 \log_{10}(2) \approx 3.01$.[3] If the spectral density doubles for each doubling of frequency, we say the spectrum rises at 3 dB/oct.

---

[3]For quantities that are not powers, or powerlike in having units of x^2, the decibel is defined to be 20 times the logarithm of the ratio. A doubling of voltage, for example, is then approximately 6 dB.

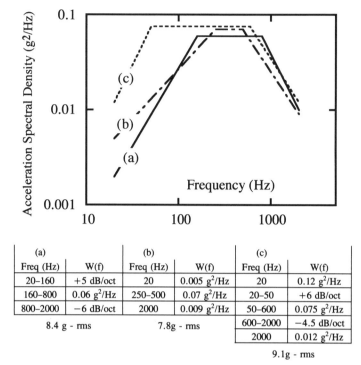

(a)		(b)		(c)	
Freq (Hz)	W(f)	Freq (Hz)	W(f)	Freq (Hz)	W(f)
20–160	+5 dB/oct	20	0.005 g²/Hz	20	0.12 g²/Hz
160–800	0.06 g²/Hz	250–500	0.07 g²/Hz	20–50	+6 dB/oct
800–2000	−6 dB/oct	2000	0.009 g²/Hz	50–600	0.075 g²/Hz
8.4 g - rms		7.8g - rms		600–2000	−4.5 dB/oct
				2000	0.012 g²/Hz
				9.1g - rms	

Figure 5.5 Examples of contemporary random vibration specification in aerospace industry.

Some examples of contemporary random vibration specifications are shown on log-log plots in Figure 5.5. The following formulas are convenient for calculating the area under the rising, falling, and flat portions of such spectra. Let C be a slope parameter

$$C = \frac{\log_{10}[W_X(f_2)/W_X(f_1)]}{\log_{10}(f_2/f_1)} = \frac{\text{dB/oct}}{10 \log_{10} 2} \approx \frac{\text{dB/oct}}{3} \qquad (5.27)$$

For $W_X(f)$ rising from f_1 to f_2, reaching $W_X(f_2) = W$,

$$A_{\text{rising}} = \frac{W}{C + 1}\left[f_2 - f_1\left(\frac{f_1}{f_2}\right)^C\right] \qquad C > 0 \qquad (5.28)$$

For $W_X(f)$ falling from f_1 to f_2, starting from $W_X(f_1) = W$,

$$A_{\text{falling}} = \begin{cases} \dfrac{W}{C + 1}\left[f_2\left(\dfrac{f_2}{f_1}\right)^C - f_1\right] & C \neq -1, C < 0 \\[3mm] Wf_1 \ln\left(\dfrac{f_2}{f_1}\right) & C = -1 \end{cases} \qquad (5.29)$$

For $W_X(f)$ flat from f_1 to f_2, at $W_X(f) = W$,

$$A_{\text{flat}} = W(f_2 - f_1) \qquad C = 0 \tag{5.30}$$

One last note: It is common to refer to an *acceleration spectrum* or a *displacement spectrum*. These are just the spectral density functions for $X(t)$ having units of acceleration and displacement, respectively.

Example 5.2: Practical Computation of RMS. For the spectra shown in Figure 5.5, the reader may verify that the RMS accelerations are $8.4g$, $7.8g$, and $9.1g$ for spectra (a), (b), and (c), respectively. The corresponding probabilities that the acceleration observed at a random instant in time is more than $\pm 30g$ are 0.04, 0.01, and 0.10%, respectively.

5.1.4 White Noise and Bandpass Filtered Spectra

In this section, we present four idealized random processes that are commonly used in theoretical studies. Their spectral densities are illustrated in Figures 5.6–5.9.

White Noise. In the frequency domain, white noise is the simplest random process conceivable. It is defined by a spectral density function that is constant for all frequencies from $-\infty$ to $+\infty$:

$$S_X(\omega) = S_0 \qquad -\infty < \omega < +\infty \tag{5.31}$$

This spectrum is illustrated in Figure 5.6. The "white" part of its name derives from the spectrum of white light, which contains the frequencies of all colors. The "noise" part of its name derives from its autocorrelation function, which exists only in the limiting sense and is given by

$$R_X(\tau) = 2\pi S_0 \delta(\tau) \tag{5.32}$$

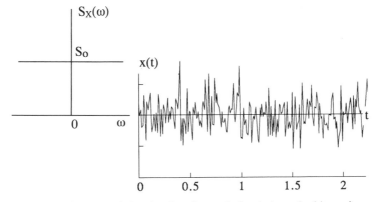

Figure 5.6 Spectral density function and simulation of white noise.

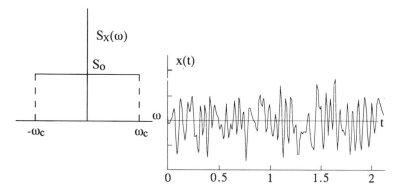

Figure 5.7 Spectral density function and simulation of low-pass noise.

That is, the autocorrelation function of white noise is a delta function at $\tau = 0$, so the process is completely uncorrelated with itself at all lags except zero. Hence, the process is "noise," completely incoherent.

The simplicity of this definition of white noise belies certain conceptual difficulties in the time domain. First of all, it is not physically realizable: Its variance is infinite, meaning that the process has infinite power.[4] Furthermore, the process can be shown (by the methods in Appendix A) to be continuous but not differentiable. One might think of this process as the limiting case of taking independent samples of random variable X at time intervals that become infinitesimally smaller; the value of $X(t + dt)$ is totally independent of $X(t)$.

White noise has many practical applications in the study of stochastic differential equations. In Chapter 8, we apply white noise as an input to a dynamic system. Also, other random processs of interest to us may be thought of as the outputs of filters that see a white-noise input.

The integral of white noise is known as the Weiner–Lévy process, after mathematicians Norbert Wiener and Paul Lévy, who developed much of the theory. It is also known as Brownian motion, after botanist Robert Brown, who described (1828) the rapid random motion of pollen grains suspended in water. This process has played an important role in the history of science and engineering. For example, in 1905, Albert Einstein published as his doctoral thesis a theoretical explanation of Brownian motion based on the statistical mechanics of atomic motion. The 1926 Nobel Prize for Physics was awarded to Jean-Baptiste Perrin for his experimental verification of Einstein's theory.

Low-Pass Noise. Suppose white noise is passed through a low-pass filter. The output is a random process that has uniform spectral density for all frequencies below its cutoff frequency ω_c:

---

[4]The sample realization shown in Figure 5.6 is from a simulation having a high, but finite, upper cutoff frequency and a finite variance.

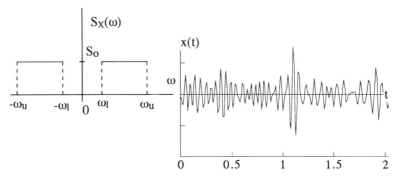

Figure 5.8 Spectral density function and simulation of bandpass noise.

$$S_X(\omega) = \begin{cases} S_0 & |\omega| < \omega_c \\ S_0/2 & |\omega| = \omega_c \\ 0 & |\omega| > \omega_c \end{cases} \qquad (5.33)$$

This spectrum is illustrated in Figure 5.7. The corresponding autocorrelation function is given by

$$R_X(\tau) = \sigma_X^2 \frac{\sin(\omega_c\tau)}{\omega_c\tau} \qquad (5.34)$$

where $\sigma_X^2 = 2\omega_c S_0$. We have chosen to write σ_X^2 to emphasize the fact that the area under the spectral density function is equal to the mean-square of the process.

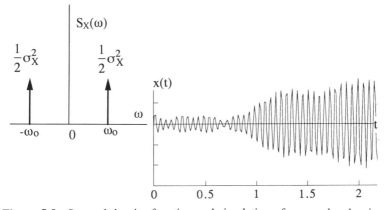

Figure 5.9 Spectral density function and simulation of narrow-band noise.

Bandpass Noise. Filtering white noise through a bandpass filter produces a random process that has uniform spectral density for all frequencies between the filter's upper and lower cutoff frequencies, ω_u and ω_l, respectively:

$$S_X(\omega) = \begin{cases} S_0 & \omega_l < |\omega| < \omega_u \\ S_0/2 & |\omega| = \omega_l, \omega_u \\ 0 & \text{otherwise} \end{cases} \tag{5.35}$$

This spectrum is illustrated in Figure 5.8. To derive the autocorrelation function, one may think of this process as the difference of two independent low-pass processes: subtracting a process with a cutoff of ω_l from one with a cutoff of ω_u. The corresponding autocorrelation function is

$$R_X(\tau) = \sigma_X^2 \frac{\sin\left(\frac{1}{2}\Delta\omega\,\tau\right)}{\frac{1}{2}\Delta\omega\,\tau} \cos(\omega_0\tau) \tag{5.36}$$

where $\sigma_X^2 = 2(\omega_u - \omega_l)S_0$, $\Delta\omega = \omega_u - \omega_l$, and $\omega_0 = \frac{1}{2}(\omega_u + \omega_l)$. Notice that the autocorrelation function consists of a cosine function within a slowly varying sinc envelope. The envelope variations are made slower if the cutoff frequencies are moved closer together.

Narrow-Band Noise. Taking the limit of a bandpass process, as the upper and lower cutoff frequencies move closer together, results in a narrow-band spectrum centered on ω_0. The spectral density function is impulsive:

$$S_X(\omega) = \sigma_X^2[\tfrac{1}{2}\delta(\omega + \omega_0) + \tfrac{1}{2}\delta(\omega - \omega_0)] \tag{5.37}$$

This spectrum is illustrated in Figure 5.9. The corresponding autocorrelation function is a cosine:

$$R_X(\tau) = \sigma_X^2\cos(\omega_0\tau) \tag{5.38}$$

The reader will recall this to be the autocorrelation function of a single harmonic process. The sample functions are sine functions with random phase.[5] The process is stationary but not ergodic.[6]

We will find this representation of a process useful later when describing the dynamic response of systems to random inputs. Again, this is another mathematical approximation to physical reality that gives useful results. It is not reality itself.

---

[5]The sample realization shown in Figure 5.9 is from a simulation having a small, but finite, bandwidth.

[6]In general, the narrower the bandwidth of a process, the "less ergodic" it will be.

5.2 CROSS-SPECTRAL DENSITY FUNCTION

This section defines the cross-spectral density function and describes its mathematical properties. The coherence function is introduced as a nondimensional, frequency-domain measure of the linear dependence of two processes. The effects of linear transformations and of differentiation are described.

5.2.1 Definition

The cross-spectral density function is defined as the Fourier transform of the cross-correlation function. There are two cross-spectral density functions, $S_{XY}(\omega)$ and $S_{YX}(\omega)$, defined as follows:

$$S_{XY}(\omega) = \frac{1}{2\pi} \int_{-\infty}^{\infty} R_{XY}(\tau)e^{-i\omega\tau}\, d\tau \tag{5.39}$$

$$S_{YX}(\omega) = \frac{1}{2\pi} \int_{-\infty}^{\infty} R_{YX}(\tau)e^{-i\omega\tau}\, d\tau \tag{5.40}$$

Unlike the spectral density function, which is real valued, the cross-spectral density functions are complex.

We note that some authors refer to $S_X(\omega)$ as the *auto-spectral density function (ASD)* to emphasize that it is the Fourier transform of the autocorrelation function. The notations $R_{XX}(\tau)$ and $S_{XX}(\omega)$ are also applied with the same rationale.

The cross-correlation functions may be recovered from the inverse transforms of the above:

$$R_{XY}(\tau) = \int_{-\infty}^{\infty} S_{XY}(\omega)e^{i\omega\tau}\, d\omega \tag{5.41}$$

$$R_{YX}(\tau) = \int_{-\infty}^{\infty} S_{YX}(\omega)e^{i\omega\tau}\, d\omega \tag{5.42}$$

Symmetry. Just as the cross-correlation functions are different, but related by $R_{XY}(\tau) = R_{YX}(-\tau)$, the cross spectra have an interesting relationship to one another. Substituting $R_{YX}(-\tau)$ for $R_{XY}(\tau)$ and changing the variable of integration by letting $u = -\tau$, we show the following:

$$S_{XY}(\omega) = \frac{1}{2\pi} \int_{-\infty}^{\infty} R_{XY}(\tau)e^{-i\omega\tau}\, d\tau = \frac{1}{2\pi} \int_{-\infty}^{\infty} R_{YX}(-\tau)e^{-i\omega\tau}\, d\tau$$

$$= \frac{1}{2\pi} \int_{-\infty}^{\infty} R_{XY}(u)e^{+i\omega u}(-du) = \frac{1}{2\pi} \int_{-\infty}^{\infty} R_{XY}(u)e^{-i(-\omega)u}\, du$$

$$= S_{YX}(-\omega) \tag{5.43}$$

So the same kind of symmetry holds for the cross spectra as for the cross correlations. We can go one step further. Since we are dealing with real-valued processes, the cross-correlation function is real and its Fourier transform has Hermitian symmetry:

$$S_{YX}(-\omega) = S_{YX}^*(\omega) \tag{5.44}$$

Thus, we have

$$S_{XY}(\omega) = S_{YX}^*(\omega) \tag{5.45}$$

That is, $S_{XY}(\omega)$ and $S_{YX}(\omega)$ are complex conjugates of each other.

Covariance. Just as the area under the spectral density function $S_X(\omega)$ is equal to the mean-square or variance of $X(t)$, the integral of the cross spectrum is equal to the expected value of the product of $X(t)Y(t)$, that is, their covariance. This can be seen by letting $\tau = 0$ in Eq. (5.41) above. Note that, because the cross-spectral density has Hermitian symmetry, the imaginary part of the function is odd and integrates to zero.

5.2.2 Coherence Function and Linear Transformations

The coherence function $\gamma_{XY}^2(\omega)$ is defined as the ratio of the squared modulus of the cross-spectral density function, $|S_{XY}(\omega)|^2$, to the auto-spectral density functions $S_X(\omega)$ and $S_Y(\omega)$:

$$\gamma_{XY}^2(\omega) = \frac{|S_{XY}(\omega)|^2}{S_X(\omega)S_Y(\omega)} = \frac{S_{XY}(\omega)S_{XY}^*(\omega)}{S_X(\omega)S_Y(\omega)} \tag{5.46}$$

The coherence function provides a nondimensional measure of the linear dependence between $X(t)$ and $Y(t)$ at each frequency, similar to the correlation coefficient. It can be shown that the coherence function has values between 0 and 1:

$$0 \leq \gamma_{XY}^2(\omega) \leq 1 \tag{5.47}$$

Values near 1 imply a linear relationship between $X(t)$ and $Y(t)$ at frequency ω. If $X(t)$ and $Y(t)$ are uncorrelated, their cross-correlation, cross-spectral density, and coherence functions are equal to zero.

We consider the effect of simple transformations of $X(t)$ on the cross-spectral density below.

Attenuation and Delay. Let $X(t)$ be delayed by δ and attenuated by a, so that $Y(t) = aX(t - \delta)$. The resulting spectral density functions are given by

$$S_Y(\omega) = a^2 S_X(\omega) \tag{5.48}$$

$$S_{XY}(\omega) = aS_X(\omega)e^{i\omega\delta} \tag{5.49}$$

The delay has the effect of shifting the phasing of $Y(t)$ relative to $X(t)$. Since the signal is delayed by the same time at all frequencies, the amount of phase shift is directly proportional to the frequency.

By inspection we see that the coherence function is equal to 1 at all frequencies: $Y(t)$ is linearly dependent on $X(t)$.

The cross-spectral density function has successfully been used to study propagation problems where the cross-correlation approach has difficulty (see Bendat and Piersol, 1980). Specifically, in dispersive propagation problems, the time delay a signal experiences is frequency dependent. Seismic waves, for example, travel faster and farther at low frequencies (long wavelengths) than at high frequencies (short wavelengths). As seen in the equation immediately above, the cross-spectral density can identify the attenuation and delay of each frequency independently.

Sum of Two Processes. Let $Z(t)$ be the sum of two processes $X(t)$ and $Y(t)$, that is, $Z(t) = X(t) + Y(t)$. Taking the transforms of Eqs. (3.37) and (3.39), the spectral densities are

$$S_{ZX}(\omega) = S_X(\omega) + S_{YX}(\omega) \tag{5.50}$$

$$S_Z(\omega) = S_X(\omega) + S_{XY}(\omega) + S_{YX}(\omega) + S_Y(\omega) \tag{5.51}$$

If $X(t)$ and $Y(t)$ are uncorrelated, these reduce to

$$S_{ZX}(\tau) = S_X(\tau) \tag{5.52}$$

$$S_Z(\omega) = S_X(\omega) + S_Y(\omega) \tag{5.53}$$

For this case, the coherence function between $Z(t)$ and $X(t)$ is given by

$$\gamma_{ZX}^2(\omega) = \frac{S_X(\omega)}{S_X(\omega) + S_Y(\omega)} \tag{5.54}$$

Example 5.3: Coherence Function for Sum of Two Processes. Figure 5.10 shows the spectral densities of two uncorrelated processes, $X(t)$ and $Y(t)$, and the coherence function between the sum, $Z(t) = X(t) + Y(t)$, and $X(t)$.

5.2.3 Derivatives of Stationary Processes

In Chapter 3, we establish the relationship between the autocorrelation function of $X(t)$ and the cross-correlation function involving its deriviative, $\dot{X}(t)$:

$$\frac{d}{d\tau} R_X(\tau) = R_{X\dot{X}}(\tau) \tag{3.41}$$

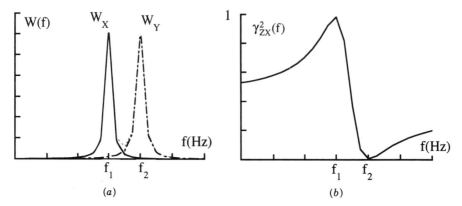

Figure 5.10 (*a*) Spectral density functions of $X(t)$ and $Y(t)$ and (*b*) coherence function of $Z(t) = X(t) + Y(t)$ with $X(t)$.

In Chapter 4, we establish the relationship between the Fourier transform of a function $g(t)$ and that of its derivative $\dot{g}(t)$:

$$\mathfrak{F}\left\{\frac{d}{dt}\,g(t)\right\} = i\omega G(\omega) \tag{4.30}$$

Hence, we can write the cross-spectral density of $X(t)$ and $\dot{X}(t)$ directly as

$$S_{X\dot{X}}(\omega) = \mathfrak{F}\{R_{X\dot{X}}(\tau)\} = \mathfrak{F}\left\{\frac{d}{d\tau}\,R_X(\tau)\right\} = i\omega S_X(\omega) \tag{5.55}$$

We can see that the cross-spectral density is purely imaginary and odd. Its integrated area is equal to zero, implying $R_{X\dot{X}}(0) = 0$. This confirms our previous finding that $X(t)$ and $\dot{X}(t)$ are uncorrelated at any given t.

From the other relationships between the auto- and cross-correlation functions of $X(t)$ and $\dot{X}(t)$, we may write the following relationships:

$$S_{\dot{X}}(\omega) = \omega^2 S_X(\omega) \tag{5.56}$$

$$S_{X\ddot{X}}(\omega) = -\omega^2 S_X(\omega) \tag{5.57}$$

$$S_{\ddot{X}}(\omega) = \omega^4 S_X(\omega) \tag{5.58}$$

Note that $X(t)$ and its "acceleration," $\ddot{X}(t)$, are negatively correlated; that is, positive values of "displacement" are correlated with negative values of "acceleration" and vice versa.

In terms of the engineering one-sided spectral density function, we may write

$$W_{\dot{X}}(f) = (2\pi)^2 f^2 W_X(f) \tag{5.59}$$

$$W_{X\dot{X}}(f) = -(2\pi)^2 f^2 W_X(f) \tag{5.60}$$

$$W_{\ddot{X}}(f) = (2\pi)^4 f^4 W_X(f) \tag{5.61}$$

Finally, the variances of the derivative processes are

$$\sigma_{\dot{X}}^2 = \int_{-\infty}^{\infty} S_{\dot{X}}(\omega)\, d\omega = \int_{-\infty}^{\infty} \omega^2 S_X(\omega)\, d\omega = (2\pi)^2 \int_0^{\infty} f^2 W_X(f)\, df \tag{5.62}$$

$$\sigma_{\ddot{X}}^2 = \int_{-\infty}^{\infty} S_{\ddot{X}}(\omega)\, d\omega = \int_{-\infty}^{\infty} \omega^4 S_X(\omega)\, d\omega = (2\pi)^4 \int_0^{\infty} f^4 W_X(f)\, df \tag{5.63}$$

PROBLEMS

5.1 Derive the autocorrelation functions for the spectra:
   a White noise
   b Low pass
   c Bandpass
   d Narrow band

5.2 Find the spectral density functions corresponding to
   a $R_X(\tau) = e^{-\alpha|\tau|}$
   b $R_X(\tau) = e^{-\alpha|\tau|} \cos(\omega_0 \tau)$

5.3 Derive the formulas given in the text for the area under a spectral density function specified in terms of decibels and octaves.

5.4 A low-pass random process $X(t)$ has a cutoff frequency of ω_c (or $2\pi f_c$). It is proposed to estimate the spectral density function of this process using Eq. (5.21) and sample records of length $T = 10/f_c$. Is T long enough? What if T is ten times longer? How long would you make it and why? *Hint:* You wish to consider the rate at which the following ratio [see Eq. (5.19)] approaches unity:

$$\frac{\int_{-T}^{T} R_X(\tau) e^{-i\omega\tau} (1 - |\tau|/T)\, d\tau}{\int_{-\infty}^{\infty} R_X(\tau) e^{-i\omega\tau}\, d\tau}$$

5.5   Let $W(t) = X(t)Y(t)$, where $X(t)$ and $Y(t)$ are uncorrelated random processes. Find the spectral density function of $W(t)$ and the cross-spectral density and coherence of $W(t)$ and $X(t)$.

5.6   *Semirandom binary transmission* (Papoulis, 1965). Let $X(t)$ be determined by periodically tossing a coin at fixed equal intervals of time T. If the coin lands heads up, $x = +1$; if tails is up, $x = -1$. The value of $X(t)$ remains constant until the next coin toss.

   a Sketch two sample realizations of $X(t)$.

   b What are the mean and variance of $X(t)$?

   c Find the autocorrelation function $R_X(t_1, t_2)$ and sketch it.

   d Is this process stationary? Ergodic? *Hint*: The ACF depends on whether or not t_1 and t_2 are in the same time interval.

5.7   *Random binary transmission* (Papoulis, 1965). Extend the previous problem by letting the sample realizations have random phasing. That is, let $Y(t) = X(t - \Theta)$, where Θ is uniformly distributed between 0 and T.

   a Sketch two sample realizations of $Y(t)$.

   b Show that the autocorrelation function is given by

$$R_Y(t_1, t_2) = \begin{cases} 1 - \dfrac{|t_1 - t_2|}{T} & |t_1 - t_2| < T \\ 0 & \text{otherwise} \end{cases}$$

   c Is this process stationary? Ergodic? *Hint*: It still depends on whether or not t_1 and t_2 are in the same time interval, but this now depends on Θ.

5.8   Find the spectral density function for the random binary transmission process in the previous problem.

5.9   At a certain time in a rocket launch, an on-board instrument experiences an average acceleration of 2.0g plus a stationary Gaussian component having the spectral density shown in Figure 5.5c. What fraction of the time does the instrument see accelerations less than zero?

5.10 A local average process $Y_T(t)$ is defined (Vanmarcke, 1984) in problem 3.7 by

$$Y_T(t) = \frac{1}{T} \int_{t - T/2}^{t + T/2} X(u) \, du$$

Show that the spectral density function of $Y_T(t)$ is given by

$$S_{Y_T}(\omega) = S_X(\omega) \left[\frac{\sin(\omega T/2)}{\omega T/2} \right]^2$$

Hint: Think of a window of length T multiplying $X(t)$ and integrating over $-\infty$ to ∞.

5.11 *Simulation by discrete Fourier transform.* The following (long) project is suggested for those students who have access to a computer program, e.g., a sophisticated spreadsheet, that has a function call to a discrete Fourier transform, e.g., through a fast Fourier transform (FFT) algorithm. (See the problems for Chapter 4 for a brief explanation.) The student may wish to defer these problems until after studying Chapter 13.

The simulation procedure is essentially the reverse of the analytical procedure suggested in Eq. (5.21). The value of n is assumed to be a power of 2.

a Create a real vector S_k of length $n/2 + 1$, $k = 0, \ldots, n/2$, in which to hold half of the two-sided spectral density function to be simulated.

   i Fill the values of S_k, $k = 0, \ldots, n/2 - 1$, with the values of the spectral density corresponding to positive frequencies.

   ii Set $S_k = 0$ for $k = 0$ and $k = n/2$. (The reason is explained in Chapter 13.)

b Create a vector Θ_k of length $n/2 + 1$, $k = 0, \ldots, n/2$, and fill it with random numbers distributed uniformly on $(0, 2\pi)$. This vector is a sample realization of the random phase angle for each frequency of the sample realization to be generated.

c Create a complex vector X_k of length n, $k = 0, \ldots, n - 1$. Set $X_k = \sqrt{S_k} \exp(i\Theta_k)$ for $k = 0, \ldots, n/2$, and set $X_{n-k} = X_k^*$ for $k = n/2 + 1, \ldots, n - 1$. This vector is a sample realization of the complex Fourier transform of the sample realization to be generated.

d Inverse Fourier transform from the frequency domain to the time domain using the inverse discrete Fourier transform. Multiply by $\sqrt{n}/(2\Delta t)$ to get x_j.

Let $n = 128$ and generate simulations of:

   a "White noise"

   b Low-pass noise with $f_c = n \, \Delta f/4$

   c Bandpass noise with $f_{upper} = n \, \Delta f/4$ and $f_{lower} = n \, \Delta f/8$

Experiment with different values of n and of the frequency cutoffs. Document your code and experiments.

6

STATISTICAL PROPERTIES OF RANDOM PROCESSES

Every sample realization of a random process is unique. Statistics such as the largest value reached in a given time interval, say $(t_1, t_2]$, vary from one realization to another and are random variables. This chapter presents the derivations of some of the most well-known of these statistics for stationary random processes, including the level-crossing problem and the distribution of peaks. Figure 6.1 illustrates these statistics.

To read the original development of the level-crossing analysis, the envelope distribution, and the peak distribution, the reader is referred to the original work of Rice (1944, 1945).

6.1 LEVEL CROSSINGS

The level-crossing problem is the most basic of the statistics we wish to cover. In this section, we introduce the concept of a first-passage problem, derive the expected value of the rate of level crossings, and specialize the general result for certain important cases. The concept of an envelope process is introduced as an approach to correct the results for the clumping phenomenon.

6.1.1 Preliminary Remarks

Figure 6.2 illustrates several sample realizations of a stationary random process $X(t)$. Pick an arbitrary time interval $(t, t + \Delta t]$ and count the number of times each sample realization crosses an arbitrary level, $x = a$. Call this number $N_a(t, t + \Delta t)$. As indicated by the figure, the number of level crossings, or "arrivals," is a random variable; each sample realization produces an obser-

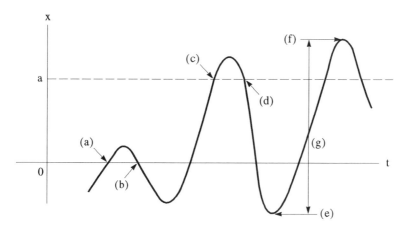

Figure 6.1 Illustration of statistics of interest: (a, b) crossing level $x = 0$ with positive and negative velocity, respectively; (c, d) crossing level $x = a$ with positive and negative velocity, respectively; (e, f) extrema, a valley and a peak, respectively; (g) rise from valley to next peak.

vation. Furthermore, we may think of $N_a(0, t)$ as a nonstationary random process counting the number of level crossings (arrivals), starting with $N_a(0, 0) = 0$ at $t = 0$[1] (Figure 6.3).

If the arrivals are independent and the arrival rate is stationary, the counting process $N_a(t)$ is a Poisson process and is completely characterized by the arrival rate λ. The expected value and variance of $N_a(t)$ are

$$\mu_{N_a}(t) = E[N_a(t)] = \lambda t \tag{6.1}$$

$$\sigma^2_{N_a}(t) = \lambda t \tag{6.2}$$

The waiting time until the first arrival, Y, is exponentially distributed and the expected time to the first arrival is

$$\mu_Y = E[Y] = \frac{1}{\lambda} \tag{6.3}$$

This suggests that we focus on calculating the expected arrival rate λ.

A close look at Figure 6.4 reveals that the Poisson process assumption of independent arrivals may not be appropriate: There are two kinds of obvious correlations between level crossings. First, the crossings almost always occur in pairs, one with positive velocity followed by another with negative velocity, that is, an up crossing and a down crossing, respectively. This kind of correlation is addressed in the next section by narrowing our focus to only the up crossings, omitting the dependent down crossings from the analysis.

[1]For a stationary process, the starting point $t = 0$ is arbitrarily located.

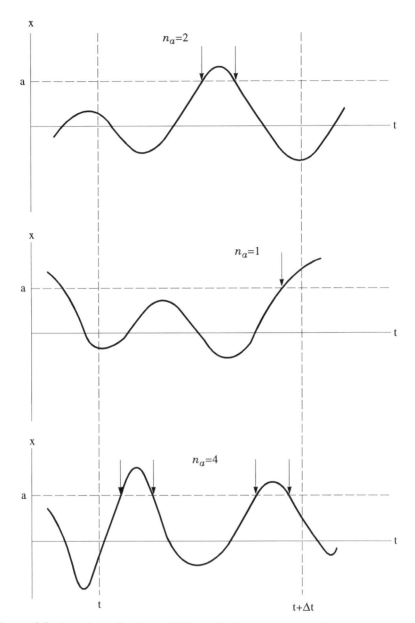

Figure 6.2 Sample realizations of $X(t)$ indicating crossings of level $x = a$ during interval $(t, t + \Delta t)$.

The second kind of obvious correlation between level crossings is apparent when the process is narrow band and it appears to be a randomly modulated sine wave. If $X(t)$ up-crosses the level $x = a$ in one cycle, it is highly likely that it will up-cross it again in the next cycle: level up crossings tend to occur in clusters or clumps. We address this kind of correlation between arrivals later

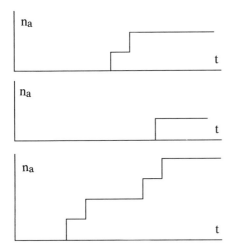

Figure 6.3 Sample realizations of $N_a(t)$ counting level crossings of processes in Figure 6.2.

in this chapter by defining another random process that up-crosses $x = a$ only once per clump. This new process is called an envelope.

The general problem of going beyond the expected value of the arrival rate and determining the full probability distribution of arrival times is known as the first-passage time problem. The solution of this difficult problem is outside the scope of this text. The interested reader is referred to Vanmarcke (1984) and Marley (1991) for good summaries. In Chapter 10, simple approaches to the design against first-passage failures are discussed.

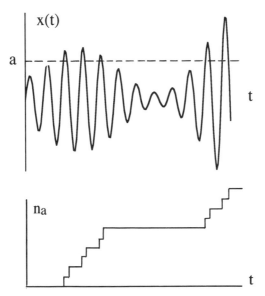

Figure 6.4 Sample realization of $X(t)$ and corresponding $N_a(t)$ showing two kinds of correlation between level crossings violating Poisson arrival assumption.

6.1.2 Derivation of Expected Rate of Level Crossing

Let $X(t)$ be a zero-mean, stationary random process. For convenience, we interpret $X(t)$ as a displacement and its derivatives $\dot{X}(t)$ and $\ddot{X}(t)$ as its velocity and acceleration, respectively. Because the process is stationary, the expected number of level crossings in a time interval $(t, t + \Delta t]$ depends only on the length of the interval, Δt:

$$E[N_a(t, t + \Delta t) = E[N_a(\Delta t)] \tag{6.4}$$

It follows that the expected number of level crossings should be proportional to Δt, for example, an interval of $2\,\Delta t$ should, on average, have twice as many crossings as in Δt. So we let ν_a be the expected rate of level crossings per unit time and write

$$E[N_a(\Delta t)] = \nu_a\,\Delta t \tag{6.5}$$

Since up crossings and down crossings may be correlated, we concentrate on only the "up crossings" with positive slope. The expected level-up-crossing rate ν_{a^+} is half the total crossing rate:

$$\nu_{a^+} = \tfrac{1}{2}\nu_a \tag{6.6}$$

We may calculate this rate from the probability that a level up crossing occurs.

Consider the limiting behavior of the counting process as the interval Δt tends to zero. If $X(t)$ is continuous and differentiable, there will be either zero or one up crossing of $x = a$ in Δt. Let A represent the event that there is an up crossing of $x = a$ in the infinitesimal time interval dt:

$$A = \{x = a \text{ is crossed with positive slope in } dt\}$$

We may express the probability distribution of the number of level up crossings in dt in terms of $P(A)$ as

$$P(N_{a^+}(dt)) = \begin{cases} P(A) & N_{a^+}(dt) = 1 \\ 1 - P(A) & N_{a^+}(dt) = 0 \\ 0 & \text{otherwise} \end{cases} \tag{6.7}$$

The expected number of level up crossings in dt is calculated from this distribution as

$$E[N_{a^+}(dt)] = (1) \cdot P(A) + (0) \cdot [1 - P(A)] = P(A) \tag{6.8}$$

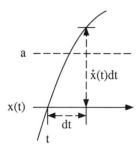

Figure 6.5 Event A, up crossing of $x = a$ in interval $(t,\ t + dt)$.

But Eq. (6.5) also calculates the expected number of level up crossings: $E[N_{a^+}(dt)] = \nu_{a^+}\ dt$. So we have the equality

$$\nu_{a^+}\ dt = P(A) \tag{6.9}$$

and we have a route to calculate ν_{a^+} in terms of $P(A)$.

The probability of event A may, in turn, be calculated directly from the conditions on $X(t)$ that produce a level up crossing. Figure 6.5 illustrates the occurrence of event A for a sample realization. In order to have an up crossing of $x = a$ in the interval $(t,\ t + dt]$, the following must be true:

- $X(t)$ must be below $x = a$ at the start of the interval:

$$X(t) < a \tag{6.10}$$

- $X(t)$ must have positive slope at the start of the interval:

$$\dot{X}(t) > 0 \tag{6.11}$$

- $X(t)$ must be above $x = a$ at the end of the interval:

$$X(t + dt) > a \tag{6.12}$$

which may be expressed in terms of the derivative $\dot{X}(t)$ as

$$X(t) + \dot{X}(t)\ dt > a \quad \text{or} \quad X(t) > a - \dot{X}(t)\ dt \tag{6.13}$$

The probability that an up crossing occurs is the probability that these three conditions are met. Writing them as a single probability statement we have

$$P(A) = P(a - \dot{X}(t)\ dt < X(t) < a \cap \dot{X}(t) > 0) \tag{6.14}$$

These conditions define a triangular area in the x–$\dot{x}$ plane, as shown in Figure 6.6. The probability of event A is calculated by integrating the joint probability

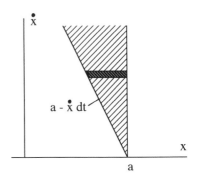

Figure 6.6   Region in x–$\dot{x}$ plane for which event
A occurs.

density function of X and $\dot{X}$ over this region:

$$P(A) = \int_0^\infty \int_{a-vdt}^a f_{X\dot{X}}(u, v)\, du\, dv \qquad (6.15)$$

As dt tends to zero, the integral over u is equal to the height of the density function evaluated at $x = a$, $f_{X\dot{X}}(a, v)$, times the width of the interval, $|v|dt$, giving us

$$P(A) = \int_0^\infty f_{X\dot{X}}(a, v)(|v|\, dt)\, dv \qquad (6.16)$$

Thus, the level-up-crossing rate may be calculated from

$$\nu_{a^+} = \int_0^\infty v f_{X\dot{X}}(a, v)\, dv \qquad (6.17)$$

This is a general result for any stationary random process, regardless of distribution or bandwidth.

6.1.3   Specializations

The joint density function of $X(t)$ and $\dot{X}(t)$ is generally unknown unless the process is Gaussian. Recall that the derivative $\dot{X}(t)$ is the limit, as h goes to zero, of a difference, $[X(t + h) - X(t)]/h$. If $X(t)$ is Gaussian, the reproductive property of the Gaussian distribution says that $\dot{X}(t)$ will also be Gaussian. If $X(t)$ is not Gaussian, the distribution of $\dot{X}(t)$ must generally be determined by convolution.

This section presents the specialization of the general level-crossing rate result for Gaussian processes and simple applications of the formula to find the rate of zero up crossings, the rate of peaking, and a simple measure of bandwidth.

Level Up Crossings. If $X(t)$ is a Gaussian process, the fact that $X(t)$ and $\dot{X}(t)$ are uncorrelated implies that they are independent.[2] Therefore, their joint density function is equal to the product of their marginal densities and Eq. (6.17) becomes

$$\nu_{a+} = \int_0^\infty v f_X(a) f_{\dot{X}}(v)\, dv = f_X(a) \int_0^\infty v f_{\dot{X}}(v)\, dv \tag{6.18}$$

Substituting the Gaussian density functions into this equation gives the result

$$\nu_{a+} = \frac{1}{2\pi} \frac{\sigma_{\dot{X}}}{\sigma_X} \exp\left(-\frac{1}{2}\frac{a^2}{\sigma_X^2}\right) \tag{6.19}$$

Notice that the level-up-crossing rate decreases as the threshold a increases. For a given σ_X, the crossing rate increases as the RMS "velocity" $\sigma_{\dot{X}}$ increases.

Zero Up Crossings. If the level of interest is zero, $a = 0$, the crossings are called zero up crossings. Substituting $a = 0$ into Eq. (6.19) for the Gaussian case yields

$$\nu_{0+} = \frac{1}{2\pi} \frac{\sigma_{\dot{X}}}{\sigma_X} \tag{6.20}$$

The units of ν_{0+} are number of zero up crossings per unit time and ν_{0+} is often thought of as the expected frequency of the process, measured in zero-crossing cycles per unit time, or hertz.

If a frequency with units of radians per unit time, ω_{0+}, is desired, the formula is given by

$$\omega_{0+} = 2\pi\nu_{0+} = \frac{\sigma_{\dot{X}}}{\sigma_X} \tag{6.21}$$

Recall that the mean-square displacement and velocity can be related to the moments of the spectral density function as

$$\sigma_X^2 = \int_{-\infty}^\infty S_X(\omega)\, d\omega = \int_0^\infty W_X(f)\, df \tag{6.22}$$

$$\sigma_{\dot{X}}^2 = \int_{-\infty}^\infty \omega^2 S_X(\omega)\, d\omega = (2\pi)^2 \int_0^\infty f^2 W_X(f)\, df \tag{6.23}$$

---

[2]Recall that this is true of the joint Gaussian distribution but is not true in general.

Thus, the zero-up-crossing frequency may be conveniently written in terms of the moments of the spectral density function as

$$\nu_{0+} = \sqrt{\frac{\displaystyle\int_0^\infty f^2 W_X(f)\, df}{\displaystyle\int_0^\infty W_X(f)\, df}} \tag{6.24}$$

or

$$\omega_{0+} = \sqrt{\frac{\displaystyle\int_{-\infty}^\infty \omega^2 S_X(\omega)\, d\omega}{\displaystyle\int_{-\infty}^\infty S_X(\omega)\, d\omega}} \tag{6.25}$$

Peak Frequency. The same level-crossing analysis can be performed on the "velocity" process $\dot{X}(t)$. A zero down crossing of $\dot{X}(t)$, that is, a change in velocity from positive to negative, corresponds to the occurrence of a peak in $X(t)$. The *peak frequency*, ν_p may be shown to be

$$\nu_p = \int_{-\infty}^0 -w f_{\dot{X}\ddot{X}}(0, w)\, dw \tag{6.26}$$

In the Gaussian case, this becomes

$$\nu_p = \frac{1}{2\pi} \frac{\sigma_{\ddot{X}}}{\sigma_{\dot{X}}} \tag{6.27}$$

$$\omega_p = 2\pi\nu_p = \frac{\sigma_{\ddot{X}}}{\sigma_{\dot{X}}} \tag{6.28}$$

In terms of the moments of the spectral density function we have

$$\nu_p = \sqrt{\frac{\displaystyle\int_0^\infty f^4 W_X(f)\, df}{\displaystyle\int_0^\infty f^2 W_X(f)\, df}} \tag{6.29}$$

$$\omega_p = \sqrt{\frac{\displaystyle\int_{-\infty}^\infty \omega^4 S_X(\omega)\, d\omega}{\displaystyle\int_{-\infty}^\infty \omega^2 S_X(\omega)\, d\omega}} \tag{6.30}$$

Bandwidth and Irregularity. Suppose $X(t)$ is a narrow-band Gaussian process with spectral density function in the limit given by

$$S_x(\omega) = \sigma_X^2[\tfrac{1}{2}\delta(\omega + \omega_m) + \tfrac{1}{2}\delta(\omega - \omega_m)] \tag{6.31}$$

where ω_m is the midband, or nominal, frequency of the process. Substituting this spectral density into the equations above gives

$$\omega_{0+} = \omega_m \tag{6.32}$$

$$\omega_p = \omega_m \tag{6.33}$$

For a theoretical narrow-band process, the zero-up-crossing and peak frequencies are exactly the same as the nominal frequency of the process.

For non-narrow-band processes, the ratio of the zero-up-crossing frequency to the peak frequency gives a measure of the "bandwidth" of the process. The ratio is known as the irregularity factor α:

$$\alpha = \frac{\nu_{0+}}{\nu_p} = \frac{\omega_{0+}}{\omega_p} \tag{6.34}$$

One may think of this as the ratio of the expected number of zero up crossings in a given time interval Δt to the expected number of peaks in the same interval:

$$\alpha = \frac{E[N_{0+}(\Delta t)]}{E[N_p(\Delta t)]} \tag{6.35}$$

For a narrow-band process, $\alpha = 1$: There is exactly one peak for every zero up crossing. For wider bandwidths, more than one peak might occur between zero up crossings, causing α to decrease. The theoretical limiting case is $\alpha = 0$, which corresponds to an infinite number of peaks for every zero up crossing. One might imagine this as a high-frequency dithering of a process as illustrated in Figure 6.7.

Sometimes a factor that is more in keeping with the notion of "width" is used. The *spectral width parameter* ϵ is given by

$$\epsilon = \sqrt{1 - \alpha^2} \tag{6.36}$$

So defined, $\epsilon = 0$ corresponds to a narrow-band process, and as the bandwidth increases, so does ϵ.

Further insight into the bandwidth of a process can be gained in the Gaussian case. If $X(t)$ is Gaussian, we have

$$\alpha = \frac{\sigma_{\dot{X}}^2}{\sigma_X \sigma_{\ddot{X}}} \tag{6.37}$$

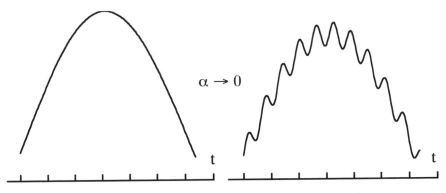

Figure 6.7 Limiting case of irregularity factor $\alpha \to 0$.

Note that the correlation between the displacement and acceleration is given by

$$E[X(t)\ddot{X}(t)] = \sigma_{X\ddot{X}} = -\sigma_{\dot{X}}^2 \tag{6.38}$$

This gives

$$\alpha = \frac{-\sigma_{X\ddot{X}}}{\sigma_X \sigma_{\ddot{X}}} = -\rho_{X\ddot{X}} \tag{6.39}$$

Hence, for a Gaussian process, the irregularity factor is equal to minus the correlation coefficient between the displacement and acceleration. The more strongly $X(t)$ and $\ddot{X}(t)$ are correlated, the more regular the process will be.

Example 6.1: Results for Gaussian Bandpass Noise. Consider Gaussian bandpass noise defined by the one-sided spectrum

$$W_X(f) = \begin{cases} W_0 & f_a < f < f_b \\ 0 & \text{otherwise} \end{cases}$$

and let $f_a = \beta f_b$. The formulas for the statistics of this random process may be shown to be

$$\sigma_X^2 = W_0(f_b - f_a) = W_0 f_b(1 - \beta)$$

$$\sigma_{\dot{X}}^2 = \frac{(2\pi)^2}{3} W_0(f_b^3 - f_a^3) = \frac{(2\pi)^2}{3} W_0 f_b^3(1 - \beta^3)$$

$$\sigma_{\ddot{X}}^2 = \frac{(2\pi)^4}{5} W_0(f_b^5 - f_a^5) = \frac{(2\pi)^4}{5} W_0 f_b^5(1 - \beta^5)$$

$$\nu_a = \frac{1}{\sqrt{3}} \sqrt{\frac{f_b^3 - f_a^3}{f_b - f_a}} \exp\left(-\frac{1}{2}\frac{a^2}{\sigma_x^2}\right)$$

$$= f_b \frac{1}{\sqrt{3}} \sqrt{\frac{1 - \beta^3}{1 - \beta}} \exp\left(-\frac{1}{2}\frac{a^2}{\sigma_x^2}\right)$$

$$\nu_{0+} = \frac{1}{\sqrt{3}} \sqrt{\frac{f_b^3 - f_a^3}{f_b - f_a}} = f_b \frac{1}{\sqrt{3}} \sqrt{\frac{1 - \beta^3}{1 - \beta}}$$

$$\nu_p = \sqrt{\frac{3}{5}} \sqrt{\frac{f_b^5 - f_a^5}{f_b^3 - f_a^3}} = f_b \sqrt{\frac{3}{5}} \sqrt{\frac{1 - \beta^5}{1 - \beta^3}}$$

$$\alpha = \frac{\sqrt{5}}{3} \frac{f_b^3 - f_a^3}{\sqrt{(f_b - f_a)(f_b^5 - f_a^5)}} = \frac{\sqrt{5}}{3} \frac{1 - \beta^3}{\sqrt{(1 - \beta)(1 - \beta^5)}}$$

For the sake of a numerical example, let $W_0 = 0.01$, $f_b = 100$, and $a = 3\sigma_X$. The results for various values of β are as follows:

β	σ_X	$\nu_{a=3\sigma_X}$	ν_{0+}	ν_p	α
0	1.00	0.641	57.735	77.460	0.745
0.1	0.949	0.676	60.828	77.498	0.785
0.2	0.894	0.714	64.291	77.759	0.827
0.3	0.837	0.756	68.069	78.432	0.868
0.4	0.775	0.801	72.111	79.653	0.905
0.5	0.707	0.848	76.376	81.504	0.937
0.6	0.623	0.898	80.829	84.012	0.962
0.7	0.548	0.949	85.440	87.164	0.980
0.8	0.447	1.002	90.185	90.919	0.992
0.9	0.316	1.056	95.044	95.219	0.998

Notice that all crossing frequencies increase as the average frequency increases due to shifting the lower cutoff frequency to the right.

6.1.4 Rice's Narrow-Band Envelope

As mentioned in our preliminary remarks, level up crossings tend to occur in clusters or clumps if the process is narrow band (Figure 6.4). We intuitively expect the waiting times to have a mixed distribution: Most arrivals occur in clusters approximately one cycle period apart, but between clusters there may be a substantial amount of time. The waiting times are obviously not exponentially distributed.

For engineering design we are often mainly interested in the first time $X(t)$ crosses the level $x = a$. Subsequent crossings after the first are less important.

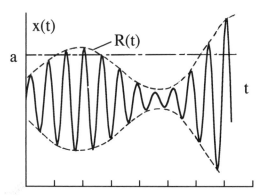

Figure 6.8 Envelope random process $R(t)$.

To address this, we define a random process called an "envelope," such as shown in Figure 6.8. The envelope first crosses $x = a$ at about the same time as $X(t)$ but remains above $x = a$ while the clustered arrivals occur. So our problem is now to find the up-crossing rate of the envelope.

The following simple envelope is introduced by S. O. Rice (1944, 1945)[3] in his landmark paper on stochastic processes. Let $X(t)$ be a stationary narrow-band Gaussian random process. Figure 6.8 suggests that we may think of $X(t)$ as a randomly modulated harmonic with a random phase angle such that

$$X(t) = R(t) \cos [\omega_m t + \Theta(t)] \tag{6.40}$$

Random process $R(t)$ is the envelope and $R(t)$ and $\Theta(t)$ vary slowly in comparison with $X(t)$. We may restrict $R(t)$ to only positive values and $\Theta(t)$ to $(0, 2\pi)$ without loss of generality.

To apply the level-up-crossing formula [Eq. (6.17)], we require the joint density function of $R(t)$ and its derivative $\dot{R}(t)$. The important intermediate results in obtaining this distribution are summarized below.

Another representation of $X(t)$ is

$$X(t) = C(t) \cos (\omega_m t) - S(t) \sin (\omega_m t) \tag{6.41}$$

where $C(t)$ and $S(t)$ are independent and identically distributed Gaussian processes with mean equal to zero and variance equal to σ_X^2. The processes $R(t)$, $\Theta(t)$, $C(t)$, and $S(t)$ are related by

$$C(t) = R(t) \cos [\Theta(t)] \tag{6.42}$$

$$S(t) = R(t) \sin [\Theta(t)] \tag{6.43}$$

---

[3]This paper is one of the most frequently cited in the random vibrations literature and was motivated by the study of telephone transmission lines.

Their derivatives are related by

$$\dot{C}(t) = \dot{R}(t) \cos [\Theta(t)] - R(t) \sin [\Theta(t)]\dot{\Theta}(t) \tag{6.44}$$

$$\dot{S}(t) = \dot{R}(t) \sin [\Theta(t)] + R(t) \cos [\Theta(t)]\dot{\Theta}(t) \tag{6.45}$$

Rice shows that if the one-sided spectral density function $W_X(f)$ is symmetric about the midband frequency ω_m, then $C(t)$, $S(t)$, and their derivatives $\dot{C}(t)$ and $\dot{S}(t)$ are all independent.[4] Let σ_1^2 be the common variance of $\dot{C}(t)$ and $\dot{S}(t)$, which we will determine later. The joint probability density function of $C(t)$, $S(t)$, $\dot{C}(t)$, and $\dot{S}(t)$ is then

$$f_{CS\dot{C}\dot{S}}(c, s, \dot{c}, \dot{s}) = \frac{1}{4\pi^2 \sigma_X^2 \sigma_1^2} \exp \left[-\frac{1}{2} \left(\frac{c^2 + s^2}{\sigma_X^2} + \frac{\dot{c}^2 + \dot{s}^2}{\sigma_1^2} \right) \right]$$

$$c, s > 0 \qquad -\infty < \dot{c}, \dot{s} < \infty \tag{6.46}$$

By transformation, the joint probability density function of $R(t)$, $\Theta(t)$, $\dot{R}(t)$, and $\dot{\Theta}(t)$ is

$$f_{R\Theta\dot{R}\dot{\Theta}}(r, \theta, \dot{r}, \dot{\theta}) = \frac{r^2}{4\pi^2 \sigma_X^2 \sigma_1^2} \exp \left[-\frac{1}{2} \left(\frac{r^2}{\sigma_X^2} + \frac{\dot{r}^2 + r^2\dot{\theta}^2}{\sigma_1^2} \right) \right]$$

$$r > 0 \qquad 0 < \theta \le 2\pi \qquad -\infty < \dot{r} < \infty$$

$$-\infty < \dot{\theta} < \infty \tag{6.47}$$

The reader may verify that envelope $R(t)$ has a Rayleigh distribution and the phase $\Theta(t)$ has a uniform distribution.[5] The joint probability density function of $R(t)$ and $\dot{R}(t)$ is

$$f_{R\dot{R}}(r, \dot{r}) = \frac{r}{\sqrt{2\pi}\sigma_X^2 \sigma_1} \exp \left[-\frac{1}{2} \left(\frac{r^2}{\sigma_X^2} + \frac{\dot{r}^2}{\sigma_1^2} \right) \right]$$

$$r > 0 \qquad -\infty < \dot{r} < \infty \tag{6.48}$$

The joint density function is the product of the two marginal densities:

$$f_R(r) = \frac{r}{\sigma_X^2} \exp \left(-\frac{1}{2} \frac{r^2}{\sigma_X^2} \right) \qquad r > 0 \tag{6.49}$$

[4]By assumption, we know that $C(t)$ and $S(t)$ are independent. Since the process is stationary, each is uncorrelated with its own derivative. Symmetry is required of $W_X(f)$ in order to assure that $C(t)$ and $\dot{S}(t)$ are uncorrelated, and likewise $S(t)$ and $\dot{C}(t)$. For Gaussian random variables, uncorrelated implies independent.
[5]This should be expected from the Box–Muller transformation discussed in Chapter 2.

$$f_{\dot{R}}(\dot{r}) = \frac{1}{\sqrt{2\pi}\sigma_1} \exp\left(-\frac{1}{2}\frac{\dot{r}^2}{\sigma_1^2}\right) \qquad -\infty < \dot{r} < \infty \qquad (6.50)$$

So $R(t)$ and $\dot{R}(t)$ are independent.

Next, we determine σ_1^2, the common variance of $\dot{C}(t)$ and $\dot{S}(t)$. Differentiating Eq. (6.41), the derivative of $X(t)$ is

$$\dot{X}(t) = \dot{C}(t) \cos(\omega_m t) - C(t)\omega_m \sin(\omega_m t)$$
$$- \dot{S}(t) \sin(\omega_m t) - S(t)\omega_m \cos(\omega_m t) \qquad (6.51)$$

We see that $\dot{X}(t)$ is a linear combination of four independent random processes. Its variance is the sum of the four variances and reduces to

$$\sigma_{\dot{X}}^2 = \sigma_1^2 + \omega_m^2\sigma_X^2 \qquad (6.52)$$

Solving for σ_1^2, we have

$$\sigma_1^2 = \sigma_{\dot{X}}^2 - \omega_m^2\sigma_X^2 \qquad (6.53)$$

Substituting $\sigma_{\dot{X}}^2 = \sigma_X^2\omega_{0+}^2$, we have

$$\sigma_1^2 = \sigma_X^2(\omega_{0+}^2 - \omega_m^2) \qquad (6.54)$$

This may also be written as a moment of the spectral density function taken about ω_m:

$$\sigma_1^2 = \int_{-\infty}^{\infty} (\omega^2 - \omega_m^2)S_X(\omega)\, d\omega \qquad (6.55)$$

We see that as $X(t)$ tends to its theoretical narrow-band limit, σ_1^2 tends to zero. This means that each sample realization of the envelope of a theoretical narrow-band process does not vary and $X(t)$ is a pure sine wave without random variation.

The expected rate of level up crossing for the envelope is found by substituting the joint density function of $R(t)$ and $\dot{R}(t)$ [Eq. (6.48)] into the general formula for the up-crossing rate [Eq. (6.17)]:

$$\nu_{R=a^+} = \int_0^{\infty} \dot{r}f_{R\dot{R}}(a, \dot{r})\, d\dot{r}$$
$$= \int_0^{\infty} \dot{r}\frac{r}{\sqrt{2\pi}\sigma_X^2\sigma_1} \exp\left[-\frac{1}{2}\left(\frac{r^2}{\sigma_X^2} + \frac{\dot{r}^2}{\sigma_1^2}\right)\right] d\dot{r} \qquad (6.56)$$

Evaluating the integral yields

$$\nu_{R=a^+} = \frac{1}{\sqrt{2\pi}}\frac{a\sigma_1}{\sigma_X^2} \exp\left(-\frac{1}{2}\frac{a^2}{\sigma_X^2}\right) \qquad (6.57)$$

By eliminating σ_1, we have the following equation for the expected level-up-crossing rate of the envelope:

$$\nu_{R=a^+} = \frac{1}{\sqrt{2\pi}} \frac{a}{\sigma_X} \sqrt{\omega_{0^+}^2 - \omega_m^2} \exp\left(-\frac{1}{2}\frac{a^2}{\sigma_X^2}\right) \tag{6.58}$$

The meaning of this result can best be seen by comparing it with the expected rate of level up crossings of the original Gaussian process $X(t)$. For almost every level up crossing of the envelope, there will be one or more crossings by $X(t)$. R. H. Lyon (1961) calls the ratio of the two expected crossing rates the *average clump size* $\langle cs(a) \rangle$:

$$\langle cs(a) \rangle = \frac{\nu_{X=a^+}}{\nu_{R=a^+}} \tag{6.59}$$

Substituting the equations for these two crossing rates [Eqs. (6.19) and (6.58)], we get

$$\langle cs(a) \rangle = \frac{1}{\sqrt{2\pi}} \frac{\sigma_X}{a} \frac{\omega_{0^+}}{\sqrt{\omega_{0^+}^2 - \omega_m^2}} = \frac{1}{\sqrt{2\pi}} \frac{\sigma_X}{a} \frac{1}{\sqrt{1 - (\omega_m/\omega_{0^+})^2}} \tag{6.60}$$

As seen in the example below, the average clump size may be quite large.

The clumping phenomenon has particular impact on first-passage time problems. Taking the expected envelope crossing rate instead of the original process crossing rate extends the expected time to the first crossing by a factor equal to the average clump size:

$$E[\text{time of first crossing of } x = a \text{ by } R(t)]$$

$$= \langle cs(a) \rangle E[\text{time of first crossing of } x = a \text{ by } X(t)] \tag{6.61}$$

This may be a significant increase in waiting time.

Example 6.2 Narrow-Band Envelope Up-Crossing Rate for Bandpass Noise.
Let us continue the case of bandpass noise from the previous example using narrow-band-like values of β. The envelope level-up-crossing rate and average clump size may be shown to be

$$\nu_{R=a^+} = \sqrt{2\pi} \frac{a}{\sigma_X} \sqrt{f_0^2 - f_m^2} \exp\left(-\frac{1}{2}\frac{a^2}{\sigma_X^2}\right)$$

where $f_0 = \nu_{0^+}$ and $f_m = \frac{1}{2}(f_a + f_b) = \frac{1}{2}f_b(\beta + 1)$.

For the sake of a numerical example, again let $W_0 = 0.01$, $f_b = 100$, and $a = 3\sigma_X$ (see Example 6.1). Let $\beta = 0.8, 0.9$, so that $\alpha = 0.992, 0.998$, respectively. The results are as follows:

β	α	f_0	f_m	$\nu_{a=3\sigma_X}$	$\nu_{R=a=3\sigma_X}$	$\langle cs(a = 3\sigma_X)\rangle$
0.8	0.992	90.185	90.000	1.002	0.482	2.077
0.9	0.998	95.044	95.000	1.056	0.241	4.378

Notice that, in this example, the envelope level-up-crossing rate decreases, rather than increases, as the average frequency increases because the bandwidth decreases. The envelope crossing rate is dependent on the bandwidth. Narrower bandwidth implies less variation of the envelope.

The average clump size for $\beta = 0.9$ and $a = 3\sigma_X$ is 4.378. That is, for each envelope up crossing of $x = 3\sigma_X$, on average there are over four peaks of $X(t)$ above that level. Viewed another way, the expected waiting time for the first crossing of the envelope over $x = 3\sigma_X$ is over four times longer than the expected waiting time for the first crossing of $X(t)$ itself. The average clump size is very sensitive to the bandwidth for $\alpha > 0.99$.

6.2 DISTRIBUTIONS OF EXTREMA

The probability distributions of the extrema of a random process, that is, its "peaks and valleys," are of particular interest in engineering design problems. This section presents a simple approach to deriving the distribution of the peaks of a narrow-band process and then a more general solution for the peaks of a wide-band process. An approximate distribution for the height of the rise between a valley and the next peak and associated joint distributions are presented at the end of the section.

6.2.1 Simple Approach to Distribution of Peak

The distribution of the peak values in a random process, say Z, is of classical interest in engineering design problems. For example, if $X(t)$ is a random loading on a structure, what is the peak load? For such problems, the interest is usually in determining a single value to represent the "largest" load over the design life. This issue is discussed at length in Chapter 10.

Cyclic loading is also of interest in design for metal fatigue. Our interest in this case is in the height H of the rise in a random process from a "valley" to the next "peak." Finding the exact distribution of these rises is a more difficult problem than the distribution of peaks. However, both the peak distribution and the rise distribution are known exactly for the theoretical narrow-band process. We derive these distributions next.

The approach followed in this article is due to Alan Powell (1958). As done earlier in this chapter, let event A be the event that $x = z$ is crossed with positive slope in dt. The probability of event A is equal to the expected level-up-crossing rate times dt:

$$\nu_{z^+} \, dt = P(A) \tag{6.62}$$

Consider a second level, a little higher than the first, say at $x = z + \Delta z$. There are slightly fewer crossings of this level, because it is higher. Intuitively, the difference between the two crossing rates must be equal to the expected rate at which peaks occur in the interval $(z, z + \Delta z]$:

$$E[\text{rate of peaking in } (z, z + \Delta z]] = \nu_{z^+} - \nu_{z + \Delta z^+} \qquad (6.63)$$

The expected rate of peaking in this interval may also be expressed as a fraction of the total rate of peaking. This is given by the expected rate of peaking in any interval ν_p times the probability that the peak is in $(z, z + \Delta z]$:

$$E[\text{rate of peaking in } (z, z + \Delta z]]$$
$$= E[\text{total rate of peaking}] \cdot P(\text{peak in } (z, z + \Delta z]) \qquad (6.64)$$

Thus, the probability that a peak occurs in $(z, z + \Delta z]$ is the ratio of the expected rate of peaking in $(z, z + \Delta z]$ to the expected total rate of peaking.

Let the interval width Δz shrink to zero. The limit of the difference in Eq. (6.63) becomes

$$\lim_{\Delta z \to 0} \{\nu_{z^+} - \nu_{z + \Delta z^+}\} = -\frac{d}{dz} \nu_{z^+} \, dz \qquad (6.65)$$

Also, the probability in Eq. (6.64) becomes

$$\lim_{\Delta z \to 0} P(\text{peak in } (z, z + \Delta z]) = f_Z(z) \, dz \qquad (6.66)$$

where random variable Z is the value of the peak. Combining equations for the expected rate of peaking, we have

$$f_Z(z) dz = \frac{-\dfrac{d}{dz} \nu_{z^+} \, dz}{\nu_p} \qquad (6.67)$$

Eliminating the differentials yields

$$f_Z(z) = \frac{-\dfrac{d}{dz} \nu_{z^+}}{\nu_p} \qquad (6.68)$$

If the zero-mean random process is narrow band, the peaking frequency may be replaced by the zero-up-crossing frequency:

$$f_Z(z) = \frac{-\dfrac{d}{dz} \nu_{z^+}}{\nu_{0^+}} \qquad (6.69)$$

The implication is that every zero up crossing corresponds to exactly one peak greater than zero, $z > 0$. This equation is applicable to zero-mean narrow-band random processes with arbitrary distribution.

In the special case of Gaussian processes, we may substitute our previous results for the level-up-crossing rate and zero-up-crossing rate, Eqs. (6.19) and (6.20), respectively. This gives us the probability density function of peaks, $f_Z(z)$, for a zero-mean, narrow-band Gaussian process as

$$f_Z(z) = \frac{z}{\sigma_X^2} \exp\left(-\frac{1}{2}\frac{z^2}{\sigma_X^2}\right) \qquad z > 0 \qquad (6.70)$$

This is recognized as a Rayleigh distribution.

For a zero-mean narrow-band process, the height of the rise, H, from a valley to the next peak is twice the height of the peak. In the Gaussian special case, the distribution of rises is found by change of variable, $H = 2Z$, to be

$$f_H(h) = \frac{h}{(2\sigma_X)^2} \exp\left(-\frac{1}{2}\frac{h^2}{(2\sigma_X)^2}\right) \qquad h > 0 \qquad (6.71)$$

Example 6.3: Probability of Peak, Narrow-Band Case. The probability that a peak Z_{nb} in a zero-mean, narrow-band Gaussian random process $X(t)$ is greater than $3\sigma_X$ is $P(Z_{nb} > 3\sigma_X) = 0.0111$. Note that this is greater than the probability that at an arbitrary time $X(t)$ itself is greater than $3\sigma_X$: $P(X(t) > 3\sigma_X) = 0.0013$. The two probabilities should not be confused.

6.2.2 General Approach to Distribution of Peak

Let us reanalyze the above as a conditional probability. Specifically, we would like to calculate the probability that a peak chosen at random from all the peaks of $X(t)$ would be equal to Z. This is a conditional probability because it is understood that the peak values have been preselected for us, so that every sample we take is a peak. Let C be the conditioning event that we randomly sample $X(t)$ in an arbitrary time interval of length dt, and $X(t)$ is a peak. Let B be the event that a peak chosen at random from the population of peaks is equal to Z. We need to calculate the conditional probability:

$$P(B|C) = \frac{P(B \cap C)}{P(C)} \qquad (6.72)$$

$$P(\text{peak equals } Z | X(t) \text{ is a peak}) = \frac{P(\text{peak equals } Z \cap X(t) \text{ is a peak})}{P(X(t) \text{ is a peak})}$$

$$(6.73)$$

We must precisely specify what these events are in terms of the displacement $X(t)$, its velocity $\dot{X}(t)$, and its acceleration $\ddot{X}(t)$. The conditions to have a peak of any magnitude are the same as for a zero down crossing of $\dot{X}(t)$, namely:

- $\dot{X}(t)$ must be above zero at the start of the interval:

$$\dot{X}(t) > 0 \tag{6.74}$$

- $\dot{X}(t)$ must have negative slope at the start of the interval:

$$\ddot{X}(t) < 0 \tag{6.75}$$

- $\dot{X}(t)$ must be below zero at the end of the interval:

$$\dot{X}(t + dt) < 0 \tag{6.76}$$

which may be expressed in terms of $\ddot{X}(t)$ as:

$$\dot{X}(t) + \ddot{X}(t)\, dt < 0 \quad \text{or} \quad \dot{X}(t) < 0 - \ddot{X}(t)\, dt \tag{6.77}$$

The probability of event C, that is, of having a peak of any magnitude, may be written by combining these statements as

$$P(C) = P(0 < \dot{X}(t) < 0 - \ddot{X}(t)\, dt \cap \ddot{X}(t) < 0) \tag{6.78}$$

There are no conditions on the value of $X(t)$ itself in this statement.

Event B is a subset of event C, adding the additional constraint that $X(t)$ is equal to Z, or more precisely, $z < X(t) \leq z + dz$. Thus, we write the joint probability of B and C as

$$P(B \cap C) = P(z < X(t) < z + dz,\, 0 < \dot{X}(t) < 0 - \ddot{X}(t)\, dt,\, \ddot{X}(t) < 0) \tag{6.79}$$

Let the joint probability density function of $X(t)$, $\dot{X}(t)$, and $\ddot{X}(t)$ be $f_{X\dot{X}\ddot{X}}(u, v, w)$. The conditional probability of Eq. (6.73) may be written in terms of the integrals of $f_{X\dot{X}\ddot{X}}(u, v, w)$ over the regions defined in Eqs. (6.78) and (6.79):

$$f_Z(z)\, dz = \frac{\displaystyle\int_{-\infty}^{0} \int_{0}^{0-wdt} \int_{z}^{z+dz} f_{X\dot{X}\ddot{X}}(u, v, w)\, du\, dv\, dw}{\displaystyle\int_{-\infty}^{0} \int_{0}^{0-wdt} \int_{-\infty}^{\infty} f_{X\dot{X}\ddot{X}}(u, v, w)\, du\, dv\, dw} \tag{6.80}$$

$$= \frac{\displaystyle\int_{-\infty}^{0} dz\, (-w\, dt) f_{X\dot{X}\ddot{X}}(z, 0, w)\, dw}{\displaystyle\int_{-\infty}^{0} (-w\, dt) f_{\dot{X}\ddot{X}}(0, w)\, dw} \tag{6.81}$$

The denominator is recognized to be the peaking frequency ν_p [Eq. (6.26)]. Canceling differentials yields

$$f_Z(z) = \frac{\displaystyle\int_{-\infty}^{0} - w f_{X\dot{X}\ddot{X}}(z, 0, w) \, dw}{\nu_p} \tag{6.82}$$

This is applicable to zero-mean random processes with arbitrary bandwidth and arbitrary distribution.

If $X(t)$ is Gaussian, the result of the integrations may be expressed, after lengthy manipulations, as

$$f_Z(z) = (1 - \alpha^2) \frac{1}{\sqrt{2\pi(1 - \alpha^2)} \, \sigma_X} \exp\left(-\frac{1}{2} \frac{z^2}{(1 - \alpha^2)\sigma_X^2}\right)$$

$$+ \alpha\Phi\left(\frac{\alpha}{\sqrt{1 - \alpha^2}} \frac{z}{\sigma_X}\right) \frac{z}{\sigma_X^2} \exp\left(-\frac{1}{2} \frac{z^2}{\sigma_X^2}\right) \qquad -\infty < z < \infty \tag{6.83}$$

where α is the irregularity factor and $\Phi(\cdot)$ is the cumulative of the standard normal distribution. This result was first derived by S. O. Rice and the distribution is sometimes called the Rice distribution. Note that the first term of the sum is a normal density function and the second term is like a Rayleigh density function. Each term is weighted by a factor dependent on the irregularity factor. In the limiting case of a narrow-band process ($\alpha = 1$), the first term vanishes and the factors modifying the Rayleigh density are equal to 1. Thus, this equation reduces to the Rayleigh distribution, consistent with the previous section. Figure 6.9 shows this distribution for various values of the irregularity factor α.

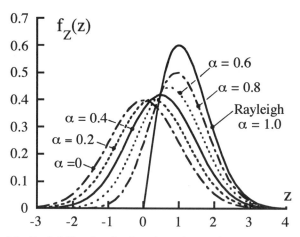

Figure 6.9 Probability density function of peak for different values of α.

Notice that peaks occur below $x = 0$ if the process is not perfectly narrow band. The fraction of peaks that are negative increases as bandwidth increases (α decreases). For the theoretical limit $\alpha = 0$, the peak distribution is Gaussian and identical to the marginal distribution of $X(t)$; half the peaks are negative. This situation is not physically realizable.

We note that the Rayleigh density function has greater values in the right tail than Rice's density functions for all values of α. This means that the probability of a peak being greater than a given value, $P(Z > z)$, is largest when calculated with the Rayleigh distribution. For most design purposes this is conservative, and the bandwidth effects are often ignored in practice.

Note that the height of the rise, H, from a valley to the next peak is not twice the height of the peak if the process is not narrow band. That is, the Rice distribution cannot be transformed to the distribution of rises. The following section addresses this problem.

Example 6.4: Probability of Peak, Wide-Band Case. Suppose $\alpha = 0.745$, corresponding to the case of low-pass noise. The probability that a peak Z in a zero-mean Gaussian process with this irregularity is greater than $3\sigma_X$ is $P(Z > 3\sigma_X) = 0.0083$. Compare this with the previous calculation for a narrow-band process: $P(Z_{nb} > 3\sigma_X) = 0.0111$. The probability for the narrow-band case is slightly larger. The reader might notice that the ratio of these two probabilities is equal to α. In general, for large values of n,

$$\lim_{n \to \infty} \frac{P(Z > n\sigma_X)}{P(Z_{nb} > n\sigma_X)} = \alpha$$

The proof is left to the reader as a problem.

6.2.3 Approximate Distribution for Height of Rise or Fall

A rise H is defined as the difference between $X(t)$ at a valley and $X(t)$ at the next peak. A fall is the difference between a peak and the next valley. If the process is stationary, the distributions of rise and fall are symmetric.

The rises and falls of a random process are of interest in oceanography, as they correspond to the heights of waves in the ocean. They are also of interest in engineering design for metal fatigue, as fatigue damage is better related to cyclic stress ranges than to stress peaks.

While an exact distribution for the peaks of a Gaussian random process is known, an exact distribution for the rises is unknown. The critical difficulty is in determining the time of the next peak. This problem is a first-passage time problem: Given that the velocity of $X(t)$ is zero at $t = 0$, when does the next zero crossing occur? As stated earlier, this distribution is unknown in general.

Despite the difficulty in finding the complete distribution, the exact value of the average rise μ_H is easily obtained (Rice and Beer, 1965). The average distance traveled by $X(t)$ in any direction during time interval Δt is equal to the average absolute velocity times Δt. The same distance is also equal to the average height of the rises and falls times the number of rises and falls occurring in Δt, which is equal to twice the number of peaks, or $2\nu_p \, \Delta t$. Equating the two calculations of the average distance and solving for μ_H gives

$$\mu_H = \frac{E[|\dot{X}(t)|]}{2\nu_p} \tag{6.84}$$

For a Gaussian process, the average height is

$$\mu_H = \alpha\sqrt{2\pi}\sigma_X \tag{6.85}$$

where α is the irregularity factor. The average height decreases as bandwidth increases.

There are several approximations for the distribution of H available in the literature (Rice, 1964; Yang, 1974; Tayfun, 1981; Madsen et al., 1986). Empirical evidence (such as Forristall, 1984) indicates that the distribution is close to Rayleigh or Weibull. The approximation presented below finds a Rayleigh distribution dependent on the irregularity factor α (Ortiz, 1985). The derivation sidesteps the first-passage problem by assuming that the trajectory of $X(t)$ between a valley and the next peak can be determined entirely by the conditions existing at the time of the valley; the arrival time of the next peak is not needed. The same result is derived by Krenk (1978) using envelopes.

The trajectory of $X(t)$ between a valley and the next peak is assumed to be sinusoidal. This is reasonable for processes that are not too irregular. For convenience, suppose the valley occurs at $t = 0$ and let the trajectory (or shape) function be

$$\Psi(t) = -\frac{H}{2}\cos(\omega_p t) \tag{6.86}$$

The frequency of the harmonic is assumed to be ω_p, the frequency associated with peaking, and $H > 0$ is the height of the rise. The derivatives of the trajectory function are

$$\dot{\Psi}(t) = \omega_p(\tfrac{1}{2}H)\sin(\omega_p t) \tag{6.87}$$

$$\ddot{\Psi}(t) = \omega_p^2\,(\tfrac{1}{2}H)\cos(\omega_p t) = A\cos(\omega_p t) \tag{6.88}$$

At the valley, we know the conditional distribution of $X(t)$ is $f_Z(-z)$ (Rice's distribution) and $\dot{X}(t)$ is equal to zero. The distribution of $\ddot{X}(t)$ at the valley can be found using the same techniques applied earlier in this chapter. Namely, define the conditions for $X(t)$ to be at a valley and calculate the distribution of $\ddot{X}(t)$ given this condition. Let A be the random variable representing this acceleration. The reader may verify the following general equations:

$$f_A(a)\,da = \frac{\displaystyle\int_a^{a+da}\int_{0-wdt}^0\int_{-\infty}^\infty f_{X\dot{X}\ddot{X}}(u,\,v,\,w)\,du\,dv\,dw}{\displaystyle\int_0^\infty\int_{0-wdt}^0\int_{-\infty}^\infty f_{X\dot{X}\ddot{X}}(u,\,v,\,w)\,du\,dv\,dw} \qquad a > 0 \quad (6.89)$$

$$f_A(a) = \frac{a f_{X\ddot{X}}(0,\,a)}{\nu_p} \qquad a > 0 \quad (6.90)$$

For the special case of a Gaussian process $X(t)$ the result is

$$f_A(a) = \frac{a}{\sigma_{\ddot{X}}^2}\exp\left(-\frac{1}{2}\frac{a^2}{\sigma_{\ddot{X}}^2}\right) \qquad a > 0 \qquad (6.91)$$

This is recognized as a Rayleigh distribution with $\sigma_{\ddot{X}}$ as the parameter.

The distribution of the rises H is found by a change of variable, $H = 2A/\omega_p^2$:

$$f_H(h) = \frac{h}{\theta_H^2}\exp\left(-\frac{1}{2}\frac{h^2}{\theta_H^2}\right) \qquad h > 0 \qquad (6.92)$$

where $\theta_H = 2\sigma_{\ddot{X}}/\omega_p^2$. The parameter may also be written as $\theta_H = 2\sigma_X\alpha$, so the distribution can be written as

$$f_H(h) = \frac{h}{(2\sigma_X\alpha)^2}\exp\left(-\frac{1}{2}\frac{h^2}{(2\sigma_X\alpha)^2}\right) \qquad h > 0 \qquad (6.93)$$

Note the similarity with the distribution for the narrow-band case [Eq. (6.71)]. As the narrow-band limit is approached ($\alpha = 1$), the two distributions agree. Also, the expected value is equal to the exact average for all values of α.

The trajectory approximation approach may be extended to derive the joint distribution of the height of a rise, H, with a valley, V, a peak, P, and the midpoint between the adjoining valley and peak, $M = \frac{1}{2}(P + V)$. The joint distribution of neighboring peaks and valleys may also be found. Once the first joint distribution has been found, all the others may be derived by change of variables.

The following results are derived by Perng (1989), who uses them to count the fatigue cycles that rise above a floor level set by a crack opening stress. Notice that each joint density function is the product of a Gaussian term and a Rayleigh term. The symbol $\phi(\cdot)$ is the standard normal density function.

- For the height of a rise, H, and the valley, V, from which is starts,

$$f_{HV}(h, v) = \frac{1}{\sqrt{2\pi(1 - \alpha^2)}\sigma_X} \exp\left[-\frac{1}{2}\frac{(v + \frac{1}{2}h)^2}{(1 - \alpha^2)\sigma_X^2}\right]$$

$$\cdot \frac{h}{4\alpha^2\sigma_X^2} \exp\left[-\frac{1}{2}\frac{h^2}{4\alpha^2\sigma_X^2}\right]$$

$$= \frac{1}{\sqrt{1 - \alpha^2}\sigma_X} \phi\left(\frac{v + \frac{1}{2}h}{\sqrt{1 - \alpha^2}\sigma_X}\right)$$

$$\cdot \frac{h}{4\alpha^2\sigma_X^2} \exp\left[-\frac{1}{2}\frac{h^2}{4\alpha^2\sigma_X^2}\right]$$

$$0 < h < \infty \qquad -\infty < v < \infty \tag{6.94}$$

- For the height of a fall, H, and the peak, P, from which it starts,

$$f_{HP}(h, p) = \frac{1}{\sqrt{2\pi(1 - \alpha^2)}\sigma_X} \exp\left[-\frac{1}{2}\frac{(p - \frac{1}{2}h)^2}{(1 - \alpha^2)\sigma_X^2}\right]$$

$$\cdot \frac{h}{4\alpha^2\sigma_X^2} \exp\left[-\frac{1}{2}\frac{h^2}{4\alpha^2\sigma_X^2}\right]$$

$$= \frac{1}{\sqrt{1 - \alpha^2}\sigma_X} \phi\left(\frac{p - \frac{1}{2}h}{\sqrt{1 - \alpha^2}\sigma_X}\right)$$

$$\cdot \frac{h}{4\alpha^2\sigma_X^2} \exp\left[-\frac{1}{2}\frac{h^2}{4\alpha^2\sigma_X^2}\right]$$

$$-\infty < p < \infty \quad 0 < h < \infty \tag{6.95}$$

- For the height of a rise, H, and its midpoint, M,

$$f_{HM}(h, m) = \frac{1}{\sqrt{2\pi(1 - \alpha^2)}\sigma_X} \exp\left[-\frac{1}{2}\frac{m^2}{(1 - \alpha^2)\sigma_X^2}\right]$$

$$\cdot \frac{h}{4\alpha^2\sigma_X^2} \exp\left[-\frac{1}{2}\frac{h^2}{4\alpha^2\sigma_X^2}\right]$$

$$= \frac{1}{\sqrt{1 - \alpha^2}\sigma_X} \phi\left(\frac{m}{\sqrt{1 - \alpha^2}\sigma_X}\right)$$

$$\cdot \frac{h}{4\alpha^2\sigma_X^2} \exp\left[-\frac{1}{2}\frac{h^2}{4\alpha^2\sigma_X^2}\right]$$

$$0 < h < \infty \qquad -\infty < m < \infty \tag{6.96}$$

Note that H and M are independent because their joint probability density function is factorable.

- For a neighboring peak–valley (P–V) pair,

$$f_{PV}(p, v) = \frac{1}{\sqrt{2\pi(1 - \alpha^2)}\sigma_X} \exp\left(-\frac{1}{2}\frac{(p + v)^2}{4(1 - \alpha^2)\sigma_X^2}\right)$$

$$\cdot \frac{p - v}{4\alpha^2\sigma_X^2} \exp\left(-\frac{1}{2}\frac{(p - v)^2}{4\alpha^2\sigma_X^2}\right)$$

$$= \frac{1}{\sqrt{1 - \alpha^2}\sigma_X} \phi\left(\frac{p + v}{2\sqrt{1 - \alpha^2}\sigma_X}\right)$$

$$\cdot \frac{p - v}{4\alpha^2\sigma_X^2} \exp\left(-\frac{1}{2}\frac{(p - v)^2}{4\alpha^2\sigma_X^2}\right)$$

$$-\infty < v < p < \infty \qquad (6.97)$$

PROBLEMS

6.1 Derive the results given in the text for the peak frequency of a random process and for the Gaussian specialization.

6.2 Show that the velocity of a Gaussian process $X(t)$ has a Rayleigh distribution when $X(t)$ is at a zero up crossing.

6.3 Let $X(t)$ be the sum of two narrow-band processes, i.e., the one-sided spectral density consists of two impulses:

$$W_X(f) = \sigma_1^2\delta(f - f_1) + \sigma_2^2\delta(f - f_2)$$

Sketch this spectral density function. Find the zero-crossing rate, peaking rate, and irregularity factor.

6.4 Let the two-sided spectral density function of $X(t)$ be given by

$$S_X(\omega) = \frac{1}{2}\left\{\frac{1}{\sqrt{2\pi}\lambda} \exp\left[-\frac{1}{2}\left(\frac{\omega + \omega_0}{\lambda}\right)^2\right]\right\}$$

$$+ \frac{1}{2}\left\{\frac{1}{\sqrt{2\pi}\lambda} \exp\left[-\frac{1}{2}\left(\frac{\omega - \omega_0}{\lambda}\right)^2\right]\right\}$$

Sketch this spectral density function. Notice that this is a Gaussian function with a center frequency of ω_0. Find the zero-crossing rate, peaking rate, and irregularity factor. *Hint:* The moments can be found easily knowing the central moments of a Gaussian probability density function, which are given in Eq. (2.48).

6.5   Suppose $X(t)$ is a narrow-band Gaussian process. What is the value of the peak with a 1% probability of being exceeded?

6.6   In the ocean, the significant wave height is the term applied to the average height of the one-third highest waves. If the waves follow a narrow-band Gaussian process, what is the relationship between the significant wave height H_s and the variance of the underlying process?

6.7   At a certain time in a rocket launch, an on-board instrument experiences an average acceleration of $2.0g$ plus a stationary Gaussian component having the spectral density shown in Figure 5.5c. What is the probability that a single peak acceleration will be greater than $10g$? Less than zero? (You may integrate numerically.)

6.8   Use the environment specified in Figure 5.5c (You may integrate numerically):

   a What is the probability that a peak exceeds $\pm 30g$? (Assume the mean is zero.)

   b Suppose that for the purposes of a conservative design, the spectral density is doubled, e.g., $0.075 \ g^2/\text{Hz}$ is doubled to $0.150 \ g^2/\text{Hz}$. Now what would be the probability that a peak exceeds $\pm 30g$?

   c What would be the value of the peak in the conservative design environment that has the same probability of being exceeded as $\pm 30g$ did originally?

   d Someone might say that, "with this approach, the factor of safety is 2." Is this true? How else would one measure the conservatism?

6.9   (Lin, 1967). The joint probability density function of $X(t)$ and its derivatives is given by a joint uniform distribution:

$$f_{X\dot{X}\ddot{X}}(x, \dot{x}, \ddot{x}) = \frac{1}{8abc} \quad -a < x < a \quad -b < \dot{x} < b \quad -c < \ddot{x} < c$$

Find the zero-crossing rate, peaking rate, and irregularity factor. Find the probability distribution of peaks.

6.10 (Lin, 1967). The joint probability density function of a narrow-band process, $X(t)$, and its first derivative is given by a joint Laplace distribution:

$$f_{X\dot{X}}(x, \dot{x}) = \frac{1}{4\alpha\beta} \exp\left(-\frac{|x|}{\alpha} - \frac{|\dot{x}|}{\beta}\right) \quad -\infty < x < \infty \quad -\infty < \dot{x} < \infty$$

where α and β are positive. Show that an approximate probability density function for the peak magnitude is

$$f_Z(z) = \frac{1}{\alpha} \exp\left(-\frac{z}{\alpha}\right) \quad z \geq 0$$

6.11 Show that, for large peaks, the probability of exceedance, $P(Z > z_0)$, given by Rice's distribution of peaks is approximately α times the probability found from the Rayleigh distribution.

6.12 *Simulation of a narrow-band process.* The following (long) project may be performed using the discrete Fourier transform simulation code developed in problem 5.11. For $f_0 = 100$ and $\zeta = 0.02$, the following spectral density is approximately narrow band:

$$W_X(f) = \left\{ \left[1 - \left(\frac{f}{f_0} \right)^2 \right]^2 + \left(2\zeta \frac{f}{f_0} \right)^2 \right\}^{-1}$$

Generate a simulation. Experiment with the parameters of the simulation, such as n, f_0, and ζ.

a Plot the histogram of the simulation and compare against the theoretical Gaussian density. Why should the simulation be Gaussian anyway?

b Plot the histogram of the peaks and compare against the Rayleigh density for peaks. Is the match to the theoretical as close as for the marginal in (a)?

c Plot the histogram of the rises and compare against the Rayleigh density for rises. Is the match to the theoretical as close as for the marginal? The peaks?

d Plot a sample realization of the process. Pick a relatively high level, say 2σ, and mark every crossing of that level. Is clumping apparent?

e Sketch the "envelope" of the same sample realization and mark the level crossings of the envelope.

7

VIBRATION OF SINGLE-DEGREE-OF-FREEDOM SYSTEMS

7.1 FREE VIBRATION OF SINGLE-DEGREE-OF-FREEDOM SYSTEM

7.1.1 Background

Reviewed in this chapter are fundamental concepts and definitions associated with linear elastic vibrations of single-degree-of-freedom systems. A single-degree-of-freedom system requires only one coordinate to describe its position at any instant of time. Examples of single-degree-of-freedom systems are shown in Figure 7.1.

The fundamentals of vibration theory are widely documented in the literature. It is suggested that the reader consult any of the numerous books on elementary dynamics or vibrations (e.g., Thomson, 1981; Weaver, Timoshenko, and Young, 1990; Rao, 1990; Inman, 1994; James et al., 1994; Steidel, 1979).

7.1.2 Harmonic Motion

Shown in Figure 7.2 is a vector **OP** of length A that rotates counterclockwise with constant angular velocity ω. At any time t the angle that **OP** makes with the horizontal is $\theta = \omega t$. Let x be the projection of **OP** on the vertical axis. Then

$$x = A \sin \omega t \qquad (7.1)$$

Here, x, a function of time, is plotted versus ωt, as shown. A particle that experiences this motion is said to have harmonic motion.

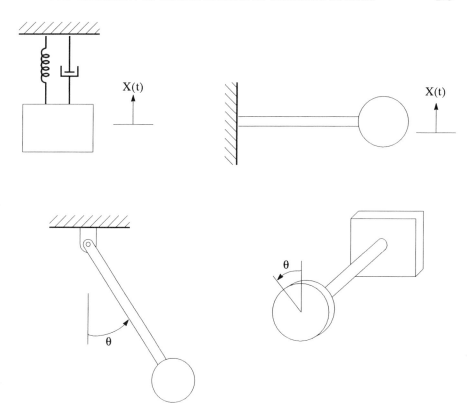

Figure 7.1 Examples of single-degree-of-freedom systems.

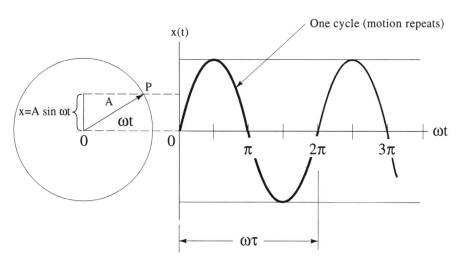

Figure 7.2 Harmonic motion.

Now consider the units of θ. Let C be the circumference of the circle shown in Figure 7.2. Thus, $C = 2\pi A$. Or we can write $C = A\theta$, where $\theta = 2\pi$ for one revolution. Thus defined, θ is said to be in *radians* and is equivalent to $360°$. Therefore, one radian is approximately equal to $58.3°$. And, in general, for any arc length, $s = A\theta$, where θ is in radians. It follows that ω in Figure 7.2 and Eq. (7.1) would be in radians per second.

Some basic definitions of harmonic motion:

1. *Amplitude A* is as shown in Figure 7.2.
2. *Range 2A* is the peak-to-peak displacement.
3. The *period* is the time for the motion to repeat (the value of τ in Figure 7.2). Note that

$$\omega\tau = 2\pi \qquad (7.2)$$

4. *Cycle* is the motion in one period, as shown in Figure 7.2
5. *Frequency* is the number of cycles per time. The most common unit of time used in vibration analysis is seconds. Cycles per second is called *hertz*. In terms of the period, the frequency is

$$f = \frac{1}{\tau} \qquad (7.3)$$

The frequency f is related to ω from Eqs. (7.2) and (7.3):

$$f = \frac{\omega}{2\pi} \qquad (7.4)$$

$$\omega = 2\pi f \qquad (7.5)$$

6. The *velocity* of a particle is given by

$$v = \frac{dx}{dt} = \dot{x} \qquad (7.6)$$

7. The *acceleration* of a particle is given by

$$a = \frac{dv}{dt} = \frac{d^2x}{dt^2} = \ddot{x} \qquad (7.7)$$

Thus, displacement, velocity, and acceleration have the following relationships in harmonic motion:

$$x = A \sin \omega t$$
$$v = \dot{x} = A\omega \cos \omega t \qquad (7.8)$$
$$a = \ddot{x} = -A\omega^2 \sin \omega t$$

7.1.3 Free Vibration of Undamped Single-Degree-of-Freedom System

Consider the single-degree-of-freedom spring–mass system, as shown in Figure 7.3. The spring is originally in the unstretched position, as shown. It is assumed that the spring obeys Hooke's law. The force in the spring is proportional to displacement with the proportionality constant (spring constant) equal to k.

When the mass m (weight $= w$) is applied, the spring will deflect to a static equilibrium position, δ_{st}. If the mass is perturbed and allowed to move dynamically, the displacement x, measured from the equilibrium position, will be a function of time. Here, $x(t)$ is the absolute motion of the mass. To determine the position as a function of time, the equations of motion are employed; the free-body diagrams are drawn as shown in Figure 7.3. Note that x is measured positive downward.

Applying Newton's second law,

$$W - k(x + \delta_{st}) = m\ddot{x} \qquad (7.9)$$

But from the static condition, note that $W = k\delta_{st}$. Thus, the equation of motion becomes

$$m\ddot{x} + kx = 0 \qquad (7.10)$$

Generations of engineering students have solved this problem as undergraduates in a differential equations course. The general solution is $x(t) = Ce^{st}$. Substituting into the equation of motion and noting Euler's formula, $e^{\pm i\alpha} = \cos \alpha \pm i \sin \alpha$, where i is the imaginary unit, it is shown that

$$x(t) = A \cos \omega_n t + B \sin \omega_n t \qquad (7.11)$$

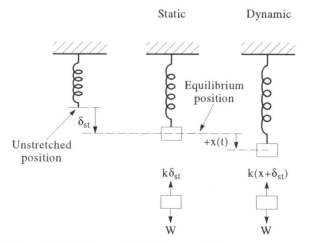

Figure 7.3 Undamped single-degree-of-freedom system.

where A and B are constants of integration and

$$\omega_n = \sqrt{\frac{k}{m}} \quad \text{(rad/sec)} \tag{7.12}$$

Here, ω_n defines the *natural frequency* of the mass. This is the frequency at which the mass will move regardless of the amplitude of motion as long as the spring in the system continues to obey Hooke's law. The natural frequency in hertz is

$$f_n = \frac{1}{2\pi} \sqrt{\frac{k}{m}} \quad \text{(Hz)} \tag{7.13}$$

The initial conditions $x = x_0$ at $t = 0$ and $\dot{x} = \dot{x}_0$ at $t = 0$ are used to evaluate the constants of integration A and B. When substituted into Eq. (7.11), it follows that

$$A = x_0 \quad \text{and} \quad B = \frac{\dot{x}_0}{\omega_n} \tag{7.14}$$

7.1.4   Damped Free Vibration of Single-Degree-of-Freedom System

Consider the spring–mass system with an energy-dissipating mechanism described by the damping force, as shown in Figure 7.4. (See Sun and Lu, 1995, for a comprehensive treatment of damping.) It is assumed that the damping force F_D is proportional to the velocity of the mass, as shown; the *damping coefficient* is c. When Newton's second law is applied, this model for the damping force leads to a linear differential equation,

$$m\ddot{x} + c\dot{x} + kx = 0 \tag{7.15}$$

Assuming a solution $x(t) = Ce^{st}$ for this homogeneous linear differential equation leads to the *characteristic equation*

$$ms^2 + cs + k = 0 \tag{7.16}$$

The roots are

$$s_{1,2} = -\frac{c}{2m} \pm \sqrt{\left(\frac{c}{2m}\right)^2 - \frac{k}{m}} \tag{7.17}$$

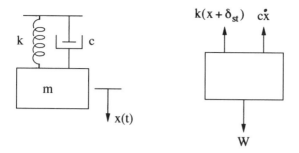

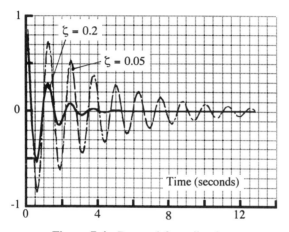

Figure 7.4 Damped free vibration.

Define the *critical damping coefficient* c_c as that value of c that makes the radical equal to zero,

$$c_c = 2m \sqrt{\frac{k}{m}} = 2m\omega_n \tag{7.18}$$

Define the *damping factor* as

$$\zeta = \frac{c}{c_c} \tag{7.19}$$

The damping factor is commonly used to characterize damping in structural systems. A review of damping factors in spacecraft structures was provided by Richard (1990).

Consider three cases of damping in which the mass has x_0 and $\dot{x}_0$.

1. *Heavy Damping:* $c > c_c$. The roots are both real. The solution to the differential equation is

$$x(t) = Ae^{s_1 t} + Be^{s_2 t} \tag{7.20}$$

where A and B are the constant of integration. Both s_1 and s_2 will be negative. Thus, given an initial displacement, the mass will decay to its equilibrium position without vibratory motion.

2. *Critical Damping:* $c = c_c$. Both roots are equal and the general solution is

$$x(t) = (A + Bt)e^{-\omega_n t} \tag{7.21}$$

The motion is again not vibratory and decays to the equilibrium position.

3. *Light Damping:* $0 < c < c_c$. The roots are complex. It is easily shown, using Euler's formula, that the general solution is

$$x(t) = [A \cos(\omega_d t + \phi)]e^{-\zeta\omega_n t} \tag{7.22}$$

where A and ϕ are the constants of integration. The damped natural frequency ω_d is given by

$$\omega_d = \omega_n \sqrt{1 - \zeta^2} \tag{7.23}$$

An illustration of damped free vibrations for the case where $\phi = 0$ is provided in Figure 7.4.

7.2 FORCED VIBRATION

7.2.1 Force-Excited System: Harmonic Excitation

Some single-degree-of-freedom systems with an external force are shown in Figure 7.5. Force can be applied both as an external force $F(t)$ or as a base motion $y(t)$, as shown. The coordinate $x(t)$ is the absolute motion of the mass. The forces W and $k\delta_{st}$ are ignored in the free-body diagrams as we know they will add to zero in the equation of motion.

Consider the force-excited system of Figure 7.5, where the applied force is harmonic, $F(t) = F_0 \sin \omega t$. Applying Newton's second law, the equation of motion becomes

$$m\ddot{x} + c\dot{x} + kx = F_0 \sin \omega t \tag{7.24}$$

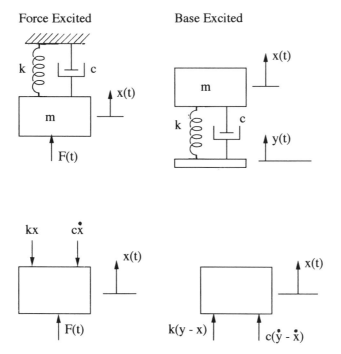

Figure 7.5 Free-body diagrams of forced vibration.

The general solution for this second-order nonhomogeneous linear differential equation is

$$x(t) = x_h(t) + x_p(t) \tag{7.25}$$

where x_h is the complementary solution or solution to the homogeneous equation [Eq. (7.15)]. But this solution dies out soon (see Figure 7.4). Our interest focuses on x_p, the particular solution. In vibration theory, the particular solution is also called the steady-state solution.

Using complex algebra, let the harmonic force be

$$F(t) = F_0 e^{i\omega t} \tag{7.26}$$

where F_0 is a real constant and i is the imaginary unit. Assume that the response has the same frequency as the force, but is, in general, out of phase with the force,

$$x(t) = X_0 e^{i(\omega t + \phi)} \tag{7.27}$$

$$= \tilde{X} e^{i\omega t}$$

where X_0 is the amplitude of the displacement and $\tilde{X}$ is the complex displacement,

$$\tilde{X} = X_0 e^{i\phi} \tag{7.28}$$

Substituting into the differential equation of motion,

$$(-m\omega^2 + ic\omega + k)\tilde{X}e^{i\omega t} = F_0 e^{i\omega t} \tag{7.29}$$

Define the *transfer function* (or frequency response function) $H(\omega)$ as the complex displacement due to a force of unit magnitude ($F_0 = 1$). Thus,

$$H(\omega) = \frac{1}{(k - m\omega^2) + ic\omega} \tag{7.30}$$

Rationalizing, the transfer function becomes

$$H(\omega) = \frac{(k - m\omega^2) - ic\omega}{(k - m\omega^2)^2 + (c\omega)^2} \tag{7.31}$$

This is also the ratio between the complex displacement response and the complex input forcing function.

Define the *gain function* as the modulus of the transfer function

$$|H(\omega)| = \sqrt{H(\omega)H^*(\omega)} = \sqrt{(\mathrm{Re}\ H)^2 + (\mathrm{Im}\ H)^2} \tag{7.32}$$

where H^* is the complex conjugate. For the force-excited system under consideration,

$$|H(\omega)| = \frac{1}{\sqrt{(k - m\omega^2)^2 + (c\omega)^2}} \tag{7.33}$$

The gain function is the amplitude of the displacement for $F_0 = 1$. Thus,

$$\frac{X_0}{F_0} = |H(\omega)| \tag{7.34}$$

It is convenient to develop a nondimensional form of the gain function. First, define the *frequency ratio*

$$r = \frac{\omega}{\omega_n} \tag{7.35}$$

Multiplying Eq. (7.33) by k and employing the definitions for ζ, c_c, and ω_n, it is easily shown that

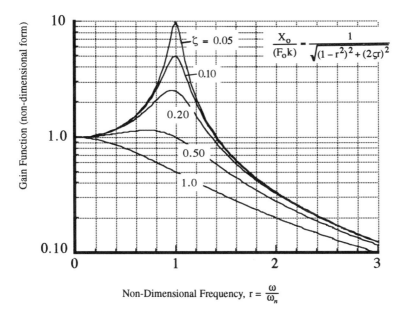

Figure 7.6   Gain function for force-excited system.

$$\frac{X_0}{F_0/k} = \frac{1}{\sqrt{(1 - r^2)^2 + (2\zeta r)^2}} \tag{7.36}$$

The nondimensional gain function of Eq. (7.36) is shown in Figure 7.6. A fact that creates difficulty for designers is that the response can become "large" when r is close to 1 or when ω is close to ω_n. This condition is called *resonance*.

Transfer functions expressing the velocity and acceleration responses can be written, based on Eq. (7.27), by multiplying $H(\omega)$ in Eq. (7.30) by $i\omega$ and $(i\omega)^2 = -\omega^2$, respectively. The resulting gain function for velocity output would be derived by multiplying both sides of Eq. (7.33) or (7.36) by ω and noting that ωX_0 is the amplitude of velocity. Similarly, the gain function for acceleration can be obtained by multiplying both sides by ω^2.

In summary, the gain function defines (a) the system and (b) the input and output.

7.2.2 Base-Excited System: Absolute Motion

Consider the base-excited system of Figure 7.5. The goal of analysis will be to determine the absolute response $x(t)$ (typically acceleration or displacement of the mass) given the base motion $y(t)$. This problem is useful when the mass, for example, is a delicate electronic instrument, and it is required to design a vibration isolation system that will limit the acceleration response of the mass.

From the free-body diagram of Figure 7.5, application of Newton's second law leads directly to the differential equation

$$m\ddot{x} + c\dot{x} + kx = ky + c\dot{y} \tag{7.37}$$

Assume that the base motion is harmonic,

$$y(t) = Y_0 e^{i\omega t} \tag{7.38}$$

and assume that the response will be harmonic,

$$x(t) = \tilde{X} e^{i\omega t} \tag{7.39}$$

where $\tilde{X}$ is the complex response, as described above [Eq. (7.28)]. The transfer function and the gain function are derived in the same manner as for the force-excited system. The transfer function is

$$H(\omega) = \frac{k + ic\omega}{(k - m\omega^2) + ic\omega} \tag{7.40}$$

The gain function is

$$\frac{X_0}{Y_0} = |H(\omega)| = \sqrt{\frac{k^2 + (c\omega)^2}{(k - m\omega^2)^2 + (c\omega)^2}} \tag{7.41}$$

In nondimensional form,

$$\frac{X_0}{Y_0} = \sqrt{\frac{1 + (2\zeta r)^2}{(1 - r^2)^2 + (2\zeta r)^2}} \tag{7.42}$$

The gain function for absolute displacement for the base-excited system is shown in Figure 7.7.

Equations (7.41) and (7.42) can also be interpreted as the gain functions for acceleration output given acceleration input. In fact, this is a more common use of these forms. And again, note that the transfer function for velocity and acceleration responses can be derived by multiplying Eq. (7.40) by $i\omega$ and $-\omega^2$, respectively.

7.2.3 Base-Excited System: Relative Motion

Consider the free-body diagram of Figure 7.5. Now the response variable under consideration will be the relative displacement

$$z(t) = x(t) - y(t) \tag{7.43}$$

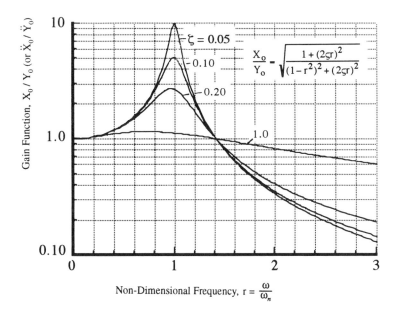

Figure 7.7 Gain function for base-excited system (absolute displacement).

In the model, the spring represents a structural element. The stresses in that element will be proportional to z. Thus, this problem would be relevant to designers of structures subjected to base motions, for example, earthquakes.

Letting $z = x - y$ in the equation of motion leads directly to

$$m\ddot{z} + \dot{z} + kz = -m\ddot{y} \tag{7.44}$$

Assuming that the base motion is harmonic,

$$y(t) = Y_0 e^{i\omega t} \tag{7.45}$$

And assuming that the response is also harmonic [see Eq. (7.28)],

$$z(t) = \tilde{Z} e^{i\omega t} \tag{7.46}$$

Following the procedures as described above, the transfer function is

$$H(\omega) = \frac{m\omega^2}{(k - m\omega^2) + ic\omega} \tag{7.47}$$

and the gain function is

$$\frac{Z_0}{Y_0} = |H(\omega)| = \frac{m\omega^2}{\sqrt{(k - m\omega^2)^2 + (c\omega)^2}} \tag{7.48}$$

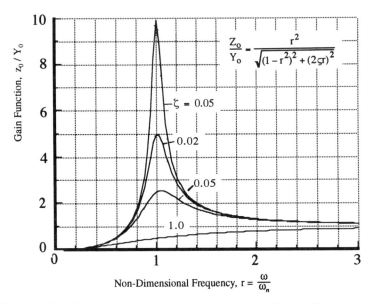

Figure 7.8 Gain function for base-excited system (relative displacement).

In nondimensional form,

$$\frac{Z_0}{Y_0} = \frac{r^2}{\sqrt{(1 - r^2)^2 + (2\zeta r)^2}} \tag{7.49}$$

The gain function for relative motion for the base-excited system is shown in Figure 7.8.

Again, note that the transfer function for relative velocity and acceleration responses can be derived by multiplying Eq. (7.47) by $i\omega$ and $-\omega^2$, respectively. The gain functions for velocity and acceleration responses can be obtained by multiplying both sides of Eqs. (7.48) or (7.49) by ω and ω^2, respectively.

Gain functions are shown in Tables 7.1 and 7.2 for various combinations of input and response for both force- and base-excited systems.

Example 7.1. A fixed bottom offshore structure is subjected to oscillatory storm waves. In a first approximation, it is estimated that the waves produce a harmonic force $F(t)$ having amplitude $F_0 = 27.5$ kips. The period of these waves is $\tau = 8$ sec.

The structure is modeled as having a lumped mass of 110 tons concentrated in the deck. The weight of the structure itself is assumed to be negligible. The natural period of the structure was measured as being $\tau_n = 4.0$ sec. It is assumed that the damping factor is $\zeta = 5\%$. It is required to determine the steady-state amplitude of the response of the structure.

Table 7.1 Gain Functions for Force-Excited Systems

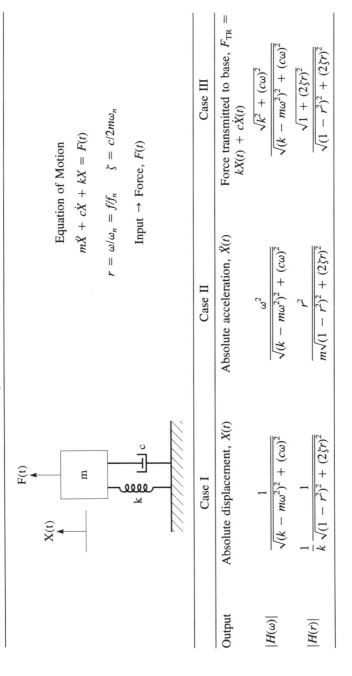

Equation of Motion

$$m\ddot{X} + c\dot{X} + kX = F(t)$$

$$r = \omega/\omega_n = f/f_n \quad \zeta = c/2m\omega_n$$

Input → Force, $F(t)$

Output	Case I Absolute displacement, $X(t)$	Case II Absolute acceleration, $\ddot{X}(t)$	Case III Force transmitted to base, $F_{TR} = kX(t) + c\dot{X}(t)$
$\lvert H(\omega) \rvert$	$\dfrac{1}{\sqrt{(k - m\omega^2)^2 + (c\omega)^2}}$	$\dfrac{\omega^2}{\sqrt{(k - m\omega^2)^2 + (c\omega)^2}}$	$\dfrac{\sqrt{k^2 + (c\omega)^2}}{\sqrt{(k - m\omega^2)^2 + (c\omega)^2}}$
$\lvert H(r) \rvert$	$\dfrac{1}{k}\dfrac{1}{\sqrt{(1 - r^2)^2 + (2\zeta r)^2}}$	$\dfrac{r^2}{m\sqrt{(1 - r^2)^2 + (2\zeta r)^2}}$	$\dfrac{\sqrt{1 + (2\zeta r)^2}}{\sqrt{(1 - r^2)^2 + (2\zeta r)^2}}$

Table 7.2 Gain Functions for Base-Excited Systems

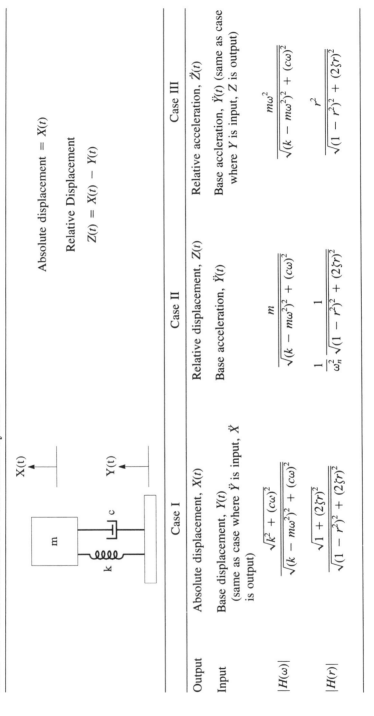

Absolute displacement $= X(t)$

Relative Displacement

$$Z(t) = X(t) - Y(t)$$

	Case I	Case II	Case III
Output	Absolute displacement, $X(t)$	Relative displacement, $Z(t)$	Relative acceleration, $\ddot{Z}(t)$
Input	Base displacement, $Y(t)$ (same as case where $\ddot{Y}$ is input, $\ddot{X}$ is output)	Base acceleration, $\ddot{Y}(t)$	Base acceleration, $\ddot{Y}(t)$ (same as case where Y is input, Z is output)
$\lvert H(\omega)\rvert$	$\dfrac{\sqrt{k^2 + (c\omega)^2}}{\sqrt{(k - m\omega^2)^2 + (c\omega)^2}}$	$\dfrac{m}{\sqrt{(k - m\omega^2)^2 + (c\omega)^2}}$	$\dfrac{m\omega^2}{\sqrt{(k - m\omega^2)^2 + (c\omega)^2}}$
$\lvert H(r)\rvert$	$\dfrac{\sqrt{1 + (2\zeta r)^2}}{\sqrt{(1 - r^2)^2 + (2\zeta r)^2}}$	$\dfrac{1}{\omega_n^2}\dfrac{1}{\sqrt{(1 - r^2)^2 + (2\zeta r)^2}}$	$\dfrac{r^2}{\sqrt{(1 - r^2)^2 + (2\zeta r)^2}}$

Solution. As modeled, this will be a force-excited system, and the response can be computed from the gain function of Eq. (7.36). The problem reduces to one of finding the frequency ratio r and the stiffness k. Because r is the ratio of the forcing frequency to the natural frequency, it follows that r will also be the ratio of the natural period to the forcing period. Thus,

$$r = \frac{T_n}{T} = \frac{4}{8} = 0.5$$

Then note that the expression for the natural frequency is

$$f_n = \frac{1}{2\pi} \sqrt{\frac{k}{W/g}}$$

To compute k, first note that the natural frequency is $f_n = 1/T_n = 0.25$ Hz. The deck weight W of 110 tons is equivalent to 220,000 lb. Using units of ft-lb-sec, $g = 32.2$ ft/sec^2, and thus $k = 16,840$ lb/ft.

Finally, substitution into Eq. (7.36) provides the solution for the amplitude of the response

$$X_0 = 2.18 \text{ ft}$$

Example 7.2. The fragility level (theoretical ultimate strength) of a 4.0-lb electronic unit is $20g$. But the base acceleration is harmonic with an amplitude of $Y_0 = 37g$ and a frequency of 53 Hz. Clearly, if the unit were hard mounted on the base, it would fail. The problem is to design a vibration isolation mount for the unit. Specifically, it is required to determine the largest allowable value of spring stiffness k. Assume that the mounts will have a damping factor of $\zeta = 0.20$.

Solution. For this base-excited system, the gain function to determine the amplitude of the absolute response of the mass is Eq. (7.42). Note that this function also applies to the ratio of the accelerations. The strategy here will be to first determine the frequency ratio. Substituting into Eq. (7.42), we obtain

$$\frac{20}{37} = \sqrt{\frac{1 + [2(.2)r]^2}{(1 - r^2)^2 + [2(.2)r]^2}}$$

from which $r = 1.80$. The response of the mass will be less than $20g$ if $r > 1.8$. Here, r is related to the natural frequency [Eq. (7.35)], and the natural frequency is related to k [Eqs. (7.12) and (7.13)]. Using Eq. (7.13) (with $g = 386$ in./sec^2), it follows that

$$f_n = 1.56\sqrt{k}$$

And the expression for r becomes

$$r = \frac{f}{f_n} = \frac{53}{1.56\sqrt{k}} > 1.8$$

from which

$$k < 356 \text{ lb/in.}$$

Note that a softer spring provides more isolation but allows more displacement.

7.3 BACKGROUND FOR RESPONSE OF SINGLE-DEGREE-OF-FREEDOM SYSTEM TO RANDOM FORCES

7.3.1 Response of Single-Degree-of-Freedom System to Impulsive Forces

The impulse response function is defined in the following through the example of the force-excited system. Note that the concept is generic in that the impulse response function can be defined for a linear system for any specified input and output.

Consider a single-degree-of-freedom system subjected to impulsive loading, as shown in Figure 7.9. The external force is

$$F(t) = F_0 \delta(t) \tag{7.50}$$

where $\delta(t)$ is the Dirac delta function, as defined in Chapter 4.

The equation of motion of the mass will be similar to Eq. (7.24) with the impulsive force of Eq. (7.50) on the right-hand side. The unit impulse is defined as $F_0 = 1$. The response $x(t)$ to the unit impulse is denoted as $h(t)$,

$$m\ddot{h} + c\dot{h} + kh = (1)\delta(t) \tag{7.51}$$

Physically speaking, for $t \approx 0$, a radical change in the system motion takes place when the short-duration high-amplitude force excites an initial motion in the system. But for $t > 0$, the response will be free vibration. Using elementary mechanics, $F(\Delta t) = m(\Delta v)$ it can be shown that the velocity of the system just after the impulse is

$$\dot{h}(0^+) = \frac{1}{m} \tag{7.52}$$

Using this and $h(0) = 0$ as the initial conditions, the free-vibration response is

$$h(t) = \frac{1}{m\omega_d} e^{-\zeta\omega_n t} \sin \omega_d t \qquad t > 0 \tag{7.53}$$

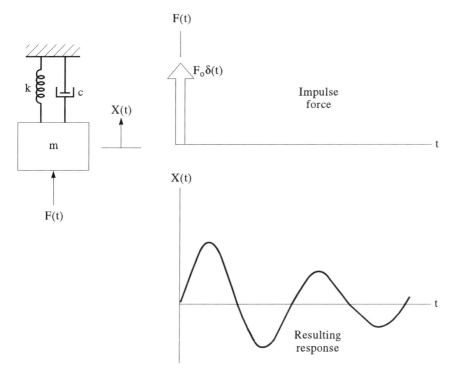

Figure 7.9 Impulse force on a system.

Here, $h(t)$ is known as the force-excited absolute displacement response, impulse response function of the single-degree-of-freedom system. Note that $h(t)$ characterizes a system just like the transfer function $H(\omega)$ does. The velocity and acceleration impulse response functions can also be obtained as derivatives of $h(t)$.

7.3.2 Response of Single-Degree-of-Freedom System to Arbitrary Loading

For a linear system, the impulse response function can be used to derive the response of a system under an arbitrary loading history. Consider the force shown in Figure 7.10. The impulse during $\Delta\tau$ is $F(\tau)\,\Delta\tau$. The response to this impulse at any time $t > \tau$ is approximately $[F(\tau)\,\Delta\tau]h(t - \tau)$. Then the response at t is the sum of the responses due to a sequence of impulses. In the limit as $\Delta\tau \to 0$,

$$x(t) = \int_{-\infty}^{t} F(\tau)h(t - \tau)\,d\tau \tag{7.54}$$

where the input $F(t)$ is accounted for as $t \to -\infty$, for example, $F(t)$ could be defined as zero for $t < 0$. The expression for $x(t)$ is called the convolution

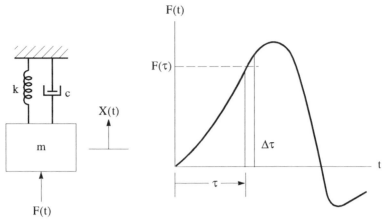

Figure 7.10 Arbitrary force on a system.

integral. Note that $h(t - \tau) = 0$ when $\tau > t$. Thus, we can expand the limits to the interval $(-\infty, \infty)$:

$$x(t) = \int_{-\infty}^{\infty} F(\tau)h(t - \tau)\, d\tau \tag{7.55}$$

Another useful form is obtained by letting $\theta = t - \tau$:

$$x(t) = \int_{-\infty}^{\infty} F(t - \theta)h(\theta)\, d\theta \tag{7.56}$$

7.3.3   Relationship Between $h(t)$ and $H(\omega)$

An important result from Fourier transform theory is that $h(t)$ and $H(\omega)$ form a Fourier transform pair. This relationship is useful in deriving responses of dynamic systems to random vibration inputs. Let

$$F(t) = e^{i\omega t} \tag{7.57}$$

Then,

$$
\begin{aligned}
x(t) &= H(\omega)e^{i\omega t} \\
&= \int_{-\infty}^{\infty} h(t - \tau)e^{i\omega\tau}\, d\tau \\
&= \int_{-\infty}^{\infty} h(\theta)e^{i\omega(t - \theta)}\, d\theta \\
&= e^{i\omega t} \int_{-\infty}^{\infty} h(\theta)e^{-i\omega\theta}\, d\theta \tag{7.58}
\end{aligned}
$$

which implies that

$$H(\omega) = \int_{-\infty}^{\infty} h(\theta)e^{-i\omega\theta} \, d\theta \qquad (7.59)$$

and therefore

$$h(t) = \frac{1}{2\pi} \int_{-\infty}^{\infty} H(\omega)e^{i\omega t} \, d\omega \qquad (7.60)$$

Thus, $h(t)$ and $H(\omega)$ form a Fourier transform pair. Note reversal of $(1/2\pi)$. See Sec. 4.1.2.

Based on the developments in Chapter 4 concerning the Fourier transform of a convolution, the expression for the response to an arbitrary input in Eq. (7.55), and representation in Eqs. (7.59) and (7.60), it is clear that we can also express the response to an arbitrary input as

$$x(t) = \frac{1}{2\pi} \int_{-\infty}^{\infty} F(\omega)H(\omega)e^{i\omega t} \, d\omega \qquad (7.61)$$

where $F(\omega)$ is the Fourier transform of $F(t)$. This expression is useful for the analysis or numerical computation of system response or as the basis for random vibration computations.

7.3.4 Relationship Between $X(\omega)$ and $F(\omega)$

The relationship between the Fourier transforms of $x(t)$ and $F(t)$ is used in Chapters 8 and 9 to derive responses of dynamic systems to random vibration input.

Take the Fourier transform of both sides of Eq. (7.56),

$$X(\omega) = \frac{1}{2\pi} \int_{-\infty}^{\infty} \left[\int_{-\infty}^{\infty} F(t-\theta)h(\theta) \, d\theta \right] e^{-i\omega t} \, dt \qquad (7.62)$$

Let $\tau = t - \theta$, $dt = d\tau$:

$$X(\omega) = \frac{1}{2\pi} \int_{-\infty}^{\infty} \left[\int_{-\infty}^{\infty} F(\tau)h(\theta) \, d\theta \right] e^{-i\omega(\tau+\theta)} \, d\tau \qquad (7.63)$$

Rearranging,

$$X(\omega) = \int_{-\infty}^{\infty} h(\theta)e^{-i\omega\theta} \, d\theta \, \frac{1}{2\pi} \int_{-\infty}^{\infty} F(\tau)e^{-i\omega\tau} \, d\tau \qquad (7.64)$$

Thus, from Eq. (7.59) and the basic relationship of the Fourier transform Eq. (4.13), it follows that

$$X(\omega) = H(\omega)F(\omega) \qquad (7.65)$$

PROBLEMS

7.1 The system shown in Figure P7.1 is a model of an instrument that will be launched using the space shuttle. A "twang" test is performed, and the response of the mass is measured. Assume that the weight of the structure is negligible.

  a What is the natural period τ_n?

  b What is the natural frequency in hertz, f_n?

  c What is the natural frequency in radians per second, ω_n?

  d What is the mass in lb-sec^2/in.? ($g = 386$ in./sec^2)

  e What is the damping factor ζ?

  f What is the damping coefficient c?

  g What is the stiffness k?

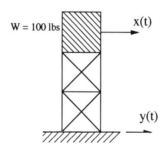

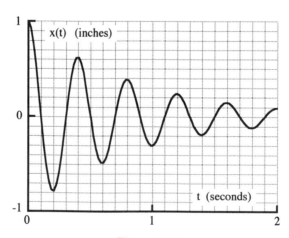

Figure P7.1

7.2 Consider problem 7.1. The lateral base acceleration is $0.60g$ (0 to peak) applied at a frequency of 2.0 Hz. From the decay record, the natural frequency is 2.5 Hz and the damping factor is 0.064.

   a Compute the gain function for absolute acceleration.

   b What is the amplitude of the absolute acceleration (in g's)?

   c Compute the gain function for relative displacement.

   d What is the amplitude of relative displacement (in inches)?

   e What is the amplitude of the shearing force in the structure?

7.3 The structural system shown in Figure P7.3 is attached to a spacecraft. The weight of the mass is 12 kg, and it is assumed that the structure is weightless. In order to determine the system characteristics, a "twang" test is employed. The base is fixed, and an optical device reads to displacement, as shown in the figure.

   a Determine the damping factor ζ and the equivalent spring stiffness k of the system.

   b The system is excited by a vertical base acceleration that is harmonic with amplitude A. A displacement measuring device on the mass records an absolute displacement amplitude of 6.1 mm at an excitation frequency of 6 Hz. What is the base acceleration amplitude A in g's?

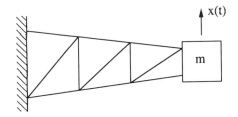

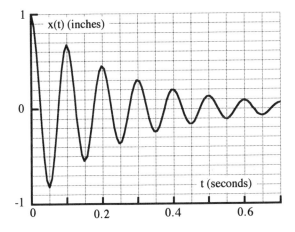

Figure P7.3

7.4 It is required to design the isolator system for an electronic package that is mounted in an aircraft. The fragility (failure) level of the 4.4-kg unit is $25g$. But the base excitation is assumed to be harmonic, having an amplitude of $55g$ and frequency of 40 Hz. The damping factor for the vibration mount is assumed to be 0.15. There will be four vibration mounts, one at each corner of the unit. Specify the requirements on the stiffness for each mount in newtons per millimeter.

7.5 A small single-story structure is modeled as a single-degree-of-freedom system with a rigid floor of weight 6000 lb and columns having lateral stiffness k. The natural period of the structure is observed to be 1.15 sec. Earthquake ground motion is modeled as a harmonic acceleration process of amplitude $0.25g$. The period of ground motion is 1.80 sec. What is the amplitude of the displacement of the floor relative to the ground? Assume a damping factor of 0.08.

8

RESPONSE OF SINGLE-DEGREE-OF-FREEDOM LINEAR SYSTEMS TO RANDOM ENVIRONMENTS

8.1 RESPONSE TO STATIONARY RANDOM FORCES

8.1.1 Preliminary Remarks

Consider the single-degree-of-freedom system shown in Figure 8.1. The force applied is a weakly stationary random process $F(t)$ $t > 0$. The process has known statistical properties: mean μ_F, variance σ_F^2, autocorrelation $R_F(\tau)$, spectral density $S_F(\omega)$, and marginal density function. The goal of this chapter is to describe the statistical properties of the response random process $X(t)$. General references include Crandall and Mark (1963), Lin (1967), Newland (1984), Clough and Penzien (1975), Schuëller and Shinozuka (1987), Elishakoff (1983), Spanos (1983, 1987), and Weaver et al. (1990).

Throughout this chapter, for purposes of simplicity, it is assumed that the input is force, denoted as F, and the output is displacement, denoted as X. The reader should interpret the input and output in a generic sense and apply mathematical intuition, plus the results presented in Chapter 7, to construct expressions for responses for other cases of input and output.

The point of departure in considering each of these will be the convolution integral, which we interpret as the connection between a particular sample realization of the random loading and the response to that loading. Noting that $F(t - \theta) = 0$ for $\theta > t$, it follows from Eq. (7.56) that,

$$X(t) = \int_{-\infty}^{t} F(t - \theta)h(\theta)\,d\theta \qquad t \geq 0 \tag{8.1}$$

Equation (8.1) will be used in the following to express the moments of the response random process. It is assumed that the input force random process

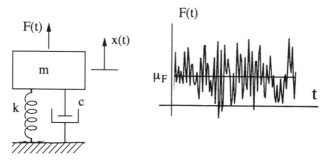

Figure 8.1 Single-degree-of-freedom system excited by random force.

$F(t)$ is zero for $t < 0$ and that it has constant mean μ_F for $t \geq 0$ and autocorrelation function $R_F(t, s) = R_F(s - t)$ for $t, s \geq 0$. We also assume that the response has zero initial conditions at $t = 0$, that is, $x(0) = \dot{x}(0) = 0$.

8.1.2 Mean of Response Process

We assume that the response random process has a well-defined finite mean. When it does, we write [see Eq. (8.1)]

$$\mu_X(t) = E[X(t)]$$

$$= E\left[\int_{-\infty}^{t} F(t - \theta)h(\theta) \, d\theta\right]$$

$$= \int_{-\infty}^{t} h(\theta)\mu_F \, d\theta$$

$$= \mu_F \int_{-\infty}^{t} h(\theta) \, d\theta \qquad t \geq 0 \qquad (8.2)$$

The response mean is finite as long as the system is stable, that is, $h(t)$ does not diverge as $t \to \infty$. As t increases, $\mu_X(t)$ clearly approaches a limiting value. In fact, depending on the system damping, this limit is quickly approached in a relatively small number of response cycles. Figure 8.2 shows how the expected value of the response approaches the limit μ_F/k as the response time increases in multiples of the natural period of the linear, single-degree-of-freedom system. It is clear that, in terms of the mean, the response approaches a stationary state following a relatively small number of cycles.

From Eq. (7.59), the integral of Eq. (8.2) approaches $H(0)$ as $t \to \infty$. Thus, the mean response is

$$\mu_X = \mu_F H(0) \qquad (8.3)$$

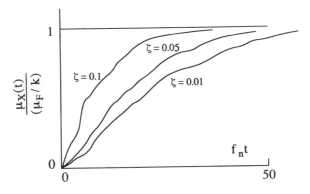

Response time measured in cycles
of the single degree of freedom system

Figure 8.2 Transient solution. Convergence of mean response of linear single-degree-of-freedom system to stationary limit; force modeled as white noise.

In words, the mean response equals the static response to the mean load. Note that when the load has a zero mean, the response also has a zero mean. This is a very important result as it implies that static and dynamic analyses can be treated separately. As an example, consider the force-excited system. The transfer function [Eq. (7.30)] is

$$H(\omega) = \frac{1}{(k - m\omega^2) + ic\omega} \qquad (8.4)$$

Noting that $H(0) = 1/k$, it follows that the stationary state mean response is

$$\mu_X = \frac{\mu_F}{k} \qquad (8.5)$$

This is a simple statement of Hooke's law. The dynamic response can be treated separately and added to the static response μ_X.

When the system initial conditions are not $x(0) = \dot{x}(0) = 0$, the homogeneous response to the initial conditions can be computed using the results of Section 7.1.4 and added to the result in Eq. (8.2).

8.1.3 Autocorrelation of Response Process

The autocorrelation function is fundamental in describing the response of the linear, single-degree-of-freedom system. We assume in the following that the response autocorrelation exists in a well-defined probabilistic sense, for example, in a mean-square sense, as described in Appendix A on convergence. When it does, we write, using the definition of R_X and Eq. (8.1),

$R_X(t, s) = E[X(t)X(s)]$

$$= E\left[\left\{\int_{-\infty}^{t} h(u)F(t - u)\ du\right\}\left\{\int_{-\infty}^{s} h(v)\ F(s - v)\ dv\right\}\right]$$

$$= E\left[\int_{-\infty}^{t} du \int_{-\infty}^{s} dv\ h(u)h(v)F(t - u)F(s - v)\right]$$

$$= \int_{-\infty}^{t} du \int_{-\infty}^{s} dv\ h(u)h(v)R_F(t - u, s - v)\right] \qquad t, s \geq 0 \quad (8.6)$$

When the excitation force is a nonstationary random process, Eq. (8.6) must be used to evaluate the response autocorrelation. The expression can be evaluated in closed form for a variety of simple inputs, but it must normally be evaluated numerically. When only the mean square of the response is sought, the expression becomes

$E[X^2(t)] = R_X(t, t)$

$$= \int_{-\infty}^{t} du \int_{-\infty}^{t} dv\ h(u)h(v)R_F(t - u, t - v) \qquad t \geq 0 \quad (8.7)$$

This expression is a clear indicator of the severity of the response random process as a function of time. The above expressions are the starting point of more advanced studies on the nonstationary response of structures. They also prove useful in the modal analysis of linear multi-degree-of-freedom structures, to be developed in the following chapter.

When the excitation force becomes a stationary random process for $t \geq 0$, we write $R_F(t, s) = R_F(s - t)$ for $t, s \geq 0$, and Eq. (8.6) simplifies to

$$R_X(t, s) = \int_{-\infty}^{t} du \int_{-\infty}^{s} dv\ h(u)h(v)R_F(s - t - (v - u)) \qquad t, s \geq 0$$

$$(8.8)$$

For t, s relatively small, $R_X(t, s)$ is a function of both t and s, even though the integrand depends only on $s - t$. The reason is that dependence on t and s also enters through the limits on the integral. However, if the system is stable, that is, if $h(t)$ does not diverge as $t \rightarrow \infty$, then as $t, s \rightarrow \infty$, $R_X(t, s)$ approaches a simple dependence on $s - t$. In practice, the time that the response requires to approach a stationary state is indicated by Figure 8.2.

The mean square of the single-degree-of-freedom system response is obtained, as a function of time, by evaluating Eq. (8.8) at $s = t$. It is

$$\sigma_X^2(t) = R_X(t, t) = \int_{-\infty}^{t} du \int_{-\infty}^{t} dv\ h(u)h(v)R_F(u - v) \qquad t, s \geq 0 \quad (8.9)$$

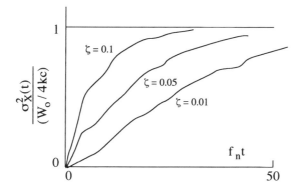

Response time measured in cycles
of the single degree of freedom system

Figure 8.3 Transient solution. Convergence of mean square response of linear single-degree-of-freedom system to stationary limit; force modeled as white noise.

This quantity approaches a limit [the quantity given in Eq. (8.11)] at $t \to \infty$. In fact, the rate at which $\sigma_X^2(t)$ approaches its constant limit as t increases is an indication of how quickly the response approaches a stationary state, and this is a function of system damping. Figure 8.3 shows how quickly some systems with light damping approach their stationary mean-square limit with increasing time.

As the response approaches a stationary state (when $t, s \to \infty$), we let $\tau = s - t$, and write

$$R_X(\tau) = \int_{-\infty}^{\infty} du \int_{-\infty}^{\infty} dv \, h(u)h(v)R_F \left[\tau - (v - u)\right] \qquad (8.10)$$

As for the nonstationary case [Eq. (8.6)], this expression can be evaluated in closed form for a variety of simple, weakly stationary inputs. It can easily be numerically evaluated in the general case.

The mean-square value of the weakly stationary response is obtained by evaluating Eq. (8.10) at $\tau = 0$. This is

$$E[X^2(t)] = R_X(0) = \int_{-\infty}^{\infty} du \int_{-\infty}^{\infty} dv \, h(u)h(v)R_F(u - v) \qquad (8.11)$$

and the result is not a function of time. This formula can be used to evaluate the response mean square, but if this is all the analyst wishes to obtain, a better approach is to use the frequency-domain equations to be developed in the following sections.

8.1.4 Spectral Density of Response Process

The power spectral density function is the Fourier transform of the autocorrelation function.

$$
\begin{aligned}
S_X(\omega) &= \frac{1}{2\pi} \int_{-\infty}^{\infty} R_X(\tau) e^{-i\omega\tau} \, d\tau \\[6pt]
&= \frac{1}{2\pi} \int_{-\infty}^{\infty} \left[\int_{-\infty}^{\infty} \int_{-\infty}^{\infty} h(u)h(v) R_F(\tau + u - v) \, du \, dv \right] e^{-i\omega\tau} \, d\tau \\[6pt]
&= \int_{-\infty}^{\infty} \int_{-\infty}^{\infty} h(u)h(v) \left[\frac{1}{2\pi} \int_{-\infty}^{\infty} R_F(\tau + u - v) e^{-i\omega\tau} \, d\tau \right] du \, dv
\end{aligned}
$$

$$(8.12)$$

Let $\theta = \tau + u - v$, $\tau = \theta - u + v$:

$$
\begin{aligned}
S_X(\omega) &= \int_{-\infty}^{\infty} \int_{-\infty}^{\infty} h(u)h(v) \left[\frac{1}{2\pi} \int_{-\infty}^{\infty} R_F(\theta) e^{-i\omega\theta} \, d\theta \right] e^{+i\omega u} e^{-i\omega v} \, du \, dv \\[6pt]
&= \left[\int_{-\infty}^{\infty} h(u) e^{i\omega u} \, du \right] \left[\int_{-\infty}^{\infty} h(v) e^{-i\omega v} \, dv \right] \left[\frac{1}{2\pi} \int_{-\infty}^{\infty} R_F(\theta) e^{-i\omega\theta} \, d\theta \right] \\[6pt]
&= H(-\omega) H(\omega) S_F(\omega)
\end{aligned}
$$

$$(8.13)$$

According to our symmetry relationships, because $h(t)$ is real, its Fourier transform has Hermitian symmetry (even real part and odd imaginary part). Thus,

$$
H(-\omega) = H^*(\omega) \tag{8.14}
$$

and it follows that

$$
S_X(\omega) = |H(\omega)|^2 S_F(\omega) \tag{8.15}
$$

In words, the spectral density of the response equals the spectral density of the loading times the square of the gain function (the modulus squared of the transfer function). This result is extremely important!

Note that while *double* integration is needed to establish the autocorrelation function, *no* integration at all is needed to establish $S_X(\omega)$. You can get $R_X(\tau)$ as the inverse Fourier transform of $S_X(\omega)$ if you need it.

Now the variance of the response may be calculated as the area under the power spectral density function,

$$
\sigma_X^2 = \int_{-\infty}^{\infty} S_X(\omega) \, d\omega
$$

$$\sigma_X^2 = \int_{-\infty}^{\infty} |H(\omega)|^2 \, S_F(\omega) \, d\omega \qquad (8.16)$$

In terms of the engineering spectral density of the response,

$$W_X(f) = |H(f)|^2 \, W_F(f) \qquad (8.17)$$

where f is frequency in hertz. The variance is

$$\sigma_X^2 = \int_0^{\infty} |H(f)|^2 \, W_F(f) \, df \qquad (8.18)$$

8.1.5 Distribution of Response Process

If $F(t)$ is a Gaussian process, and if the system is linear, then the response will also be Gaussian. Consider the response obtained from the impulse response function [Eq. (7.54)],

$$X(t) = \int_{-\infty}^{t} F(\tau)h(t - \tau) \, d\tau \qquad (8.19)$$

From the discussion in Chapter 2 on the reproductive property of the normal distribution, we know that $Z = \Sigma \, X_i$ is normal if X_i is normal and the X_i's are correlated. Thus, if we consider the integral as a sum, it follows that if $F(\tau)$ is normally distributed, $X(t)$, a sum of normally distributed random variables, is also normally distributed.

8.2 WHITE-NOISE PROCESS AS A MODEL FOR FORCE

8.2.1 Definition of Process

Formally, the white-noise random process is defined by the spectral density function,

$$S_F(\omega) = S_0 \qquad -\infty < \omega < \infty \qquad (8.20)$$

as shown in Figure 8.4. The autocorrelation function is the Fourier transform of $S_X(\omega)$ (see Chapter 4),

$$R_X(\tau) = 2\pi S_0 \delta(\tau) \qquad (8.21)$$

Clearly, white noise is physically impossible. A consequence of Eq. (8.20), the unbounded spectral density function, is that $\sigma_X \to \infty$. A consequence of

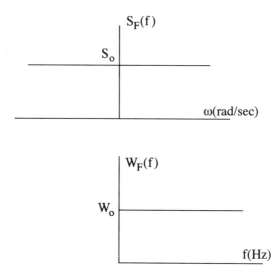

Figure 8.4 Spectral density function for white-noise process.

Eq. (8.21) is that dependency exists in the process only as $\tau \to 0$. While the white-noise process is only an idealization, it is useful in dynamic analyses.

8.2.2 Response of Force-Excited System to White Noise

For certain systems, the response to a white-noise force is finite. As an example, consider the force-excited system, the transfer function of which is given in Eq. (8.4). Assume the process has zero mean. If the force is white noise, $S_F(\omega) = S_0$ and the spectral density function of the response is

$$S_X(\omega) = |H(\omega)|^2 \, S_F(\omega) = \left| \frac{1}{(k - m\omega^2) + ic\omega} \right|^2 S_0$$

$$= \frac{S_0}{(k - m\omega^2)^2 + (c\omega)^2} \tag{8.22}$$

The response spectral density function has the same shape as the square of the gain function. The variance of the response is

$$\sigma_X^2 = \int_{-\infty}^{\infty} S_X(\omega) \, d\omega \tag{8.23}$$

If we use the complex form,

$$\sigma_X^2 = \int_{-\infty}^{\infty} \left| \frac{1}{(k - m\omega^2) + ic\omega} \right|^2 S_0 \, d\omega \tag{8.24}$$

The definite integral may be calculated by the method of residues (Lin, 1967). For our purposes, we may use the tabulated results of Appendix B. For our case, $n = 2$,

$$I_2 = \int_{-\infty}^{\infty} |H_2(\omega)|^2 \, d\omega \tag{8.25}$$

$$H_2(\omega) = \frac{B_0 + i\omega \, B_1}{A_0 + i\omega \, A_1 - \omega^2 \, A_2} = \frac{1}{k + i\omega c - m\omega^2} \tag{8.26}$$

Thus

$$B_0 = 1 \quad B_1 = 0 \quad A_0 = k \quad A_1 = c \quad A_2 = m \tag{8.27}$$

and

$$\begin{aligned} I_2 &= \frac{\pi(A_0 \, B_1^2 + A_2 \, B_0^2)}{A_0 A_1 A_2} \\ &= \frac{\pi \, (0 + m)}{kcm} \\ &= \frac{\pi}{kc} \end{aligned} \tag{8.28}$$

Finally,

$$\sigma_X^2 = S_0 \frac{\pi}{kc} \tag{8.29}$$

Or, in terms of the engineering spectral density function, W_0, it is easily shown that

$$\sigma_X^2 = \frac{W_0}{4 \, kc} = \frac{0.785 f_n W_0}{k^2 \zeta} \tag{8.30}$$

where f_n is the natural frequency, k is the spring constant, and ζ is the damping factor.

The variance obtained by integrating the response spectral density function for various cases of force- and base-excited motion is summarized in Table 8.1.

In general, when the transfer function for a measure of the response diminishes slower than ω^{-2} as $\omega \to \infty$, the integral for the mean-square value of the response will not be finite. In such a case, white noise will not be a valid approximation for the input.

Table 8.1   Mean-Square Response to White Noise

W_0, Input Spectral Density of[a]	Mean-Square Response of[a]	Formula
Force-Excited System		
$F(t)$	$X(t)$	$\sigma_X^2 = 0.785 f_n W_0/\zeta k^2$
$F(t)$	$\dot{X}(t)$	b
$F(t)$	$F_{TR}(t)$	$\sigma_{FT}^2 = 0.785 f_n W_0(1 + 4\zeta^2)/\zeta$
Base-Excited System		
$Y(t)$	$X(t)$	$\sigma_X^2 = 0.785 f_n W_0(1 + 4\zeta^2)/\zeta$
$\ddot{Y}(t)$	$\ddot{X}(t)$	$\sigma_{\ddot X}^2 = 0.785 f_n W_0(1 + 4\zeta^2)/\zeta$
$\ddot{Y}(t)$	$Z(t)$	$\sigma_Z^2 = W_0/(1984 \zeta f_n^3)$
$\ddot{Y}(t)$	$\ddot{Z}(t)$	b

[a]$F(t)$ = force on system; $X(t)$ = absolute displacement of mass; $Z(t)$ = displacement of mass relative to ground; $Y(t)$ = base displacement, F_{TR} = force transmitted to base.
[b]Response not finite.

8.2.3   Engineering Significance of White-Noise Process

Consider a force-excited system. The force $F(t)$ is a zero-mean stationary process having a spectral density function of $W_F(f)$. The exact mean-square response is

$$\sigma_X^2(\text{exact}) = \int_0^\infty |H(\omega)|^2 \, W_F(f) \, df \qquad (8.31)$$

Define $W_0 = W_F(f_n)$. Under certain conditions, a good approximation to the exact mean-square response is given by Eq. (8.30),

$$\sigma_X^2(\text{exact}) \cong \sigma_X^2(\text{approx}) \qquad (8.32)$$

providing that:

1. $W_F(f)$ is relatively smooth in the neighborhood of f_n.
2. ζ is small ($\zeta \leq 0.20$).

The white-noise approximation is illustrated in Figure 8.5. It is suggested from the figure that the reason that the approximation works is that the integral [Eq. (8.31)] will be dominated by the values of the two functions in a narrow neighborhood of f_n as long as the two conditions above apply.

The significance of the white-noise approximation is that closed-form expressions for σ_X^2 are available for many cases (Table 8.1), and it is generally *much easier* to use the white-noise model than to have to perform a numerical inte-

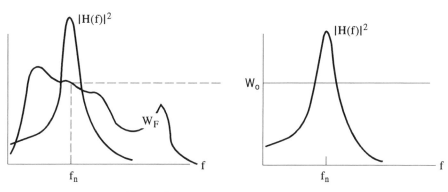

Figure 8.5 White-noise approximation.

gration of Eq. (8.31). An example of accuracy of the white-noise model is provided in Example 8.2.

8.3 EXAMPLES OF RESPONSE OF SINGLE-DEGREE-OF-FREEDOM SYSTEMS TO RANDOM FORCES

Example 8.1. The vibration environment for equipment to fly in the space shuttle is a white-noise acceleration process having a spectral density of $0.02g^2$ Hz^{-1}. An electronic unit, modeled as a single-degree-of-freedom system, to be designed for this environment has a natural frequency of 10 Hz and a damping factor of 0.05.

a. Determine the RMS acceleration response of the unit.
b. If the environment is applied for 25 sec, what is the median peak response in g's? Consider both peaks and troughs.

Solution
(a) The mean-square acceleration response of the unit is (see Table 8.1)

$$\sigma_{\ddot{X}}^2 = \frac{0.785 f_n W_0 (1 + 4\zeta^2)}{\zeta}$$

$$= \frac{0.785(10)(0.02)[1 + 4(0.05)^2]}{0.05} = 3.17$$

And the RMS value is

$$\sigma_{\ddot{X}} = 1.78g$$

(b) The spectral density function of the response, shown in Figure 8.6, has the same shape as the gain function. From examination of the figure, it is clear

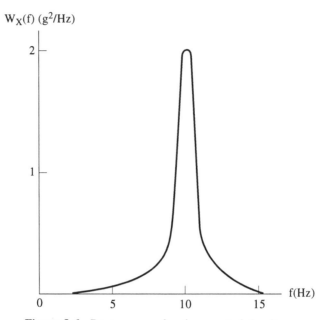

Figure 8.6 Response acceleration spectral density.

that the response acceleration process is narrow band at a center frequency of f_n. The number of peaks and troughs is

$$n = 2f_n T$$

$$= 2(10)(25) = 500$$

Assuming that the peaks and troughs are independent and that the process is narrow-band Gaussian, the distribution of the peak process is Rayleigh and the distribution function of the largest of 500 peaks is

$$F_V(v) = [F_v(v)]^{500} = \left\{ 1 - \exp\left[-\frac{1}{2}\left(\frac{v}{\sigma_{\ddot{X}}} \right)^2 \right] \right\}^{500}$$

Using the definition of the median,

$$0.50 = \left\{ 1 - \exp\left[-\frac{1}{2}\left(\frac{\tilde{V}}{1.78} \right)^2 \right] \right\}^{500}$$

Thus, the median of the largest peak is estimated as

$$\tilde{V} = 6.46g$$

Example 8.2. This example illustrates that the white-noise approximation to an RMS response provides a "good" estimate of the exact form even when the conditions are stretched.

Consider a small electronic element weighing 0.80 oz (22.7 g) that is mounted on a circuit board. The circuit, in turn, is subjected to a severe acoustic environment. The spectral density of the force due to acoustical pressure on the element has been established and is given in Figure 8.7a. Results from a shaker test indicates that the element resonates at 300 Hz and that the damping is 20% of critical. It is important that the amplitude of vibration of the element is kept less than 1.25 mm. Determine the probability of exceeding this design specification.

Solution. The RMS displacement will be computed using both the exact form and the white-noise approximation. Then the percentage of time that the displacement process exceeds 1.25 mm will be computed.

From Eq. (7.36), the gain function for the force-excited system is

$$|H(f)| = \frac{X_0}{F_0} = \frac{1}{k} \frac{1}{\sqrt{(1 - r^2)^2 + (2\zeta r)^2}}$$

where $r = f/f_n$. Hence, $f_n = 300$ and $\zeta = 0.20$. To determine the spring stiffness, the form for natural frequency is used,

$$f_n = \frac{1}{2\pi} \sqrt{\frac{k}{m}}$$

$$300 = \frac{1}{2\pi} \sqrt{\frac{k}{0.0227}}$$

And

$$k = 80.65 \text{ kN/m} = 80.65 \text{ N/mm}$$

The gain function squared is plotted in Figure 8.7b. The spectral density of the response is

$$W_X(f) = |H(f)|^2 W_F(f)$$

and $W_X(f)$ is plotted in Figure 8.7c. By numerical analysis, the mean-squared response is computed by the area under the spectral density function as $\sigma_X^2 = 0.0529$ mm^2. The RMS response (exact) is

$$\sigma_X = 0.231 \text{ mm (exact)}$$

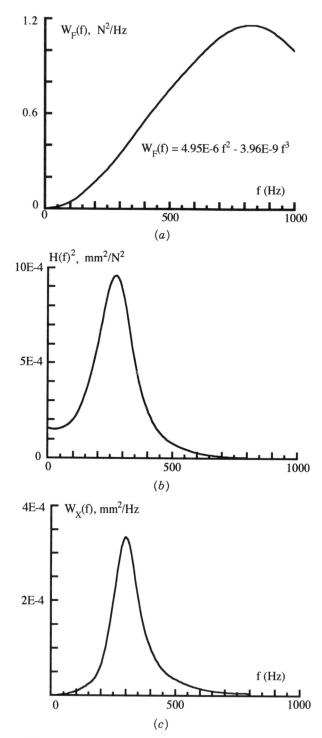

Figure 8.7 (*a*) Force spectral density. (*b*) Gain function squared. (*c*) Single degree of freedom of response.

Noting that $W_0 = W_F(300) = 0.3385$ N²/Hz and $k = 80.65$ N/mm, the white-noise approximation to the mean-squared response is (see Table 8.1)

$$\sigma_X^2 = 0.785 \frac{f_n W_0}{\zeta k^2}$$

$$= \frac{0.785(300)\,(0.3385)}{(0.20)\,(80.65)^2} = 0.0614 \text{ mm}^2$$

and the RMS response is

$$\sigma_X = 0.248 \text{ mm (approximate)}$$

When comparing the exact with the approximate RMS response, the white-noise form produced a result that was only 7.4% high, even though the damping factor was relatively large, and the spectral content was concentrated away from resonance.

Assuming that the displacement process X is Gaussian, the percent of time that $|X|$ exceeds 1.25 mm is

$$P[|X| > 1.25] = 2P[X > 1.25]$$

$$= 2P\left[\frac{X}{\sigma_X} > \frac{1.25}{0.231}\right] = 2[1 - \Phi(5.41)]$$

$$\approx 3 \times 10^{-8}$$

Example 8.3: Vehicle Problem. Consider a single-degree-of-freedom ground vehicle, as shown in Figure 8.8. As the vehicle travels with constant velocity v, road roughness provides a random base excitation to the system. The goal of the analysis is to determine the spectral density (and RMS) of the vertical acceleration of the vehicle.

Clearly the spectral density of the base displacement Y will be a function of velocity v. But σ_Y will be independent of v. Taking advantage of this fact, a normalized spectral density that describes road roughness but is independent

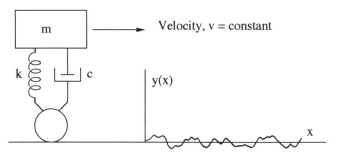

Figure 8.8 Vehicle moving on rough road surface.

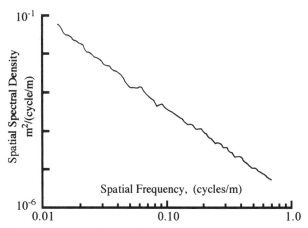

Figure 8.9 Typical spectral density of highway road profile.

of v can be constructed. Define spatial spectral density as $W_S(f_S)$, where f_S is spatial frequency. Here, W_S has units of m²/(cycle/m) or in.²/(cycle/ft); f_S has units of cycles/m or cycles/ft. A typical example of a spatial spectral density function representing a road surface is shown in Figure 8.9.

To get a better understanding of the spatial spectral density, imagine a test where a device to measure vertical displacement Y moves over a road surface at a constant velocity v over a road surface. Instrumentation measures $Y(t)$ and analysis provides the spectral density of the vertical displacement $W_Y(f)$. Then make a change of scale,

$$W_S(f_S) = vW_Y(f)$$

$$f_S = \frac{f}{v}$$

So defined, W_S defines the road roughness and ensures that σ_Y is independent of v.

Now to determine the acceleration of the mass. Given $W_S(f_S)$ and v, $W_Y(f)$ can be computed. The gain function, where base displacement is the input and absolute acceleration of the mass is the output, can be constructed from Eq. (7.42). Noting that acceleration amplitude is $\omega^2 X_0$,

$$|H(f)| = \frac{(2\pi f)^2 X_0}{Y_0} = (2\pi f)^2 \sqrt{\frac{1 + (2\zeta r)^2}{(1 - r^2)^2 + (2\zeta r)^2}}$$

Thus, the spectral density of the acceleration of the mass will be

$$W_{\ddot{X}}(f) = |H(f)|^2 W_Y(f)$$

where H and W_Y are defined as described above.

PROBLEMS

8.1 A single-degree-of-freedom spring–mass system having mass m, spring stiffness k, and damping factor ζ is subjected to a white-noise base acceleration at level W_0. It is required to determine the RMS force in the spring. This can be done two ways: (1) compute the RMS absolute acceleration and then apply a quasi-static dynamic load or (2) compute the relative displacement and use that value to calculate the RMS spring force. Perform both analyses and explain any differences in the results.

8.2 Peak lateral ground acceleration of a "strong" earthquake has been modeled as a stationary white-noise process having a level of 0.052 $(m^2/sec^4)/Hz$. A steel structure is modeled as a single-degree-of-freedom system (mass on top of a weightless structure, e.g., water tank). The system has a mass of 5000 kg. The natural frequency is 1.8 Hz; there is 5% critical damping. Compute the RMS of the absolute acceleration of the mass and the RMS of the force in the structure.

8.3 Consider a force $F(t)$ applied to the 3.7-kg mass, as shown in Figure P8.3. Here, $F(t)$ is white noise at a level of 10 N^2/Hz. It is required to design the steel rod (find the minimum b) so that the strength $R = 350$ MPa will be greater than three times the RMS stress. The damping factor is 0.10. The modulus of elasticity of steel is 207 GPa. Determine the minimum b in centimeters and the resulting natural frequency of the system.

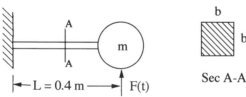

b

Sec A-A

Figure P8.3

8.4 A structural system to be launched by the space shuttle is described in problems 7.1 and 7.2. Mass $m = 0.259$ lb-sec^2/in., $k = 63.9$ lb/in., and $\zeta = 0.064$. The basic design environment is a white-noise base acceleration of $0.02g^2$ Hz^{-1}. Assume that this vibration level occurs for a 10-sec period.

a Compute the RMS of the absolute acceleration of the mass (in g's).

b The response of m will be assumed to be a narrow-band Gaussian process. Also assume that all peaks (and valleys) will be independent. What is your estimate of the median maximum acceleration (in g's) experienced by m?

c Compute the RMS of the relative displacement (in inches). Note $g = 386$ in./sec^2.

d What is your estimate of the median maximum shearing force in the structure (in pounds)? (See assumption of part b.)

8.5   An electronic instrument having mass of 0.82 kg is mounted close to a high-energy noise source in a high-performance aircraft. Using the vibration mounts selected, the natural frequency of the unit is 27.4 Hz. The damping factor is $\zeta = 0.12$. Environmental test specifications require that the design base acceleration is a white noise at a level of $0.04g^2$ Hz^{-1}. It is required to estimate the peak acceleration of the unit and the peak relative displacement. It is known (see Chapter 10) that for a reasonably long record (more than 1000 peaks), the expected maximum peak in a narrow-band process will be roughly four times the RMS. Using this algorithm, approximate the peaks of the absolute acceleration and relative displacement of the mass.

8.6   The system shown in Figure P8.6 is subjected to a random base acceleration process having the spectral density function, as shown. The natural frequency of the system is 15 Hz. It is required to determine the RMS of the relative response. (*Hint:* Use the white-noise approximation.) To determine the damping, a free vibration decay test was performed. It was observed that the peak displacement decayed from 0.80 to 0.10 in. in five cycles.

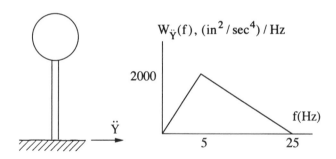

Figure P8.6

8.7   The structure shown in Figure P8.7 experiences a zero-mean base acceleration modeled as white noise at a level of $0.05g^2$ Hz^{-1}. The material is aluminum having a modulus of elasticity of $E = 10 \times 10^3$ ksi. The damping factor in the system is $\zeta = 0.05$. The weight of the mass is 4.5 lb.

a Determine the RMS response of the mass relative to the base.

b What percentage of the time will the relative response exceed 0.15 in.?

c What is the RMS of the absolute acceleration of the mass?

d Given that the vibration is applied for 20 sec, what is the probability that the absolute acceleration of the mass exceeds 15g (i.e., the probability that the largest peak is greater than 15g)?

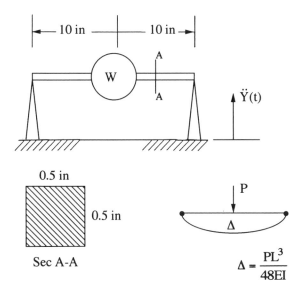

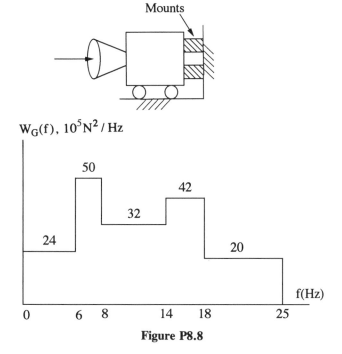

Figure P8.7

8.8 Shown in Figure P8.8 is a sketch of a rocket engine test stand. The stand is mounted to the wall with four vibration mounts, each one having a stiffness of 92.5 kN/m. The damping factor is estimated to be 0.20.

Figure P8.8

The mass of the engine is 55 kg. The engine is a rough-burning engine and it is anticipated that the force will be

$$F(t) = F_0 + G(t)$$

where $F_0 = 8.9$ kN. The spectral density of $G(t)$ is given in the figure. Here, $G(t)$ is assumed to be a Gaussian process.

a The mounts will "bottom out" at a deflection of 10 cm. Compute the probability that the instantaneous amplitude of the system will exceed 10 cm.

b The response will be approximately a narrow-band process. Compute the probability that a given peak will exceed 10 cm.

8.9   Shown in Figure P8.9 is a component of a spacecraft that is experiencing a quasi-static acceleration of $8g$ during engine burn. There is also a base vibration environment, a white noise of $0.04g^2$ Hz^{-1}, that is applied for only 25 sec. The weight of the electronic unit is 12 lb. The structural element is a thin tube having a wall thickness $t = 0.05$ in. The design requirement is that the maximum acceleration experienced by the unit during the 25 sec of engine operation is $20g$. Assume that the maximum acceleration corresponds to four times the RMS dynamic component.

a Design the unit, i.e., find the maximum required diameter. Do not be concerned here with member strength. Assume a damping factor of 0.05.

b Compute the probability that the maximum acceleration during engine fire exceeds $20g$.

Note: Force deflection for a cantilever beam is $\Delta = FL^3/3EI$ and modulus of elasticity $E = 10 \times 10^3$ ksi for aluminum.

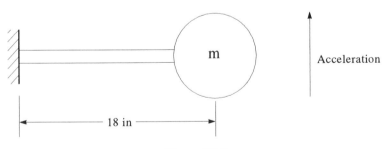

Figure P8.9

8.10   A single-wheeled vehicle weighing 1275 lb (578 kg) moves with a velocity of 40 mph on a road surface defined by the spectral density function of Figure 8.7. The fundamental frequency of the vehicle is 1.5 Hz and the damping factor is 0.25. Determine the RMS acceleration of the mass (in g's). *Hint:* Smooth the spectral density curve. Perform a ''crude'' numerical analysis.

8.11 The system shown in Figure P8.11 moves in the horizontal plane. Here, $F(t)$ is considered the input and $\theta(t)$ is the output. The force spectral density function is

$$S(\omega) = A\omega^2 \quad \text{for} \quad -\infty < \omega < \infty$$

a Derive the frequency response function. *Hint:* Write the equation of motion of rotation of a rigid body about a fixed axis (O). The moment of inertia of the rod about O is I_0. Assume small displacements.

b Derive the impulse response function.

c Determine the mean-square response.

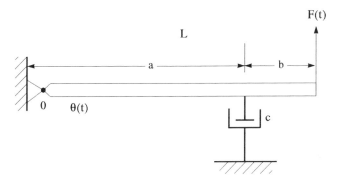

Figure P8.11

9

RANDOM VIBRATION OF MULTI-DEGREE-OF-FREEDOM SYSTEMS

9.1 EQUATIONS OF MOTION FOR MULTI-DEGREE-OF-FREEDOM SYSTEM

9.1.1 Introduction

A multi-degree-of-freedom (MDOF) system is defined as a system that requires more than one coordinate to describe its motion. An example of an MDOF system in Figure 9.1 shows forces and corresponding displacements at a node. The forces are all random processes. Note that the generalized forces can be moments and base motions as well. Similarly, the generalized displacements can include rotations. In general, all of the forces will be correlated. General references for linear MDOF systems include Hurty and Rubinstein (1964), Clough and Penzien (1975), Lin (1967), and Rao (1990), and for nonlinear systems, Spanos (1980) and Roy and Spanos (1993).

In general, this determination of statistics of the random response is analytically complex. In practice, computer methods are almost always necessary. Presented here is an overview of the topic along with several examples.

General assumptions that will be employed in this discussion include the following

1. The system is linear.
2. The force processes are
   a. Stationary
   b. Ergodic
   c. Gaussian

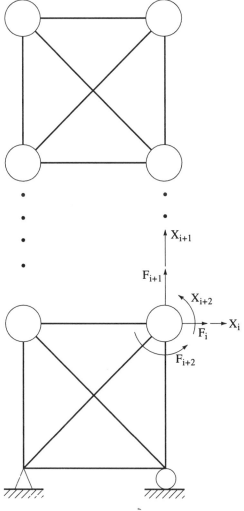

Figure 9.1 Example of MDOF system.

Two methods of approach for computing responses to MDOF systems will be discussed.

1. *Direct Method.* This method provides a direct and exact soluton to the steady-state response of systems. But while it can be employed for relatively small systems (100 degrees of freedom), computer storage requirements are so great for larger systems that it is not practical.

2. *Normal Mode Method.* This is the method of choice for analysis of large structural systems. It is a powerful method, but the solutions are only approximate.

Examples of both methods are given below.

9.1.2   Equations of Motion

Equations of motion for a single-degree-of-freedom system were developed in Chapter 7. For an n-degree-of-freedom system, the more general form of the equations of motion are (e.g., Rao, 1990)

$$[m] \{\ddot{X}\} + [c] \{\dot{X}\} + [k] [X] = \{F\} \qquad (9.1)$$
$$\underset{n \times n}{} \underset{n \times 1}{} \quad \underset{n \times n}{} \underset{n \times 1}{} \quad \underset{n \times n}{} \underset{n \times 1}{} \quad \underset{n \times 1}{}$$

where $[m]$ is the mass matrix, $[c]$ the matrix of damping coefficients, and $[k]$ the stiffness coefficient matrix. The force vector is $\{F\}$, and the response vector is $\{X\}$.

This matrix equation represents n simultaneous (and coupled) linear non-homogeneous differential equations.

Example 9.1. Consider the two-degree-of-freedom system shown in Figure 9.2. Develop the equations of motion.

Imposing the equations of motion on the free-body diagrams of each mass,

$$\Sigma F_1 = m_1 \ddot{X}_1 \; k_2(X_2 - X_1) + c_2(\dot{X}_2 - \dot{X}_1) - k_1 X_1 - c_1 \dot{X}_1 + F_1 = m_1 \ddot{X}_1$$

$$\Sigma F_2 = m_2 \ddot{X}_2 - k_2(X_2 - X_1) - c_2(\dot{X}_2 - \dot{X}_1) + F_2 = m_2 \ddot{X}_2$$

Rearranging into standard differential equation form,

$$m_1 \ddot{X}_1 + (c_1 + c_2)\dot{X}_1 + (k_1 + k_2)X_1 - c_2 \dot{X}_2 - k_2 X_2 = F_1(t)$$

$$m_2 \ddot{X}_2 - c_2 \dot{X}_1 - k_2 X_1 + c_2 \dot{X}_2 + k_2 X_2 = F_2(t)$$

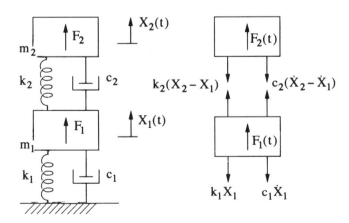

Definition of Coordinates   Free Body Diagrams

Figure 9.2   Two-degree-of-freedom system with free-body diagrams.

The set of differential equations can be expressed conveniently in matrix form as Eq. (9.1), where

$$[m] = \begin{bmatrix} m_1 & 0 \\ 0 & m_2 \end{bmatrix} \quad [c] = \begin{bmatrix} (c_1 + c_2) & -c_2 \\ -c_2 & c_2 \end{bmatrix}$$

$$[k] = \begin{bmatrix} (k_1 + k_2) & -k_2 \\ -k_2 & k_2 \end{bmatrix} \quad \{X\} = \begin{bmatrix} X_1 \\ X_2 \end{bmatrix} \quad \{F\} = \begin{bmatrix} F_1(t) \\ F_2(t) \end{bmatrix}$$

Note that the m, c, and k matrices are symmetric.

9.1.3 Transfer Function and Impulse Response Functions for System

The response of a system can be schematically described, as shown for the two-degree-of-freedom system in Figure 9.3. The response of each coordinate is linked to all of the forces, for example, coordinate 1 responds to forces at

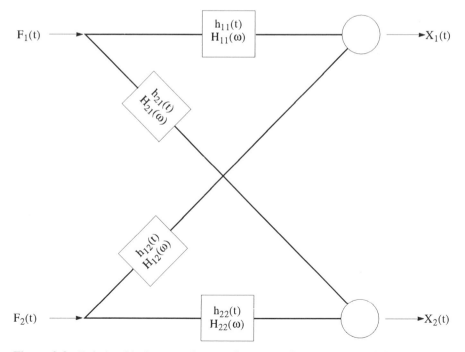

Figure 9.3 Relationship between force and response for two-degree-of-freedom system including transfer function and impulse response function.

both 1 and 2. As implied by the figure, transfer functions and impulse response functions link input to output.

In general,

$h_{ij}(t) \sim$ impulse response function for coordinate i due to force at j

$H_{ij}(\omega) \sim$ transfer function or frequency response function for coordinate i due to force at coordinate j

$H_{ij}(\omega)$, $h_{ij}(t)$ form a Fourier transform pair

We can define the transfer function and impulse response function matrices for an n-dimensional system as

$$[H(\omega)] = \begin{bmatrix} H_{11}(\omega) & H_{12}(\omega) & \cdots & H_{1n}(\omega) \\ H_{21}(\omega) & & & \\ \vdots & & & \vdots \\ H_{n1}(\omega) & \cdots & \cdots & H_{nn}(\omega) \end{bmatrix} \tag{9.2}$$

$$[h(t)] = \begin{bmatrix} h_{11}(t) & h_{12}(t) & \cdots & h_{1n}(t) \\ h_{21}(t) & & & \\ \vdots & & & \vdots \\ h_{n1}(t) & \cdots & \cdots & h_{nn}(t) \end{bmatrix} \tag{9.3}$$

To construct $[H(\omega)]$, the following procedure can be employed. First, assume the input is harmonic,

$$\{F(t)\} = \{F_0\}e^{i\omega t} \tag{9.4}$$

where, in general, the elements of the F_0 matrix are complex (defines phase relationships). The assumed harmonic output is

$$\{X(t)\} = \{X_0\}e^{i\omega t} \tag{9.5}$$

where the elements of the X_0 matrix are complex. Substitute $\{F\}$ and $\{X\}$ into the differential equations of motion [Eq. (9.1)] to get a relationship of the form

$$[V(\omega)]\{X_0\} = \{F_0\} \tag{9.6}$$

where

$$[V] = -\omega^2[m] + i\omega[c] + [k] \tag{9.7}$$

Sometimes $[V]$ is called the impedance matrix. Solving for $\{X_0\}$,

$$\{X_0\} = [H(\omega)] \{F_0\} \tag{9.8}$$

where the transfer function matrix is

$$[H(\omega)] = [V(\omega)]^{-1} \tag{9.9}$$

9.2 DIRECT MODEL FOR DETERMINING RESPONSE OF MDOF SYSTEM

9.2.1 Expression for Response

For simplicity, we will first focus our attention on the response of a single coordinate, $X(t)$, in a system. The response $X(t)$ is influenced by all of the forces on the system, as shown in Figure 9.4. Here, $X_i(t)$ is the component of the response of $X(t)$ due only to force $F_i(t)$. And $h_i(\tau)$ is the impulse response function of $X(t)$ due only to force $F_i(t)$.

For a linear system, $X(t)$ is just the sum of the responses, $X_i(t)$, to each

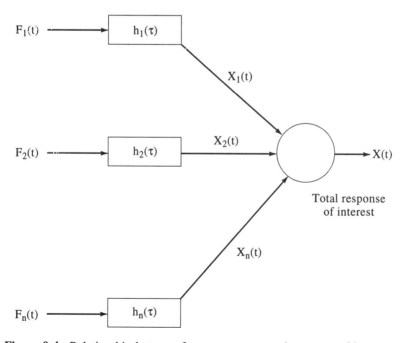

Figure 9.4 Relationship between forces on system and response of interest.

force, $F_i(t)$,

$$X(t) = \sum_{j=1}^{n} X_j(t) \tag{9.10}$$

where, from Eq. (7.56)

$$X_j(t) = \int_{-\infty}^{\infty} h_j(\tau) F_j(t - \tau) \, d\tau \tag{9.11}$$

Given that the random processes, $F_i(t)$; $-\infty < t < \infty$, $i = 1, n$ are stationary with means μ_i, cross-correlation functions $R_{F_iF_j}(\tau)$, and cross-spectral density functions $S_{F_iF_j}(\omega)$, for all ij, find the mean, auto correlation, spectral density function, and mean-square value of the response, $X_j(t)$, $j = 1, n$.

9.2.2 Mean Value of Response

First, we find the mean. By definition and direct substitution,

$$\mu_X = E[X(t)] = E\left[\sum_{i=1}^{n} \int_{-\infty}^{\infty} h_i(\tau) F_i(t - \tau) \, d\tau\right] \tag{9.12}$$

Using the property that the expected value of a sum is the sum of expected values,

$$\mu_X = \sum_{i=1}^{n} \int_{-\infty}^{\infty} h_i(\tau) \, E[F_i(t - \tau)] \, d\tau \tag{9.13}$$

By the definition of the mean of $F_i(t)$,

$$\mu_X = \sum \mu_{F_i} \int_{-\infty}^{\infty} h_i(\tau) \, d\tau \tag{9.14}$$

And because $H(\omega)$ is the Fourier transform of $h(t)$, we get, finally, [Eq. (7.59)]

$$\mu_X = \sum_{i=1}^{n} \mu_{F_i} H_i(0) \tag{9.15}$$

Note that, as in the case of a single degree-of-freedom, the mean response is simply a response to a static loading. And again, without loss of generality, we can let the mean value be zero in computing dynamic responses. This assumption is implicit in the following.

Consider now the general problem, that is, the response of all n coordinates.

The mean value of the jth coordinate is

$$\mu_{X_j} = \sum_{i=1}^{n} \mu_{F_i} H_{ji}(0) \tag{9.16}$$

or, in matrix form,

$$\{\mu_X\}_{n \times 1} = [H(0)]_{n \times n} \{\mu_F\}_{n \times 1} \tag{9.17}$$

where

$$\{\mu_X\} = \begin{bmatrix} \mu_{X_1} \\ \mu_{X_2} \\ \vdots \\ \mu_{X_n} \end{bmatrix} \qquad \{\mu_F\} = \begin{bmatrix} \mu_{F_1} \\ \mu_{F_2} \\ \vdots \\ \mu_{F_n} \end{bmatrix} \tag{9.18}$$

and again, without loss of generality, we can assume a mean of zero in considering the dynamic response.

9.2.3 Autocorrelation Function

Using the property that the expected value of a sum is the sum of expected values and the definition of the cross-correlation function, we are able to derive an expression for $R_X(\tau)$ as follows:

$$R_X(\tau) = E[X(t)X(t + \tau)] = E\left[\sum_i X_i(t) \sum_j X_j(t + \tau) \right]$$

$$= E\left[\sum_i \sum_j \int_{-\infty}^{\infty} h_i(\xi)F_i(t - \xi) \, d\xi \int_{-\infty}^{\infty} h_j(\eta)F_j(t + \tau - \eta) \, d\eta \right]$$

$$= \sum_i \sum_j \int_{-\infty}^{\infty} \int_{-\infty}^{\infty} h_i(\xi)h_j(\eta)E[F_i(t - \xi)F_j(t + \tau - \eta)] \, d\xi \, d\eta$$

$$\tag{9.19}$$

and finally,

$$R_X(\tau) = \sum_i \sum_j \int_{-\infty}^{\infty} \int_{-\infty}^{\infty} h_i(\xi)h_j(\eta)R_{F_iF_j}(\xi - \eta + \tau) \, d\xi \, d\eta \tag{9.20}$$

9.2.4 Spectral Density Function of Response

To determine the cross-spectral density function of the response, the approach will be to employ the relationships between the spectral density function of a random process, $X(t)$, and the finite Fourier transform of $X(t)$.

Define the vector of Fourier transforms of $\{X(t)\}$ and $\{F(t)\}$ as

$$\{X(\omega)\} = \begin{bmatrix} X_1(\omega) \\ X_2(\omega) \\ \vdots \\ X_n(\omega) \end{bmatrix} \qquad \{F(\omega)\} = \begin{bmatrix} F_1(\omega) \\ F_2(\omega) \\ \vdots \\ F_n(\omega) \end{bmatrix} \qquad (9.21)$$

where $X_i(\omega)$ and $F_j(\omega)$ are the Fourier transforms of $X_i(t)$ and $F_j(t)$, respectively. The cross-spectral density matrix of the force, assumed to be given, is

$$[S_F(\omega)] = \begin{bmatrix} S_{F_1} & \cdots & S_{F_1 F_n} \\ \vdots & \ddots & \vdots \\ S_{F_n F_1} & \cdots & S_{F_n} \end{bmatrix} \qquad (9.22)$$

Generalizing the results of Eq. 5.20 the cross-spectral density of $X_j(t)$ and $X_k(t)$ and $F_j(t)$ and $F_k(t)$ can be written as

$$\begin{aligned} S_{X_j X_k}(\omega) &= CE[X_j(\omega)X_k^*(\omega)] \\ S_{F_j F_k}(\omega) &= CE[F_j(\omega)F_k^*(\omega)] \end{aligned} \qquad (9.23)$$

where C is a constant.

The finite Fourier transforms of the response and the force are related by

$$\{X(\omega)\} = [H(\omega)] \{F(\omega)\} \qquad (9.24)$$

This relationship is a generalization of Eq. (7.65) for a MDOF system. The derivation directly follows that of Section 7.11 using matrix forms for X, F, and H.

Take the complex conjugate and then the transpose of Eq. (9.24) and then postmultiply:

$$\{X(\omega)\} \{X^*(\omega)\}^T = [H(\omega)] \{F(\omega)\} \{F^*(\omega)\}^T [H^*(\omega)]^T \qquad (9.25)$$

Take the expected value of both sides and multiply by a constant C:

$$CE[\{X(\omega)\} \{X^*(\omega)\}^T] = [H(\omega)]CE[\{F(\omega)\} \{F^*(\omega)\}^T] [H^*(\omega)]^T \qquad (9.26)$$

Comparing with Eq. (9.23), it follows that the spectral density function response matrix can be written as

$$[S_X(\omega)] = [H(\omega)] [S_F(\omega)] [H^*(\omega)]^T \tag{9.27}$$

The diagonal elements of $[S_X(\omega)]$ are the spectral density functions of each coordinate S_{X_i} and the off-diagonal terms $S_{X_i X_j}(\omega)$ are the cross-spectral density functions.

The mean-square response of the ith coordinate is

$$E[X_i^2(t);] = X_{ri}^2 = \int_{-\infty}^{\infty} S_{X_i}(\omega) \, d\omega \tag{9.28}$$

Here, $X_{ri}^2 = \sigma_{X_i}^2$, the variance, if the mean is equal to zero. The covariance of $X_i(t)X_j(t)$ for a time lag $\tau = 0$ is equal to (see Eq. 5.41 and 5.42)

$$\sigma_{X_i X_j}(0) = \int_{-\infty}^{\infty} S_{X_i X_j}(\omega) \, d\omega \tag{9.29}$$

9.2.5 Single Response Variable: Special Cases

There are two special cases of interest.

1. If there is a *single* input, say $F_k(t)$, then

$$S_{F_i F_j}(\omega) = \begin{cases} S_{F_k}(\omega) & \text{if } i = j = k \\ 0 & \text{otherwise} \end{cases} \tag{9.30}$$

and

$$S_X(\omega) = H_k^*(\omega)H_k(\omega)S_{F_k}(\omega) \tag{9.31}$$

or

$$S_X(\omega) = |H_k(\omega)|^2 \, S_{F_k}(\omega) \tag{9.32}$$

This is exactly the same result as the single-degree-of-freedom problem.

2. If all $F_i(t)$ are *uncorrelated*, then

$$S_{F_i F_j}(\omega) = \begin{cases} S_{F_k}(\omega) & \text{if } i = j = k = 1, 2, \ldots, n \\ 0 & \text{otherwise} \end{cases} \tag{9.33}$$

Then

$$S_X(\omega) = \sum_{k=1}^{n} |H_k(\omega)|^2 \, S_{F_k}(\omega) \tag{9.34}$$

Taking the integral with respect to ω of both sides of Eq. (9.34), the mean-square value of $X(t)$ is just the sum of the mean-squared response to each individual force. (Recall from basic probability theory that the variance of a sum of mutually independent random variables is the sum of the variances.)

$$\sigma_X^2 = \sum_{i=1}^{n} \sigma_i^2 \qquad (9.35)$$

where σ_i^2 is the variance of $X(t)$ due to $F_i(t)$ only.

Example 9.1. Example of Direct Method: Spacecraft Design Problem
Shown in Figure 9.5 is a model of a spacecraft (S/C) having a large engine in the center for attaining a synchronous orbit. When the engine fires, the mean thrust is $F_0 = 16$ kN. The engine burns rough and the thrust has a dynamic component, $F_D(t)$. The spectral density associated with F_D (a stationary Gaussian process) is shown. The two masses are $m_1 = 181$ kg and $m_2 = 362$ kg. The stiffness of the structural brackets is $k = 1.40$ MN/m. The damping coefficient is 3.50 kN-sec/m. The problems are:

1. All of the electronic gear is in the second mass and its acceleration cannot be too large.
2. The relative displacement between masses 1 and 2, proportional to the force in the structural brackets, cannot be too large.

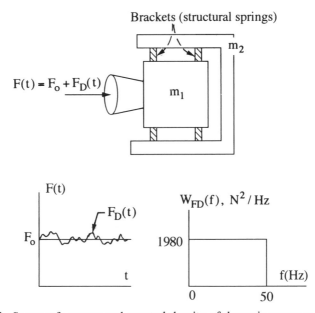

Figure 9.5 Spacecraft system and spectral density of dynamic component of force.

Solution. The goal of the analysis will be to determine the absolute acceleration $\ddot{X}_2(t)$ and the relative displacement $Z(t)$.

First, consider the mean values. From the free-body diagram of the system of Figure 9.6, we can compute the mean acceleration of the system as

$$a = \frac{F}{m} = \frac{F}{m_1 + m_2} = \frac{16{,}000}{543} = 29.4 \text{ m/sec}^2$$

or, in terms of g's,

$$G = \frac{29.4}{g} = \frac{29.4}{9.81}$$

$$= 3g$$

The mean force in spring μ_{F_s} is computed by applying the equation of motion to the free-body diagram of m_2 in Figure 9.6:

$$\mu_{F_s} = m_2 a_2$$

$$= 362(29.4) = 10.6 \text{ kN}$$

And the mean deflection of the spring is given as

$$\mu_Z = \frac{\mu_{F_s}}{k} = \frac{10.6 \times 10^3}{1.40 \times 10^6}$$

$$= 7.6 \text{ mm}$$

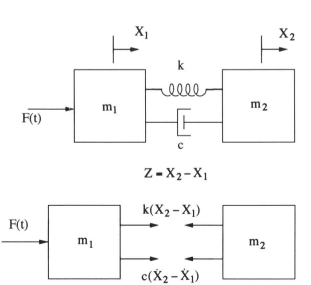

Figure 9.6 Dynamic model and free-body diagrams of spacecraft system.

Now compute the variance of the response. We can assume that the white-noise approximation is reasonable if the natural frequency is comfortably below 50 Hz (see the spectral density function of Figure 9.5). Here, the natural frequencies of the system are 0 (rigid-body mode) and 17.4 Hz (see Sections 9.3.2 and 9.3.3) so the white-noise approximation is assumed to be reasonable.

Now consider the dynamic component of the response. To obtain the response spectral density [Eq. (9.27)], we need $[H(\omega)]$ and $S_F(\omega)$. First we find $[H(\omega)]$.

The equations of motion are

$$m_1\ddot{X}_1 - k(X_2 - X_1) - c(\dot{X}_2 - \dot{X}_1) = F(t)$$

$$m_2\ddot{X}_2 + k(X_2 - X_1) + c(\dot{X}_2 - \dot{X}_1) = 0$$

But the coordinates of interest are (Z, X_2). Let

$$X_2 - X_1 = Z$$

Then the equations of motion in terms of Z and X_2 are

$$m_1\ddot{X}_2 - m_1\ddot{Z} - kZ - c\dot{Z} = F(t)$$

$$m_2\ddot{X}_2 + kZ + c\dot{Z} = 0$$

Assuming harmonic excitation, the force is

$$\{F(t)\} = \begin{bmatrix} F_0 \\ 0 \end{bmatrix} e^{i\omega t}$$

And the response is

$$\begin{bmatrix} Z(t) \\ \ddot{X}_2(t) \end{bmatrix} = \begin{bmatrix} Z_0 \\ \ddot{X}_0 \end{bmatrix} e^{i\omega t}$$

Substituting assumed force and response into the equations of motion,

$$m_1\ddot{X}_0 + m_1\omega^2 Z_0 - kZ_0 - ic\omega Z_0 = F_0$$

$$m_2\ddot{X}_0 + kZ_0 + ic\omega Z_0 = 0$$

The V matrix [Eq. (9.6)] is,

$$[V(\omega)] = \begin{bmatrix} m_1\omega^2 - k - ic\omega & m_1 \\ k + ic\omega & m_2 \end{bmatrix}$$

Finally, the transfer function is

$$[H(\omega)] = [V(\omega)]^{-1} = \frac{1}{\Delta(\omega)} \begin{bmatrix} m_2 & -m_1 \\ -k - ic\omega & m_1\omega^2 - k - ic\omega \end{bmatrix}$$

where

$$\Delta(\omega) = m_1 m_2 \omega^2 - k(m_1 + m_2) - ic\omega(m_1 + m_2)$$

Because there is only a single force applied, the spectral density function matrix for the force on the system is

$$[S_F(\omega)] = \begin{bmatrix} S_0 & 0 \\ 0 & 0 \end{bmatrix}$$

Then the spectral density function matrix for the response is given as

$$[S_X(\omega)] = [H(\omega)] [S_F] [H^*(\omega)]^T$$

Note that

$$[H^*(\omega)]^T = \frac{1}{\Delta^*(\omega)} \begin{bmatrix} m_2 & -k + ic\omega \\ -m_1 & m_1\omega^2 - k + ic\omega \end{bmatrix}$$

so that finally, the spectral density function of the response is

$$[S_X(\omega)] = \frac{S_0}{|\Delta(\omega)|^2} \begin{bmatrix} m_2^2 & -km_2 + icm_2\omega \\ -km_2 + icm_2\omega & k^2 + (c\omega)^2 \end{bmatrix}$$

Now determine the RMS of relative displacement $Z(t)$. The spectral density function of $Z(t)$ is the top term on the diagonal of the $[S_X(\omega)]$ matrix,

$$S_Z(\omega) = S_0 \frac{m_2^2}{|\Delta(\omega)|^2}$$

The mean square of Z is given by

$$\sigma_Z^2 = \int_{-\infty}^{\infty} S_Z(\omega) \, d\omega$$

Use the forms presented in Appendix B for a solution to the integral,

$$\sigma_Z^2 = \frac{\pi S_0 m_2^2}{ck(m_1 + m_2)^2}$$

But note that $S_0 = W_0/4\pi$; upon making the substitutions, it follows that

$$\sigma_Z \cong 0.212 \text{ mm}$$

Thus, the relative displacement has mean and RMS

$$\mu_Z = 7.60 \text{ mm} \qquad \sigma_Z = 0.212 \text{ mm}$$

The force in the spring has mean and RMS

$$\mu_{F_s} = k\mu_Z = 10.6 \text{ kN} \qquad \sigma_{F_s} = k\sigma_Z = 0.297 \text{ kN}$$

Now consider the absolute acceleration of mass 2. The spectral density is the second diagonal term of $[S_X(\omega)]$,

$$S_{\ddot{X}}(\omega) = S_0 \frac{k^2 + (c\omega)^2}{|\Delta(\omega)|^2}$$

$$\sigma_{\ddot{X}}^2 = \int_{-\infty}^{\infty} S_{\ddot{X}}(\omega) \, d\omega$$

Again, use the formulas in Appendix B:

$$\sigma_{\ddot{X}}^2 = \pi S_0 \left[\frac{k}{c(m_1 + m_2)^2} + \frac{c}{(m_1 + m_2)(m_1 m_2)} \right]$$

Upon making the substitutions,

$$\sigma_{\ddot{X}} = 0.852 \text{ m/s}^2$$

$$= 0.087g$$

Thus, the mean and standard deviation of $\ddot{X}_2$ are

$$\mu_{\ddot{X}_2} = 3g \qquad \sigma_{\ddot{X}_2} = 0.087g$$

Summary Comment. This exercise clearly shows that the direct method would be very difficult to execute by hand for a system larger than the one considered.

Example 9.2. Example of Direct Method: Useful Forms for Some Problems Consider the expression for the response spectral density matrix

$$[S_X] = [H] [S_F] [H*]^T$$

where

$$[H] = \begin{bmatrix} a & b \\ c & d \end{bmatrix} \qquad [S_F] = \begin{bmatrix} S_{11} & S_{12} \\ S_{21} & S_{22} \end{bmatrix}$$

for two degrees of freedom. Note that all terms will be functions of frequency ω.

Expanding $[S_X]$, the two diagonal elements are

$$S_{X_1} = aa^*S_{11} + ab^*S_{12} + a^*bS_{21} + bb^*S_{22}$$

$$S_{X_2} = cc^*S_{11} + cd^*S_{12} + c^*dS_{21} + dd^*S_{22}$$

These forms are particularly useful in establishing expressions for the mean-square response. If the inputs are white noise, then the forms of Appendix B may apply.

Consider the following special cases:

1. *Force applied at 1 only*:

$$[S_F] = \begin{bmatrix} S_{11} & 0 \\ 0 & 0 \end{bmatrix}$$

and the response at 1 is

$$\sigma_{X_1}^2 = \int_{-\infty}^{\infty} S_{11} aa^* d\omega$$

using Appendix B,

$$H_n(\omega) = \sqrt{S_{11}}\, a$$

Or if S_{11} is constant (white noise), it can be taken outside the integral.

2. *Force applied at 2 only*: The response at 1 is

$$\sigma_{X_1}^2 = \int_{-\infty}^{\infty} S_{22} bb^* d\omega$$

using

$$H_n(\omega) = \sqrt{S_{22}}\, b$$

3. *Uncorrelated forces at 1 and 2*:

$$\sigma_{X_1}^2 = \int_{-\infty}^{\infty} (S_{11} aa^* + S_{22} bb^*)\, d\omega$$

using

$$H_n^{(1)} = \sqrt{S_{11}}\, a \qquad H_n^{(2)} = \sqrt{S_{22}}\, b$$

Mathematical intuition can be applied to obtain expressions for $S_{X_i}(\omega)$ and $\sigma_{X_i}^2$ for more than two degrees of freedom. In general, a numerical integration is required to evaluate $\sigma_{X_i}^2$, but there are special cases where Appendix B can be employed.

9.3 NORMAL MODE METHOD

9.3.1 Preliminary Remarks

In the general case of a linear system, where damping is present and $F(t)$ has an arbitrary form, the equation of motion is [repeating Eq. (9.1)]

$$\underset{m \times m}{[m]} \underset{n \times 1}{\{\ddot{X}\}} + \underset{n \times n}{[c]} \underset{n \times 1}{\{\dot{X}\}} + \underset{n \times n}{[k]} \underset{n \times 1}{\{X\}} = \underset{n \times 1}{\{F\}} \qquad (9.1)$$

where m, c, and k are square mass, damping, and stiffness matrices, respectively, and X is the vector of coordinates. But these equations are n simultaneous *coupled* differential equations. In general, solutions are difficult, although digital simulation programs can be used effectively.

The normal mode approximation involves a transformation of coordinates so that the differential equations become uncoupled and easy to solve. In the following sections, the normal mode method is developed. For general reference, the reader is referred to Hurty and Rubinstein (1964) and Clough and Penzien (1975).

9.3.2 Eigenvalue Problem: Free-Vibration Solution to Equations of Motion

The first step in executing the normal mode method is to solve the eigenvalue problem. Consider free vibration (no external forces, no damping) of a system. From Eq. (9.1),

$$[m] \{\ddot{X}\} + [k]\{X\} = 0 \qquad (9.36)$$

Assume that $\{X\}$ is harmonic in free vibration,

$$\{X\} = \{\phi\}e^{i\omega t} \qquad (9.37)$$

Here ϕ_i, a scalar element of $\{\phi\}$, denotes the magnitude of the response. Equation (9.37) indicates that all masses in the system are oscillating harmonically with frequency ω and are in phase or 180° out of phase with each other. Upon substitution into Eq. (9.36),

$$[-\omega^2[m] + [k]] \{\phi\} = 0 \qquad (9.38)$$

For a nontrivial $(\{\phi\} \neq 0)$ solution, it must be that

$$|-\omega^2[m] + [k]| = 0 \qquad (9.39)$$

This determinant leads to an algebraic equation in ω^2. For an n-degree-of-freedom system, there will be n real values of ω^2 that will satisfy the *characteristic equation*. These values are called *eigenvalues*. They are the squares of natural frequencies of the system. When an ω_i is substituted into Eq. (9.38), relative values of ϕ_i are established. Such a vector is called an *eigenvector*. These n eigenvectors are the natural modes of vibration of the system. In practice, large systems are analyzed using a computer and standard eigenvalue routines are used to solve the eigenvalue problem.

9.3.3 Example of Eigenvalue Problem

Consider the system shown in Figure 9.7. When disturbed, the system can oscillate in free vibration. The equations of motion of the system are

$$m_1 \ddot{X}_1 + (k_1 + k_2)X_1 - k_2 X_2 = 0$$
$$m_2 \ddot{X}_2 - k_2 X_1 + k_2 X_2 = 0 \tag{a}$$

Assume that the system moves harmonically in free vibration,

$$X_1(t) = \phi_2 e^{i\omega t} \qquad X_2(t) = \phi_2 e^{i\omega t} \tag{b}$$

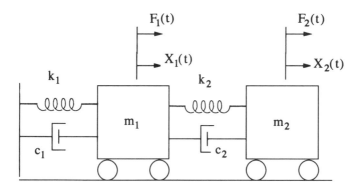

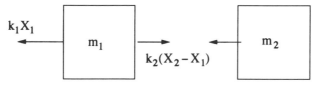

Figure 9.7 Two-degree-of-freedom system with free-body diagrams of masses in free vibration (no damping or external forces).

Substituting into the equation of motion,

$$(k_1 + k_2 - m_1\omega^2)\phi_1 - k_2\phi_2 = 0$$
$$-k_2\phi_1 + (k_2 - m_2\omega^2)\phi_2 = 0 \qquad \text{(c)}$$

For a nontrivial solution of ϕ_1 and ϕ_2, it must be that

$$\begin{vmatrix} k_1 + k_2 - m_1\omega^2 & -k_2 \\ -k_2 & k_2 - m_2\omega^2 \end{vmatrix} = 0 \qquad \text{(d)}$$

Letting $k' = k_1 + k_2$, it follows that, upon expanding the determinant,

$$m_1 m_2 \omega^4 - (m_2 k' + m_1 k_2)\omega^2 + (k' k_2 - k_2^2) = 0 \qquad \text{(e)}$$

This is called the characteristic equation. From this quadratic equation, we can find two roots: ω_1^2, ω_2^2.

Example 9.3

$$m_1 = 35 \text{ kg} \qquad m_2 = 17.5 \text{ kg}$$
$$k_1 = 8750 \text{ N/m} \qquad k_2 = 3500 \text{ N/m}$$

Then the characteristic equation becomes

$$\omega^4 - 550\omega^2 + 50{,}000 = 0 \qquad \text{(f)}$$

The two roots are

$$\omega_1^2 = 115 \qquad \qquad \omega_2^2 = 435$$
$$\omega_1 = 10.72 \text{ rad/sec} \qquad \omega_2 = 20.82 \text{ rad/sec}$$
$$f_1 = 1.71 \text{ Hz} \qquad f_2 = 3.32 \text{ Hz}$$

Note: An n-degree-of-freedom system has n natural frequencies. Here, $n = 2$. The fundamental is f_1 and often plays the most important role in dynamics analysis: $f_1 = 1.71$ Hz.

To obtain the corresponding eigenvector, substitute $\omega^2 = \omega_1^2 = 115$ into Eq. (c),

$$47\phi_{11} - 20\phi_{21} = 0$$
$$-20\phi_{11} + 8.5\phi_{21} = 0$$

Thus, ϕ_{ij} is the amplitude of displacement of the ith mass in the jth mode. Note that the two equations are identical. Here, $\phi_{21} = 2.35 \, \phi_{11}$. We can write the relative response as a vector ϕ, often called an eigenvector; and ω_1 is called an eigenvalue,

$$\phi_1 = \begin{bmatrix} \phi_{11} \\ \phi_{21} \end{bmatrix} = \begin{bmatrix} 1 \\ 2.35 \end{bmatrix}$$

This defines the first mode shape, or the configuration of the system when it is vibrating with frequency f_1. Only the ratio of ϕ_{21} to ϕ_{11} is unique. It is equally valid to write

$$\phi_1 = \begin{bmatrix} 4 \\ 9.4 \end{bmatrix} \quad \text{or} \quad \begin{bmatrix} -2 \\ -4.7 \end{bmatrix}$$

We can only obtain the *ratios* of the responses in free vibration. For the second mode, $f_2 = 3.32$ Hz, Eq. (c), with $\omega^2 = \omega_2^2 = 435$, is

$$-17\phi_{12} - 20\phi_{22} = 0$$

$$-20\phi_{12} - 23.5\phi_{22} = 0$$

which implies that $\phi_{22} = -0.850\phi_{21}$. Thus, we can write

$$\phi_2 = \begin{bmatrix} \phi_{12} \\ \phi_{22} \end{bmatrix} = \begin{bmatrix} 1 \\ -0.850 \end{bmatrix} \quad \text{or} \quad \begin{bmatrix} -1 \\ 0.850 \end{bmatrix} \quad \text{or} \quad \begin{bmatrix} -4 \\ 3.4 \end{bmatrix} \quad \text{etc.}$$

Here, ϕ_2 defines the configuration of the system when it is vibrating at $f_2 = 3.32$ Hz. The terms ϕ_1 and ϕ_2 are called *eigenvectors*.

The modal matrix is just the collection of eigenvectors,

$$[\Phi] = [\phi_1 \quad \phi_2] = \begin{bmatrix} 1 & 1 \\ 2.35 & -0.850 \end{bmatrix}$$

Note that, as a matter of convention, the ω and ϕ are ranked in ascending order. Thus, ω_1 is the lowest or fundamental frequency, ω_2 the second lowest, and so on.

9.3.4 Normal Mode Method: Orthogonality Conditions

The orthogonality conditions are the key to the normal mode method. Consider the free-vibration equation [Eq. (9.36)]. Assume harmonic motion $\{X\} = \{\phi\}e^{i\omega t}$, where $\{\phi\}$ defines the mode shape,

$$-\omega^2[m]\{\phi\} + [k]\{\phi\} = 0 \qquad (9.40)$$

But there are n values of ω and corresponding $\{\phi\}$,

$$\omega_i; \ \phi_i \equiv \{\phi\}_i \qquad i = 1, n$$

Thus,

$$\omega_i^2[m]\phi_i = [k]\phi_i \tag{9.41}$$

Consider the jth mode, ω_j, ϕ_j. Premultiply both sides by ϕ_j^T, $j \neq i$

$$\omega_i^2 \phi_j^T[m]\phi_i = \phi_j^T[k]\phi_i \tag{9.42}$$

Note the rule for transposing matrices $(ABC)^T = C^T B^T A^T$. Taking the transpose of both sides,

$$\omega_i^2 \phi_i^T[m]\phi_j = \phi_i^T[k]\phi_j \tag{9.43}$$

Consider Eq. (9.41) for the jth mode,

$$\omega_j^2[m]\phi_j = [k]\phi_j \tag{9.44}$$

Premultiply by ϕ_i^T,

$$\omega_j^2 \phi_i^T[m]\phi_j = \phi_i^T[k]\phi_j \tag{9.45}$$

Subtract Eq. (9.45) from Eq. (9.43),

$$(\omega_i^2 - \omega_j^2) \ \phi_i^T[m]\phi_j = 0 \tag{9.46}$$

Because $\omega_i \neq \omega_j$, it must be that

$$\phi_i^T[m]\phi_j = 0 \tag{9.47}$$

This equation is referred to as an orthogonality condition. Also note that

$$\phi_i^T[m]\phi_i = M_i \tag{9.48}$$

Here, M_i is called the modal mass for the ith coordinate. Because the absolute value of ϕ_i is arbitrary, so also is the value of M_i.

The same orthogonality condition applies to $[k]$. Consider Eqs. (9.43) and (9.45). Divide by ω_i^2 and ω_j^2,

$$\phi_i^T[m]\phi_j = \frac{1}{\omega_i^2} \ \phi_i^T[k]\phi_j \tag{9.49}$$

$$\phi_i^T[m]\phi_j = \frac{1}{\omega_j^2} \ \phi_i^T[k]\phi_j \tag{9.50}$$

Subtracting Eq. (9.50) from Eq. (9.49),

$$\left(\frac{1}{\omega_i^2} - \frac{1}{\omega_j^2}\right) \phi_i^T[k]\phi_j = 0 \tag{9.51}$$

Because $\omega_i \neq \omega_j$, it must be that

$$\phi_i^T[k]\phi_j = 0 \tag{9.52}$$

This is the orthogonality condition for $[k]$. Also note that

$$\phi_i[k]\phi_i = K_i \tag{9.53}$$

Here, k_i is called the modal stiffness and is also arbitrary.

Define the modal matrix as

$$[\Phi] = [\phi_1 \quad \phi_2 \quad \ldots \quad \phi_n] \tag{9.54}$$

Then, as a consequence of the orthogonality conditions, the modal mass and modal stiffness matrices, as defined below, are diagonal,

$$[M] = [\Phi]^T [m][\Phi] = \begin{bmatrix} M_1 & \ldots & \ldots & 0 \\ & M_2 & & \\ & & \ddots & \\ 0 & \ldots & \ldots & M_n \end{bmatrix} \tag{9.55}$$

$$[K] = [\Phi]^T [k][\Phi] = \begin{bmatrix} K_1 & \ldots & \ldots & 0 \\ & K_2 & & \\ & & \ddots & \\ 0 & \ldots & \ldots & K_n \end{bmatrix} \tag{9.56}$$

Orthonormalization of Eigenvectors. Recall that values of ϕ_i are relative. Therefore, we can adjust the scale on ϕ_i so that all $M_i = 1$. Let the adjusted ϕ be

$$\phi_i^N = \frac{1}{\sqrt{M_i}} \phi_i \tag{9.57}$$

Then

$$\phi_i^T [M]\phi_i = 1$$
$$[\Phi]^T [m][\Phi] = [I] \tag{9.58}$$

where $[I]$ is the identity matrix.

Consider Eq. (9.42) with $i = j$:

$$\omega_i^2 \phi_i^T [m] \phi_i = \phi_i^T [k] \phi_i \tag{9.59}$$

But from Eq. (9.58), the coefficient of ω_i^2 is unity, so that

$$\phi_i^T [k] \phi_i = \omega_i^2 \tag{9.60}$$

$$[K] = [\Phi]^T [k][\Phi] = \begin{bmatrix} \omega_1^2 & \cdots & & \cdots & 0 \\ \vdots & \omega_2^2 & & & \vdots \\ \vdots & & \ddots & & \vdots \\ 0 & \cdots & & \cdots & \omega_n^2 \end{bmatrix}$$

Example 9.4. Computing Modal Mass and Stiffness

Consider Example 9.3. The mass, stiffness, and modal matrices of eigenvectors are

$$[m] = \begin{bmatrix} m_1 & 0 \\ 0 & m_2 \end{bmatrix} = \begin{bmatrix} 35 & 0 \\ 0 & 17.5 \end{bmatrix}$$

$$[k] = \begin{bmatrix} k_1 + k_2 & -k_2 \\ -k_2 & k_2 \end{bmatrix} = \begin{bmatrix} 12,250 & -3,500 \\ -3500 & 3500 \end{bmatrix}$$

$$[\Phi] = [\phi_1 \quad \phi_2] = \begin{bmatrix} 1 & 1 \\ 2.35 & -0.850 \end{bmatrix}$$

Determine the modal mass and stiffness matrices:

$$[M] = [\Phi]^T [m][\Phi] = \begin{bmatrix} 131.6 & 0 \\ 0 & 47.6 \end{bmatrix}$$

$$[K] = [\Phi]^T [k][\Phi] = \begin{bmatrix} 15,128 & 0 \\ 0 & 20,728 \end{bmatrix}$$

Now, orthonormalize the eigenvectors so that the modal mass matrix is the unit matrix:

$$\phi_1^N = \frac{1}{\sqrt{M_1}} \phi_1 = \frac{1}{\sqrt{131.6}} \begin{bmatrix} 1 \\ 2.35 \end{bmatrix}$$

$$= \begin{bmatrix} 0.0872 \\ 0.2049 \end{bmatrix}$$

$$\phi_2^N = \frac{1}{\sqrt{M_2}} \phi_2 = \begin{bmatrix} 0.1449 \\ -0.1232 \end{bmatrix}$$

Using the normalized vectors, the normalized modal matrix is

$$[\Phi] = \begin{bmatrix} 0.0872 & 0.1449 \\ 0.2049 & -0.1232 \end{bmatrix}$$

Now the modal mass and stiffness matrices will be

$$[M] = \begin{bmatrix} 1 & 0 \\ 0 & 1 \end{bmatrix}$$

$$[K] = \begin{bmatrix} 115 & 0 \\ 0 & 435 \end{bmatrix}$$

Note that the diagonal terms of the stiffness matrix are the natural frequencies in the first and second modes, ω_1^2 and ω_2^2, respectively.

9.3.5 Normal Mode Method: Equations of Motion

Consider the case of forced vibration but with no damping,

$$[m]\{\ddot{X}\} + [k]\{X\} = \{F\} \tag{9.61}$$

We will introduce damping later. These differential equations are coupled. The essence of the normal mode method is that we are able to transform coupled (hard-to-solve) equations into uncoupled (easy-to-solve) equations.

Make a change of response variables:

$$\{X\} = [\Phi]\{q\} \tag{9.62}$$

where $\{q\}$ are the "new" variables. The transformation matrix is $[\Phi]$, the matrix of eigenvectors. Substitute into Eq. (9.61) and premultiply by $[\Phi]^T$.

$$[\Phi]^T[m][\Phi]\{\ddot{q}\} + [\Phi]^T[k][\Phi]\{q\} = [\Phi]^T\{F\} \tag{9.63}$$

From the definition of modal mass and stiffness, it follows that

$$[M]\{\ddot{q}\} + [K]\{q\} = \{Q\} \tag{9.64}$$

where $\{Q\}$ is called the modal force,

$$\{Q\} = [\Phi]^T\{F\} \tag{9.65}$$

Look what happened to the equations of motion. For $n = 2$,

$$M_1 \ddot{q}_1 + K_1 q_1 = Q_1(t)$$
$$M_2 \ddot{q}_2 + K_2 q_2 = Q_2(t) \tag{9.66}$$

The equations uncouple! They can be solved one at a time for $q_1(t)$, $q_2(t)$, ... Then in "real" coordinates, the $\{q\}$ are substituted back into Eq. (9.62). For $n = 2$,

$$\begin{bmatrix} X_1(t) \\ X_2(t) \end{bmatrix} = \begin{bmatrix} \phi_{11} & \phi_{12} \\ \phi_{21} & \phi_{22} \end{bmatrix} \begin{bmatrix} q_1(t) \\ q_2(t) \end{bmatrix} \tag{9.67}$$

The $\{q(t)\}$ are called modal coordinates. To gain a physical interpretation, consider X_1 of Eq. (9.67),

$$X_1(t) = \phi_{11} q_1(t) + \phi_{12} q_2(t) \tag{9.68}$$

Note that ϕ_{ij} is the displacement of the ith mass in the jth mode. Let $q_2 = 0$. Then

$$X_1(t) = \phi_{11} q_1(t)$$
$$X_2(t) = \phi_{21} q_1(t) \tag{9.69}$$

Note that

$$\frac{X_1}{X_2} = \frac{\phi_{11}}{\phi_{21}} \tag{9.70}$$

is the same ratio of displacement as when the system is oscillating in free vibration in the first mode. Thus, $q_1(t)$ is the participation of the first mode in the total displacement of the system. The total motion of the system can be thought of as a linear sum of the motion in all of the modes. Also note that for the ith mode, the natural frequency is ω_i [e.g., $\omega_1^2 = K_1/M_1$ in Eq. (9.66)]. Thus, $q_i \to \infty$ if the forcing frequency $\omega \to \omega_i$ for no damping. This is the condition of resonance in the ith mode. The implication here is that a system having n degrees of freedom will have n natural frequencies to be avoided if resonance is not desired.

9.3.6 Big Payoff for Using Normal Mode Method

Consider a large system, say a spacecraft structure modeled as an $n = 400$ degree of freedom system. If the force is shock or a random vibration, then the system tends to respond globally only in the first few modes, and principally the fundamental. Suppose it was assumed that only the first three modes made a significant contribution to the total response. Then

$$\{X\}_{400 \times 1} \cong [\Phi]_{400 \times 3} \{q\}_{3 \times 1} \tag{9.71}$$

Instead of solving 400 simultaneous coupled differential equations, it is now necessary to solve only three uncoupled equations for q_1, q_2, and q_3. Clearly, the normal mode method is a very powerful tool for dynamicists.

9.3.7 Introduction of Damping Into Equations of Motion

Now introduce damping to the system [Eq. (9.1)]. Making the normal mode transformation,

$$[M]\{\ddot{q}\} + [\Phi]^T[c][\Phi]\{\dot{q}\} + [K]\{q\} = \{Q\} \tag{9.72}$$

The modal damping matrix,

$$[C] = [\Phi]^T[c][\Phi] \tag{9.73}$$

C is diagonal only in special cases, one of which is,

$$[c] = a[m] + b[k] \tag{9.74}$$

This is sometimes called Rayleigh damping (see Clough and Penzien, 1975). But another scheme is generally used. First, recall that for a single-degree-of-freedom system

$$m\ddot{x} + c\dot{x} + kx = F(t)$$

$$\ddot{x} + \frac{c}{m}\dot{x} + \omega_n^2 x = \frac{F(t)}{m} \tag{9.75}$$

But

$$\frac{c}{m} = \frac{\zeta c_c}{m} = \frac{\zeta(2m\omega_n)}{m} = 2\zeta\omega_n \tag{9.76}$$

So that

$$\ddot{x} + 2\zeta\omega_n\dot{x} + \omega_n^2 x = \frac{F(t)}{m} \tag{9.77}$$

Now by analogy consider motion in the ith mode,

$$\ddot{q}_i + 2\zeta_i\omega_i\dot{q}_i + \omega_i^2 q_i = \frac{Q_i(t)}{M_i} \tag{9.78}$$

Introduce ζ_i as a damping factor in the ith mode. This approach, called modal damping is not mathematically correct generally, but it generally leads to reasonable results if damping is small.

Example 9.5.   Normal Mode Method

Consider the two-degree-of-freedom system shown in Figure 9.8 subjected to a white-noise stationary ground acceleration, $W_0 = 0.0742$ m^2/sec^4/Hz, roughly equivalent to the 1940 El Centro earthquake. This is the same system as presented in Examples 9.3 and 9.4, for which a modal analysis was performed:

$$m_1 = 35 \text{ kg} \qquad m_2 = 17.5 \text{ kg}$$
$$k_1 = 8750 \text{ N/m} \qquad k_2 = 3500 \text{ N/m}$$

The assumed damping factors for each mode are

$$\zeta_1 = 0.034 \quad \text{(first mode)}$$
$$\zeta_2 = 0.066 \quad \text{(second mode)}$$

The goal of analysis is to compute the RMS of the *relative* response of m_1 to the ground (proportional to the force in the structural member).

Let $Z_i = X_i - Y$. Substituting into the undamped differential equations of motion, if follows that

$$m_1\ddot{Z}_1 + (k_1 + k_2)Z_1 - k_2Z_2 = -m_1\ddot{Y}$$
$$m_2\ddot{Z}_2 - k_2Z_1 + k_2Z_2 = -m_2\ddot{Y}$$

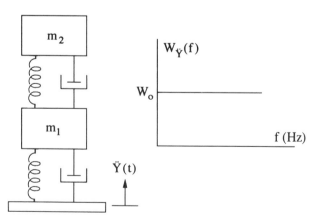

Figure 9.8   Two-degree-of-freedom system and spectral density of base acceleration.

The transformation to $[Z]$ was made so that the equations of motion were identical to those of Example 9.3. The choice of Z_2 is somewhat unconventional because it does not define the motion of m_2 relative to m_1, what one typically wants for structural design.

A summary of the analysis of the equations of motion in free vibration is presented as follows (see Example 9.3). The two natural frequencies are $\omega_1 = 10.77$ rad/sec and $\omega_2 = 20.85$ rad/sec. The modal and modal mass matries are

$$[\Phi] = \begin{bmatrix} 1 & 1 \\ 2.35 & -0.85 \end{bmatrix} \quad [M] = \begin{bmatrix} 131.6 & 0 \\ 0 & 47.6 \end{bmatrix}$$

and the transformation to modal coordinates, $\{x\} = [\Phi]\{q\}$, is

$$\begin{bmatrix} z_1 \\ z_2 \end{bmatrix} = \begin{bmatrix} 1 & 1 \\ 2.35 & -0.85 \end{bmatrix} \begin{bmatrix} q_1 \\ q_2 \end{bmatrix}$$

The modal force is

$$\{Q\} = [\Phi]^T\{F\} = \begin{bmatrix} 1 & 2.35 \\ 1 & -0.85 \end{bmatrix} \begin{bmatrix} -35\ddot{Y} \\ -17.5\ddot{Y} \end{bmatrix}$$

$$= \begin{bmatrix} -76.1\ddot{Y} \\ -20.1\ddot{Y} \end{bmatrix}$$

Following Eq. (9.78), the uncoupled equations of motion in modal coordinates are

$$\ddot{q}_1 + 2\zeta_2\omega_1\dot{q}_1 + \omega_1^2 q_1 = G_1$$

$$\ddot{q}_2 + 2\zeta_2\omega_2\dot{q}_2 + \omega_2^2 q_2 = G_2$$

where

$$G_1 = \frac{Q_1}{M_1} = \frac{-76.1\ddot{Y}}{131.6} = -0.578\ddot{Y}$$

$$G_2 = \frac{Q_2}{M_2} = \frac{-20.1\ddot{Y}}{47.6} = -0.422\ddot{Y}$$

Because Y is white noise, we can use the white-noise formulas

$$\sigma_{z_i}^2 = \frac{W_0}{1984\zeta_i f_i^3}$$

Note that, in using this form, we do have to account of the coefficients of the $\ddot{Y}$ term. Just from dimensional considerations, these terms would be squared, and we multply W_0 in the above expression. Thus,

$$\sigma_1^2 = \frac{(0.578)^2(0.0742)}{1984\ (0.034)(1.71)^3} = 7.35 \times 10^{-5}$$

$$\sigma_2^2 = \frac{(0.422)^2(0.0742)}{1984\ (0.066)(3.32)^3} = 2.76 \times 10^{-6}$$

Our only concern is with $Z_1(t)$. From the normal mode transformation,

$$Z_1 = \phi_{11}q_1 + \phi_{12}q_2$$

If we assume that $q_1(t)$ and $q_2(t)$ are independent processes, the variance of the sum is the sum of the variances, and thus the RMS of $Z_1(t)$ is

$$\sigma_{Z_1}^2 = \phi_{11}^2\sigma_1^2 + \phi_{12}^2\sigma_2^2$$

$$= (1)\ (7.35 \times 10^{-5}) + (1)\ (2.76 \times 10^{-6})$$

$$\sigma_{Z_1}^2 = 73.5 \times 10^{-6} + 2.76 \times 10^{-6}$$

Clearly, in this problem, the response appears to be dominated by the response of the system in the first mode. Finally, the RMS of $Z_1(t)$ is

$$\sigma_{Z_1} = 8.73\ \text{mm}$$

9.3.8 Normal Mode Method: Mode Combination Problem—Some Methods

The variance of the response of coordinate X_i is computed from the variance of the response of the modal coordinates as

$$\sigma_{X_i}^2 = V(X_i) = V\left(\sum_{j=1}^{J} \phi_{ij}q_j\right) \tag{9.79}$$

where J is the total number of modes considered. In the above example, we assumed that q_1 and q_2 were independent processes. In fact, there is likely to be some dependency.

It is generally thought that if two natural frequencies are "close," then there is significant dependency in q_1 and q_2 because of congruence of the gain functions. And if the two frequencies are "far apart," q_1 and q_2 will tend to be independent. No general operational definition for closeness exists.

Various methods of "combining modes" to compute the RMS response have been proposed. For example, STARDYNE has a selection of 15 mode

Modes

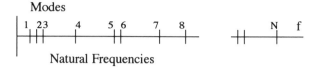

Natural Frequencies

Figure 9.9 Example of distribution of natural frequencies of MODF system.

combinations available. Figure 9.9 shows natural frequencies of a MDOF system on a frequency scale, and it illustrates how some frequencies may be close together and some have significant separation. Given that the analyst must choose which modes are close, define the following:

$$N = \text{total modes considered}$$

$$Z = \text{total number of modes considered to be close}$$

$$H = \text{number of sets of equal or close eigenvalues}$$

$$K_j = \text{number of close frequencies in any set}$$

$$\sigma_p^2 = \text{mean square value of } q_p(t).$$

$$\phi_{ip} = \text{mode vector, coordinate } i \text{ in } p\text{th mode}$$

Some of the various methods used for computing the RMS of X_i from modal responses are as follows:

1. *Square Root of the Sum of Squares (SRSS)*

$$\sigma_{X_i} = \left[\sum_{p=1}^{N} (\phi_{ip}^2 \sigma_p^2) \right]^{1/2} \tag{9.80}$$

This method assumes all modal responses are independent (see example in Section 9.3.9).

2. *Absolute Method*

$$\sigma_{X_i} = \sum_{p=1}^{N} (|\phi_{ip}\sigma_p|) \tag{9.81}$$

This method assumes all modal responses have correlation coefficients of $\rho = 1, -1$.

3. *Naval Research Laboratory (NRL) Method*

$$\sigma_{X_i} = |\phi_{11}\sigma_{q1}| + \left[\sum_{p=2}^{N} \phi_{ip}^2 \sigma_p^2 \right]^{1/2} \tag{9.82}$$

The first mode is assumed to be correlated with the sum (by SRSS) of all of the others.

4. *Close Method*

$$\sigma_{X_i} = \sum_{p=1}^{Z} (|\phi_{ip}\sigma_p^2|) + \left[\sum_{p=Z+1}^{N-Z} (\phi_{ip}^2\sigma_p^2) \right]^{1/2} \tag{9.83}$$

It is assumed that there is a group of modes that are close and another group that are separated.

5. *Modified Root Sum of Squares (MRSS)*

$$\sigma_{X_i} = \left\{ \sum_{j=1}^{H} \left[\sum_{p=1}^{K_j} (|\phi_{ip}\sigma_p|) \right]^2 + \sum_{p=Z+1}^{N-Z} (\phi_{ip}^2\sigma_p^2) \right\}^{1/2} \tag{9.84}$$

A more general form of the close method. Here, there are H groups of modes that are close. All other modes are separated (Richard et al., 1988).

6. *Combined Quadratic Combination (CQC)*

$$\sigma_{X_i} = \left(\sum_{j=1}^{N} \sum_{k=1}^{N} \sigma_{ij}\rho_{jk}\sigma_{ki} \right)^{1/2} \tag{9.85}$$

where

$$\sigma_{ij} = \phi_{ij}\sigma_j$$

$$\rho_{jk} = \frac{8\sqrt{\zeta_j\zeta_k}\,(\zeta_j + r\zeta_k)r^{3/2}}{(1-r^2)^2 + 4\zeta_j\zeta_k r(1+r^2) + 4(\zeta_j^2 + \zeta_k^2)r^2} \tag{9.86}$$

$$r = \frac{\omega_k}{\omega_j} \qquad k > i$$

For example, for two modes

$$\sigma_{X_i}^2 = \sigma_{i1}^2\rho_{11} + \sigma_{i1}\rho_{12}\sigma_{i2} + \sigma_{i2}\rho_{21}\sigma_{i1} + \sigma_{i2}^2\rho_{22} \tag{9.87}$$

or

$$\sigma_{X_i}^2 = \phi_{i1}^2\sigma_1^2 + \phi_{i2}^2\sigma_2^2 + 2\rho_{12}\phi_{i1}\phi_{i2}\sigma_1\sigma_2 \tag{9.88}$$

[see Der Kiureghian (1980) and Wilson et al. (1981).]

9.3.9 Example of Mode and Combination Algorithm Results

Shown in Figure 9.8 is a two-degree-of-freedom, base-excited system. The base motion is a stationary, Gaussian white-noise acceleration process at a level of $W_0 = 0.02g^2$ Hz^{-1} (a basic design requirement for some equipment to fly

Table 9.1 Comparison of Performance of Mode Combination Algorithms for Two Cases

	Case 1	Case 2
Data		
m_1, kg	35.0	35.0
m_2, kg	7.0	14.0
k_1, N/m	8750.0	8750.0
k_2, N/m	1750.0	3500.0
ζ_1	0.095	0.10
ζ_2	0.095	0.05
Natural frequencies		
f_1, Hz	2.02	1.84
f_2, Hz	3.14	3.43
RMS of Z_1 (mm)		
Exact	25.8	33.5
$m_2 = 0$	25.3	24.6
SRSS	24.9	26.7
ABS	30.4	33.2
CQC	25.8	27.1

on the space shuttle). It is required to determine the RMS of the displacement of m_1 relative to the base, that is, $Z_1 = X_1 - Y$.

The goal of analysis is to compare the normal mode approximation with the exact solution. The mode combination algorithms to be considered are the SRSS, ABS, and CQC methods.

Two cases are considered. System parameter values are provided in Table 9.1.

Solution. The exact solution is obtained by the direct method described in Section 9.2. The normal mode solution for the modal coordiantes q_1 and q_2 was obtained using the method illustrated in Section 9.3. Modal combination algorithms for SRSS, ABS, and CQC were employed.

Results of the analysis for the two cases are presented in Table 9.1. For reference only, the solution for $m_2 = 0$ was obtained as an approximation. It is clear from the results that no conclusion can be made relative to the relative quality of the competing mode combination methods.

9.4 COMPUTER CODES FOR ANALYZING MDOF STRUCTURAL SYSTEMS

Several commercial codes are available for performing random vibration analysis of MDOF structural systems. They include ANSYS, MSC Nastran, and STARDYNE. A brief summary of typical capabilities is given as follows:

1. Displacement, velocity, acceleration, or force power spectral densities may be used as input.
2. The power spectral density can be applied as either a base or nodal excitation.
3. Uncorrelated, partially correlated, or fully correlated multiple-input power spectral densities can be used to excite a structure.
4. The spectral analysis that follows the modal analysis can be a response spectrum or power spectral density analysis.
5. Several mode combination algorithms are available.
6. Partial correlation can be defined in one of three ways:
   a. Direct definition of input power spectral density matrices including off-diagonal cospectral and quadspectral terms.
   b. A function of distance separating the excitation points (i.e., spatial correlation
   c. In terms of a traveling wave with specified velocity, direction, and coherency properties.
7. The RMS values can be reported as displacements, velocities, or accelerations. Similarly, force velocities, force accelerations, stress velocities, and stress accelerations can also be reported.

PROBLEMS

9.1 For the system shown in Figure P9.1, the applied force $F(t)$ is a Gaussian white-noise process having a zero mean and spectral density S_0. It is required to determine the RMS of the mass m. To do this:

a Write the equations of motion.

b Derive the transfer function.

c Note the results of the example in Section 9.2.

d Determine the RMS using the tables of Appendix B.

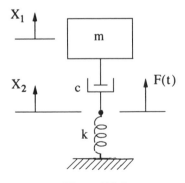

Figure P9.1

9.2 For the system shown in Figure P9.2, determine the RMS of $X_1(t)$. Here, $F(t)$ is the same as problem 9.1. See also problem 9.1 instructions.

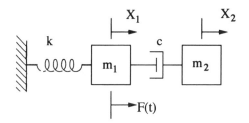

Figure P9.2

9.3 Consider the two-degree-of-freedom system shown in Figure P9.3. The system parameters are given as

$$m = 1 \qquad k = 100 \qquad c = 20$$

The force is a white-noise process at level $S_0 = 1000$. Determine the RMS of $X_1(t)$.

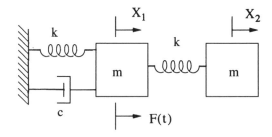

Figure P9.3

9.4 The base acceleration of the system shown in Figure P9.4 is a white-noise process having a zero mean and spectral density of S_0. Determine the RMS of the acceleration of the mass on the right.

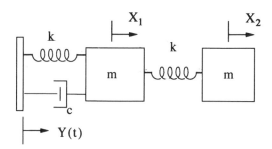

Figure P9.4

9.5 Shown in Figure P9.5 is a model of an automobile. The spectral density of the base displacement $y(t)$ is given as W_0. The ultimate goal of this problem is to determine the RMS of the acceleration of m_2. Let $m_1 = 70$ kg, $m_2 = 1080$ kg, $k_1 = 350$ kN/m, $k_2 = 90$ kN/m, and $c = 8$ kN-sec/in. Do the following:

a Draw free-body diagrams and derive the equations of motion.

b Letting X_1 and X_2 be the output, derive the transfer function matrix. For simplicity, assume that $m_1 = 0$.

c Derive the gain function for X_2 only.

d Determine the spectral density function for $\ddot{X}_2$.

e For $W_0 = 0.30$ cm^2/Hz, determine the RMS of the acceleration of m_2 in g's (use Appendix B).

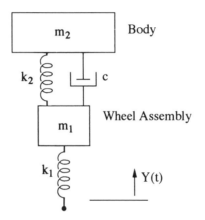

Figure P9.5

9.6 The two-degree-of-freedom system shown in Figure P9.6 is a simplified model of a structure subjected to a white-noise random force applied to mass 2 at a level of 100 lb^2/Hz. It is required to compute the RMS of the absolute displacement of both masses 1 and 2. Use the normal mode method and SRSS to combine modal responses. Assume that the damping factor in each mode is 5%. The weight of each mass is $W = 100$ lb and the lateral stiffness of the columns between each mass is $k = 250$ lb/ft.

9.7 The system shown in Figure P9.7 is subjected to a white-noise random force having a spectral density of 0.005. The system is defined by $m = 2$, $k_1 = 300$, $k_2 = 200$, and $L = 8$. The moment of inertia of the rod about point 0 is $I_0 = 3$. All of the values are in compatible units. Use the normal mode method (with SRSS) to determine the RMS of the responses, x and θ. Assume that $\zeta = 0.05$ in both modes.

9.8 Consider the system shown in Figure 9.8, with $m_1 = 0.2$, $k_1 = 50$, $m_2 = 0.08$, and $k_2 = 20$; damping in the first mode is $\zeta_1 = 0.10$ and in the

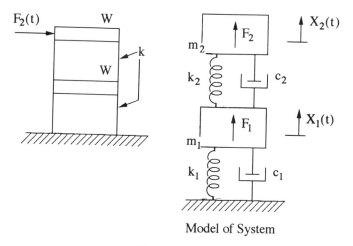

Model of System

Figure P9.6

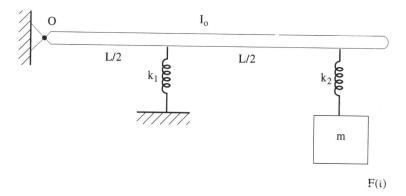

Figure P9.7

second mode $\zeta_2 = 0.05$, $W_0 = 0.02g^2$ Hz^{-1}. Using the direct method, determine the exact RMS of $Z_1 = X_1 - Y$. *Hint*: Let $Z_1 = X_1 - Y$ and $Z_2 = X_2 - X_1$. Also, the force spectral density matrix is

$$S(\omega) = S_0 \begin{bmatrix} 1 & 1 \\ 1 & 1 \end{bmatrix}$$

9.9 Consider problem 9.8.

    a Determine the natural modes and frequencies.

    b Determine the mean-square responses of the modal coordinates q_1 and q_2.

    c Determine the RMS of Z_1 by the SRSS, ABS, and CQC methods. *Note*: Let $Z_1 = X_1 - Y$, $Z_2 = X_2 - X_1$ in your analysis.

10

DESIGN TO AVOID STRUCTURAL FAILURES DUE TO RANDOM VIBRATION

Consider a single structural element subjected to a random force (or stress) process $S(t)$, as shown in Figure 10.1. The goal of design is to select an element size so the probability of failure during the intended service life is acceptably small. In this chapter, we will address elementary models for ensuring structural integrity of single components when subjected to random stress processes.

Elementary modes of failure associated with structural components include (1) yielding, (2) excessive deformation, (3) brittle fracture, (4) ductile fracture, (5) buckling (instability under compressive loads), and (6) fatigue. Discussed in this chapter are methods of designing to avoid a "level exceedance" or quasi-static failure mode that relates to items 1–5. Also, methods of fatigue life prediction under random stresses and designing to avoid fatigue are presented.

It is conventional in the design world to use capital letters to represent design values. Thus, in this chapter we frequently use upper case to denote design values, departing from exclusive use of capitals for random variables used throughout this text.

10.1 THREE-SIGMA DESIGN

10.1.1 Basic Design Criterion

The "three-sigma" design criterion is typically used to design structures where brittle fracture or fatigue are not considered to be the principal failure modes. Spacecraft structures and missiles are examples of systems that are not likely to see prolonged exposure to the design vibration environment where a fatigue failure might occur.

254

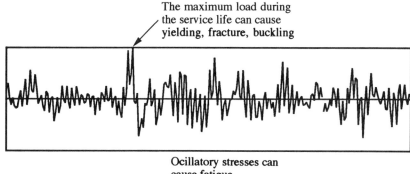

Figure 10.1 Realization of load or load effect (stress) process and possible failure modes.

Consider the structural element shown in Figure 10.2. $S(t)$ is a stationary Gaussian stress process with RMS σ_S. The strength R is deterministic. Find the minimum value of A (cross-sectional area) for a safe design. Note that the Gaussian model for $S(t)$ implies that the sample space for S is $-\infty < S < \infty$. Therefore, no matter how large you make R, it is impossible to ensure that stress S will be less than strength R for all time t with probability 1.0.

A basic requirement commonly employed in design criteria documents is that, for a safe design,

$$R \geq 3\sigma_S \tag{10.1}$$

The implication, although it is usually unstated, is that $\mu_S = 0$ and R is deterministic. Because S is Gaussian, the probability that stress will exceed three times its RMS value is

$$P\left[|S(t)| > 3\sigma_S\right] = 0.0026 \tag{10.2}$$

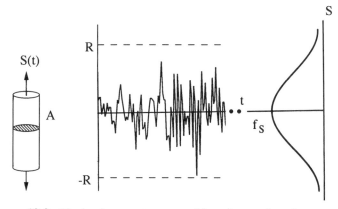

Figure 10.2 Single element structure subjected to random force process.

Thus, stress exceeds strength only about 0.3% of the time if $R = 3\sigma_S$. In reality, R is a random variable, and the design value is usually taken as a characteristically low value, for example, the lower 1% point. Therefore, the implied risk is likely to be significantly less than that of Eq. (10.2).

10.1.2 More General Statement of Three-Sigma Criterion

Consider a more general case. Assume that (1) the mean stress is not zero and (2) strength R is a random variable having a normal distribution, having mean μ_R and standard deviation σ_R. The situation is shown in Figure 10.3. Using the same probability of Eq. (10.2), we can derive a safety check criterion for this more general case. A basic criterion for a safe design can be written as

$$\mu_R \geq \xi\sigma_S \tag{10.3}$$

where the ξ factor is a function of $Q = \mu_S/\sigma_S$ and the coefficient of variation of R, $C_R = \sigma_R/\mu_R$. The function ξ has been derived by numerical analysis and the results are shown in Figure 10.4.

Example 10.1. A stationary Gaussian stress process $S(t)$ having an RMS of $\sigma_S = 70$ MPa and mean $\mu_S = 35$ MPa is applied to a component whose strength has a coefficient of variation $C_R = 10\%$. What is the requirement on mean strength?

Solution. Refer to Figure 10.4. Compute Q,

$$Q = \frac{\mu_S}{\sigma_S} = \frac{35}{70} = 0.5$$

$$C_R = 0.10$$

From Figure 10.4, $\xi = \mu_R/\sigma_S = 3.46$.

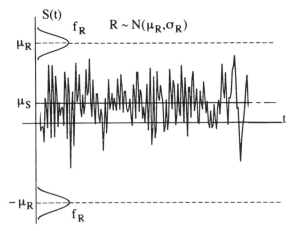

Figure 10.3 Stationary stress process with nonzero mean and random strength.

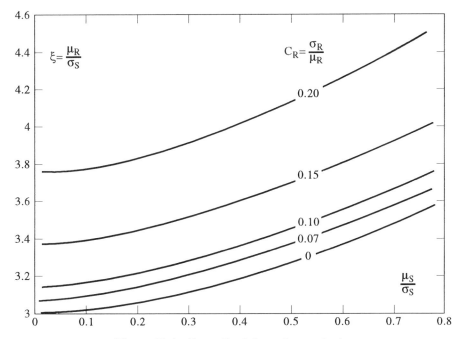

Figure 10.4 Generalized three-sigma criterion.

Thus, the mean strength must be

$$\mu_R > 3.46\sigma_S$$

for a safe design.

10.2 FIRST-PASSAGE FAILURE

10.2.1 Introductory Remarks

Failure is defined as the first time that the stress process exceeds strength R, $|S(t)| > R$. First-passage failure is illustrated in Figure 10.5; note that now R is assumed to be a function of time. The figure shows a decreasing R (deteriorating structure), but that assumption is not necessary in the analysis that follows. The goal of design is to ensure that the probability of a first-passage failure in service life T_S is acceptably small.

For first-passage failures, we define the following terms:

T = time to failure, a random variable

T_S = service life

$S(t)$ = a stationary random stress process; $0 \le t \le T_S$

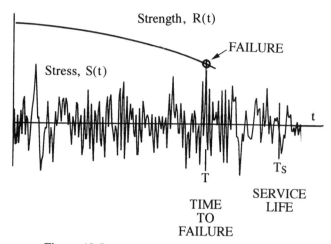

Figure 10.5   Realization of first-passage failure.

Z = peak of $S(t)$, a random variable

Y = max (Z), largest peak of $S(t)$ during T_S, a random variable

$R(t)$ = strength, in general a deterministic function of time

σ_S = RMS of $S(t)$

μ_S = mean value of $S(t)$

Throughout the discussion on first-passage failure, it is assumed that $S(t)$ is stationary.

The probability of a first-passage failure can be formulated as

$$p_f = P(T < T_S) \qquad (10.4)$$

In the special case where R is not a function of time,

$$p_f = P(Y > R) = 1 - F_Y(R) \qquad (10.5)$$

Over the years, several investigators have addressed the first-passage problem. Some of the pioneers have been Crandall, Chandiramani, and Cook (1966), Lin (1970), Yang and Shinozuka (1971), and Vanmarcke (1975). A summary of the elementary development of first-passage theory was provided by Vanmarcke (1984). A comprehensive summary of the development of elementary first-passage theory was provided by Marley (1991), and in part, the following is based upon his presentation.

In general, $S(t)$ will be wide-band non-Gaussian and nonstationary, and calculation of p_f can be a tough and tedious problem. Following are some special cases and approximations that are useful for engineering application.

10.2.2 Basic Formulation of First-Passage Problem

Assume that $R(t)$ is large relative to $S(t)$ so that crossings of $R(t)$ by $S(t)$ are rare. The implication is that it is assumed at the outset that the probability of failure should be small. Consider a small time increment Δt that occurs at time t. In this increment, it is assumed that the up crossings of $S(t)$ past $R(t)$ form a point process whose rate of arrival is described by a Poisson process. The rate of occurrence is $\nu_R^+(t)$. Thus, it follows (see Section 6.1) that

$$\text{Probability of no up crossing in } \Delta t = \exp\left[-\nu_R^+(t)\,\Delta t\right] \qquad (10.6)$$

This equation is a direct application of the Poisson distribution [Section 3.3.2] process.

Consider the service life T_S. Subdivide T_S into k small equal increments, Δt. Then the probability of no up crossing in T_S is the probability of the mutual intersection of the events of no up crossing in all Δt. Assuming that the event of an up crossing in all intervals Δt are mutually independent, it follows that

$$\text{Probability of no up crossing in } T_S = \prod_{i=1}^{k} \exp\left[-\nu_R^+(t_i)\,\Delta t\right]$$

$$= \exp\left[-\sum_{i=1}^{k} \nu_R^+(t)\,\Delta t\right] \qquad (10.7)$$

In the limit, as $\Delta t \to 0$,

$$\text{Probability of no up crossing in } T_S = \exp\left[-\int_0^{T_S} \nu_R^+(t)\,dt\right] \qquad (10.8)$$

The probability of failure is defined as the probability of the event of one or more up crossings in T_S. Thus, the probability of failure is

$$p_f = 1 - \exp\left[-\int_0^{T_S} \nu_R^+(t)\,dt\right] \qquad (10.9)$$

In the special case that $X(t)$ is a Gaussian process, it was shown in Equations 6.19 and 6.20 that

$$\nu_R^+ = \nu_0^+ \exp\left[-\frac{1}{2}\left(\frac{R(t) - \mu_S}{\sigma_S}\right)^2\right] \qquad (10.10)$$

Analysis to estimate the probability of failure now focuses on the functional form of $R(t)$ and $\nu_R^+(t)$. Further discussion is provided by Marley (1991), who provides useful approximations and examples.

An important application is the special case of a fatigue problem where crack growth is modeled by the fracture mechanics fatigue model. As the crack grows, the ultimate strength $R(t)$ becomes smaller (regardless of whether a ductile or brittle fracture model is used). Failure occurs the first time that $S(t)$ exceeds $R(t)$. This issue is addressed as an exercise (problem 10.12) at the end of the chapter.

Example 10.2. Consider the special case where R is constant and ν_R^+ is also constant. The integral of Eq. (10.9) becomes $\nu_R^+ T_S$. Consider the series expansion of exp $(\cdot)$. If p_f is to be small, then it follows that

$$p_f \approx \nu_R^+ T_S \tag{10.11}$$

Example 10.3. If $X(t)$ is Gaussian and narrow band, $\nu_0^+ = f_0$ where f_0 is the center frequency of the process. And if failure occurs for excursion of the constant boundary $+R$, then the rate of crossings will be $\nu_0^+ = f_0$. From Eqs. (10.10) and (10.11), the probability of failure can be approximated as

$$p_f \approx (f_0 T_S) \exp\left[-\frac{1}{2}\left(\frac{R - \mu_S}{\sigma_S}\right)^2\right] \tag{10.12}$$

Example 10.4. Consider the problem where the boundaries are constant and symmetric, $\pm R$, and $\mu_S = 0$. If $X(t)$ is narrow band, $\nu_0^+ = 2f_0$. From Eqs. (10.10) and (10.11), it follows that

$$p_f \approx (2f_0 T_S) \exp\left[-\frac{1}{2}\left(\frac{R}{\sigma_S}\right)^2\right] \tag{10.13}$$

As discussed by Vanmarcke (1984), the Poisson assumption for Gaussian processes is exact as R approaches infinity. But, in general, for finite R, the error will depend upon both the level of R and the bandwidth of the process. Additional detail is provided by Vanmarcke (1975, 1984).

Dependency within the peaks, obvious from examination of a realization of any narrow-band process, can be accounted for with an envelope process, as described in Section 6.1.4. As shown in Section 6.1.4, expected time to first passage of the envelope process may be significantly higher than the expected time to first passage of $S(t)$. The significance of this result is that the assumption that peaks are independent leads to estimates of the probability of failure, which err on the high side.

10.2.3 First-Passage Failure: Exact Distribution of Largest in Sample of Size n Independent Peaks

Another approach to the first-passage problem is presented as follows. Assume that $X(t)$ is a stationary zero-mean process. The strength R is not a function of time. The peaks Z_i form a random process. The distribution function of Z is

$F_Z(z)$. Assume that Z_i are mutually independent. Let B_i be the event that the ith peak is less than R. Let B be the event that Y the largest peak in a sample of size n is less than R. The event that the largest peak is less than R is equivalent to the event that all peaks are less than R, and we can write

$$B = B_1 \cap B_2 \cap \cdots \cap B_n$$
$$= \bigcap_{i=1}^{n} B_i \tag{10.14}$$

Because all B_i are independent,

$$P(B) = \prod_{i=1}^{n} P(B_i) = [P(B_i)]^n \tag{10.15}$$

But $P(B) = P(Y < R) = F_Y(R)$, where $F_Y(y)$ is the distribution function of Y. Here, $P(B_i) = P(Z \leq R) = F_Z(R)$, and it follows that,

$$F_Y(R) = [F_Z(R)]^n$$
$$P(\text{no peak exceeds } R) = P(B) = [F_Z(R)]^n \tag{10.16}$$

The probability of a first-passage failure is the probability that at least one peak exceeds R. Thus,

$$p_f = 1 - [F_Z(R)]^n = 1 - F_Y(R) \tag{10.17}$$

For a stationary Gaussian wide-band process, the PDF of Z (and therefore, F_Z) is given as the Rice distribution in Figure 6.9. However, if it can be assumed that $S(t)$ is also narrow band, then Z has a Rayleigh distribution (see

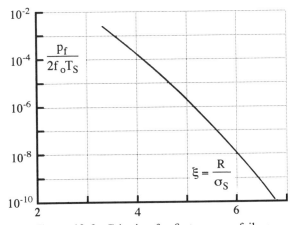

Figure 10.6 Criterion for first-passage failure.

Section 6.2.1),

$$F_Z(z) = 1 - \exp\left[-\frac{1}{2}\left(\frac{z}{\sigma_S}\right)^2\right] \qquad (10.18)$$

Then, using Eq. (10.16), the CDF of Y becomes

$$F_Y(R) = \left\{1 - \exp\left[-\frac{1}{2}\left(\frac{R}{\sigma_S}\right)^2\right]\right\}^n \qquad (10.19)$$

The mean and standard deviation of Y can be computed from $F_Y(y)$. Values are given in Table 10.1 as a function of n.

Note from Eqs. (10.17) and (10.19) that if p_f is small, then the exponential term of Eq. (10.19) is also small. Using the binomial series expansion of Eq. (10.19) and letting $n = 2f_0 T_s$, it follows that

$$p_f \approx (2f_0 T_s) \exp\left[-\frac{1}{2}\left(\frac{R}{\sigma_S}\right)^2\right] \qquad (10.20)$$

This expression is used to construct the chart of Figure 10.6. Note that this is the same result as obtained using the Poisson approximation of Section 10.2.2.

Continuing with the example, derive the requirement on strength R for a component subjected to a stationary narrow-band zero-mean Gaussian stress having an RMS $\sigma_S = 10$ ksi. The center frequency $f_0 = 5$ Hz, and the duration of application $T_S = 100$ sec. The design reliability goal $p_0 = 10^{-3}$.

Solution. The total number of peaks $n = 2f_0 T_s = 2(5)(100) = 1000$. Then, from Eq. (10.20), and Fig. 10.6.

$$R > 5.26\sigma_S = 52.6 \text{ ksi}$$

for a safe design.

Table 10.1 Mean and Standard Deviation of Largest Peak in Sample of Size n

Number of Peaks, n	Mean Value of Largest Peak, μ_Y/σ_S	Standard Deviation of Largest Peak, σ_Y/σ_S
5	2.043	0.608
10	2.365	0.507
50	2.972	0.409
100	3.198	0.383
500	3.670	0.338
1000	3.856	0.323
2000	4.032	0.310

Note that the application of a first-passage criterion can result in very conservative designs relative to the three-sigma approach. However, the first-passage criterion is very commonly used, particularly for large-scale structures subjected to extreme loadings that tend to be rare.

10.2.4 First-Passage Failure: Use of Extreme-Value Distribution to Model Distribution of Largest Peak

Consider a stationary peak process having a distribution of peak heights defined by the CDF $F_Z(z)$. In the limit, as the number of peaks $n \rightarrow \infty$, $F_Y(y)$ will approach the extreme-value distribution (EVD) for most commonly used distributions except the lognormal (see Ang and Tang, 1984). The CDF of the EVD is

$$F_Y(y) = P(Y \le y) = \exp \left\{ -\exp \left[-\alpha(y - \beta) \right] \right\} \qquad (10.21)$$

where the mean values in terms of the parameters are (see Section 2.5.6)

$$\mu_Y = \beta + \frac{0.577}{\alpha} \qquad \sigma_Y = \frac{1.283}{\alpha} \qquad (10.21a)$$

The parameters α and β can be computed from the distribution of individual peak heights, Z,

$$\alpha = n f_Z(\beta) \qquad \beta = F_Z^{-1} \left(1 - \frac{1}{n} \right) \qquad (10.22)$$

In the special case where the stress process is narrow-band Gaussian with RMS of σ_S, Z has a Rayleigh distribution, and using Eq. (10.22), it is easily shown that μ_Y and σ_Y are given by

$$\mu_Y = \left(\sqrt{2 \ln n} + \frac{0.577}{\sqrt{2 \ln n}} \right) \sigma_S$$

$$\sigma_Y = \frac{1.283 \sigma_S}{\sqrt{2 \ln n}} \qquad (10.23)$$

and the distribution parameters are

$$\alpha = \frac{\sqrt{2 \ln n}}{\sigma_S} \qquad \beta = \sqrt{2 \ln n} \; \sigma_S \qquad (10.24)$$

It is known that convergence of this asymptotic form to the exact distribution of the largest of n independent peaks is "slow." [See Fig. 2.32] For $n = 1000$, Eq. (10.23) gives

$$\mu_Y/\sigma_S = 3.872 \quad \text{(exact value} = 3.856; \text{Table 10.1)}$$

$$\sigma_Y/\sigma_S = 0.345 \quad \text{(exact value} = 0.323; \text{Table 10.1)} \qquad (10.25)$$

Example 10.5: Determination of Design Wave Height in a Storm. A "design storm" (ocean) used for design of marine structures has a length of 3 hr with a total of 1000 waves. The RMS value of the ocean wave elevation is 9.0 ft (significant wave height of 36 ft). Assume that the wave heights have a Rayleigh distribution.

The design wave is defined as the upper 95% point of the distribution of the largest value.

Solution. Assume that the largest wave follows the asymptotic distribution (EVD). From Eq. (10.23),

$$\mu_Y = 3.872\sigma = 34.85 \qquad \sigma_Y = 0.345\sigma = 3.105$$

From Eq. (10.21a), the EVD parameters are

$$\alpha = \frac{1.283}{\sigma_Y} = 0.413 \qquad \beta = \mu_Y - \frac{0.577}{\alpha} = 33.45$$

And the design value S_0 is that value associated with the 95% level,

$$0.95 = F_Y(S_0) = \exp\{-\exp[-0.413(S_0 - 33.45)]\}$$

And thus,

$$S_0 = 40.6 \text{ ft}$$

10.2.5 First-Passage Failure: Determination of Design Value Using Concept of Return Period

The return period can be used to define a design value S_0 of peak stress for use in a deterministic design procedure, that is, using safety factors. Consider a stationary peak process having a distribution of peaks defined by the CDF $F_Z(z)$. Given n peaks in the design life (frequently, in practice, chosen to be greater than the service life), define the design load S_0 as being that peak having a probability of exceedance of $1/n$,

$$P(Z > S_0) = 1 - F_Z(S_0) = \frac{1}{n} \qquad (10.26)$$

Thus, the *return period* of S_0 is $T_R = n$. Here, S_0 is a value of Z that will be exceeded once on the average every n times.

Example 10.6: Design Value Using Return Period Concept. Tests on a prototype aircraft structural section have shown that the long-term distribution of peaks of a random load process have a Weibull distribution with a mean of 10 and standard deviation of 4. The service life of the component is specified as 20,000 cycles.

The design load S_0 is specified as that value having a return period of twice the service life. Determine S_0.

Solution. Here $T_R = n = 2(20,000) = 40,000$. Then

$$P(Z > S_0) = 1 - F_Z(S_0) = \frac{1}{n}$$

$$= \frac{1}{40,000} = 2.5 \times 10^{-5}$$

The Weibull parameters are

$$\alpha = C_Z^{-1.08} = \left(\frac{4}{10}\right)^{-1.08} = 2.60 \qquad \beta = \frac{10}{\Gamma(1/\alpha + 1)} = 11.24$$

where C_Z is the coefficient of variation and $\Gamma(\cdot)$ is the gamma function. The CDF is

$$F_Z(S_0) = P(Z \le S_0) = 1 - \exp\left[-\left(\frac{S_0}{11.24}\right)^{2.69}\right]$$

It follows that

$$S_0 = 27.0$$

Thus, the peak load in the structural section will exceed 27.0 on the average once every 40,000 cycles.

As a footnote to this problem, note that the design value produced by the return period has a high probability of exceedance during the service life:

$$F_Y(S_0) = P(Y < S_0) = [F_Z(S_0)]^n = \left(1 - \frac{1}{n}\right)^n$$

for example, if $n = 20,000$, then $P(Y > S_0) = 0.632$.

10.3 FATIGUE: AN INTRODUCTION

10.3.1 Physical Process of Fatigue

The physical process of fatigue is described in Figure 10.7. Under the action of oscillatory tensile stresses of sufficient magnitude, a small crack will initiate at a point of the stress concentration. As a crude rule of thumb for metallic materials, the amplitude of the stress at the notch should exceed 50% of the ultimate strength of the material. But if there is a preexisting crack or flaw (e.g., a tool mark), stress amplitude to initiate a growing crack may be considerably less.

Once the crack is initiated, it will tend to grow in a direction orthogonal to the direction of the oscillatory tensile loads. Finally, the material having a reduced cross-sectional area experiences a quasi-static brittle or ductile fracture.

Fatigue is perhaps the most important failure mode to be considered in mechanical design. Traditionally, fatigue has been associated with metallic materials, but fatigue in other materials, such as filamentary composites and concrete, is also considered. It has been reported that fatigue accounts for 80–90% of all observed service failures in mechanical and structural systems. Fatigue failures are often catastrophic; they come without warning and may cause significant property damage as well as loss of life.

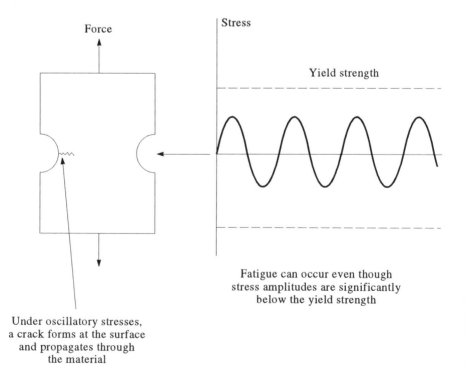

Figure 10.7 Elementary description of fatigue process.

There are several reasons for the dominance of this failure mode and the problems of designing to avoid it: (1) the fatigue process is inherently unpredictable, as evidenced by the statistical scatter in laboratory data; (2) it is often difficult to translate laboratory data of material behavior into field predictions; (3) it is extremely difficult to accurately model the mechanical environments to which the system is exposed over its entire design lifetime; and (4) environmental effects produce complex stress states at fatigue-sensitive "hot spots" in the system. In summary, material fatigue can involve a very complicated interaction of several processes.

An engineering description of material behavior in fatigue can be established by classical fatigue test, as illustrated in Figure 10.8. A specimen is subjected to a constant amplitude stress S. The process continues until the specimen breaks. Cycles to failure N are measured.

The data S versus N are typically plotted on a log-log plot, as suggested by Figure 10.8. Many data sets tend to plot a straight line in this space. Sometimes a linear scale is used for stress (e.g., MIL-HDBK 5, 1987). There is considerable variability in observed cycles to failure in almost every material and at almost every stress level, although the variability in N is typically larger at the lower stress levels. Generally, the coefficient of variation of N will range from 50 to 150%.

It will be assumed herein that the fatigue strength of a material will be defined by a single S–N curve. This curve is generally considered to be a lower bound design curve, although the median curve through the data is sometimes used. Uncertainty in fatigue behavior, as evidenced by the observed scatter in the data, can be accounted for using reliability methods. Thus, the fatigue strength will be defined by a function of the form

$$N = N(S; \mathbf{A}) \qquad (10.27)$$

where N denotes cycles to failure, S will be stress range or stress amplitude, depending upon the choice of the analyst, and $\mathbf{A}$ is a vector of parameters that are used to model the S–N curve. Reliability methods can be employed to manage the large uncertainty in fatigue design factors, including scatter in data. Such methods are used to quantify structural integrity and derive design criteria. In this text, all design factors are assumed to be deterministic. For a summary of reliability considerations, see Wirsching (1991, 1995).

10.3.2 Fatigue Strength Models

The theory and application of engineering models that are used to describe fatigue strength are documented in several references, including Nelson (1978), Fuchs and Stephens (1980), Collins (1981), Broek (1984), Almar-Naess (1985), Gurney (1979), Rolfe and Barsom (1987), Hertzberg, (1989), Dowling (1993), Winterstein (1988), and the *Fatigue Design Handbook* (SAE, 1988) and the Damage Tolerant Design Handbook (U.S. Air Force Material Laboratory, 1988). Fundamentally, there are three basic models that are used for fatigue:

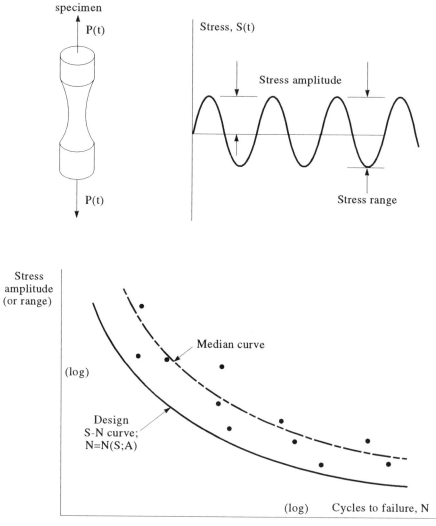

Figure 10.8 Illustration of fatigue test and how results are presented.

(1) *stress-based approach*, (2) *strain-based approach*, and (3) *fracture mechanics approach*. An excellent summary of these methods is provided by Dowling (1993).

In the stress-based approach, it is assumed that the material will remain in the elastic state. Oscillatory stress at the point of stress concentration is computed and compared to the strength, as defined by an *S–N* curve such as shown in Figure 10.8. Typically, cycles to failure *N* will exceed 10^5, and the failure is said to be *high cycle fatigue*.

For the strain-based approach, local stresses and strains are estimated at the point where fatigue cracking is most likely to occur. It is assumed that localized yielding will occur at this point. The effects of local yielding are included, and the material's cyclic stress–strain and strain life curves from smooth axial test specimens are employed. This method has sometimes been called *local strain analysis*. When random stress processes are applied to a component, the point in question will experience a random hystersis process, and a crack will form after a number of cycles N. Cycles to crack initiation are likely to be less than 10^5, and crack initiation is called *low cycle fatigue*. As described by Dowling (1993) and *Fatigue Design Handbook* (SAE, 1988), the process is quite complex but can be automated by a computer code. In this text, there will be no additional discussion of prediction of fatigue life with a strain-based approach and random stresses.

The essence of the fracture mechanics approach is that a pre-existing crack or flaw exists at the point where the fatigue crack will initiate. Flaws are inherent in many components owing to the process by which they are manufactured or fabricated, for example, in a welded joint where defects due to porosity and lack of penetration, lack of fusion, and so on, exist prior to the application of any load. Tool marks, forging laps, and inclusions are also sources of initial cracks.

The basic parameter of fracture mechanics fatigue analysis is the stress intensity factor range given by

$$\Delta K = Y(a)S\sqrt{\pi a} \tag{10.28}$$

in which S is the applied stress range, a the crack depth for a surface flaw or half-width for a penetration flaw, and $Y(a)$ the geometry correction factor, which may depend on a. The geometry factor depends on the crack size, structural geometry, and applied far-field stress. See above references for more detail. The process is described in Figure 10.9.

It has been found from experimental data that the crack growth rate da/dn (n = cycles) can be modeled as shown in Figure 10.9. A commonly used model for the central region is the Paris law (Paris, 1964),

$$\frac{da}{dn} = C(\Delta K)^m \tag{10.29}$$

in which C and m are empirical constants. These depend on such factors as the mean cycling stress and the test environment. The level of the stress intensity factor below which the crack will not propagate is shown in Figure 10.9 as the threshold stress intensity factor ΔK_{th}. In the following, ΔK_{th} is conservatively assumed to be equal to zero. On the other hand, when ΔK approaches the critical stress intensity ΔK_c, rapid crack growth and fracture will occur.

A convenient equation for cycles to failure, N, can be derived by integration of Eq. (10.29) from an initial crack length a_0 to a critical crack length or failure

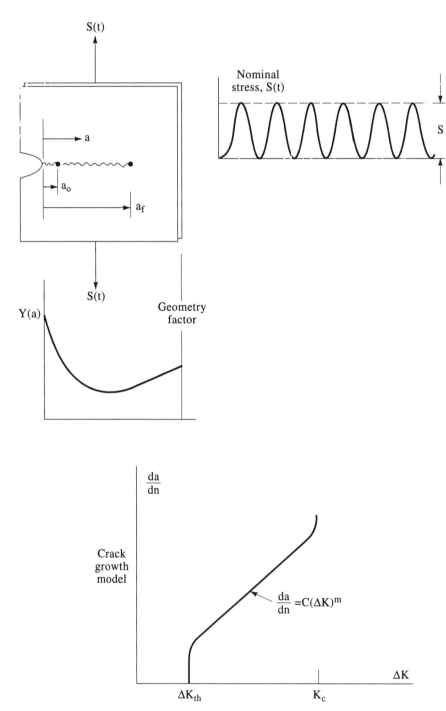

Figure 10.9 Illustration of fracture mechanics fatigue model.

crack length a_f (often, a_f is assumed on the basis of engineering judgment) and for n from 0 to N. The result of integration of Eq. (10.29) produces a cycle life

$$N = \frac{1}{S^m C \pi^{m/2}} \int_{a_0}^{a_f} \frac{da}{Y^m(a) a^{m/2}} \tag{10.30}$$

Note that this form also gives us an expression of N as a function of S and would provide an expression of fatigue strength similar to that of Figure 10.8 and Eq. (10.27). Thus, the discussions on random fatigue that follow, which employ the stress-based approach, also apply to the fracture mechanics model of Eq. (10.30) as well. As a footnote, a model that predicts random fatigue using the Paris law with a $\Delta K_{th} > 0$ has been discussed by Wirsching and Chen (1988).

More refined stochastic crack growth models have been presented by Yao et al. (1986), Sobczyk and Spencer (1991), Veers et al. (1989), Ortiz (1985), and Ortiz et al. (1988).

10.3.3 Miner's Rule

Almost all available fatigue data for design purposes is based on constant-amplitude tests, for example, as suggested in Figures 10.8 and 10.9. In practice, however, fatigue stresses are typically variable amplitude or random, as illustrated in Figure 10.10. The key issue is how to use the mountains of available constant-amplitude data to predict fatigue in a component.

Collins (1981) provides a comprehensive review of the models that have been proposed to predict fatigue life in components subject to variable-amplitude stress using constant-amplitude data to define fatigue strength. Yet the original model, a linear damage rule originally suggested by Palmgren (1924) and developed by Miner (1945), maintains its popularity principally because of its simplicity. Moreover, the engineering profession has not been convinced that any of the other more refined models will consistently outperform Miner's rule. As a result, Miner's rule is specified in almost every design code worldwide.

In the following development of Miner's rule, it is assumed that the stress process can be described by discrete events (stress cycles) and that a spectrum of amplitudes of stress cycles can be defined. Such a spectrum will lose any information on the applied sequence of stress cycles that may be important in some cases. It is also assumed that a constant-amplitude $S-N$ curve is available, and this curve is compatible with the definition of stress; that is, at this point there is no explicit consideration of the possibility of mean stress. Inclusion of mean stress into the model will be discussed later.

Illustrated in Figure 10.11 is a stress spectrum described as a sequence of constant-amplitude blocks, each block having stress amplitude S_i and the total number of applied cycles n_i. The constant-amplitude $S-N$ curve is also shown.

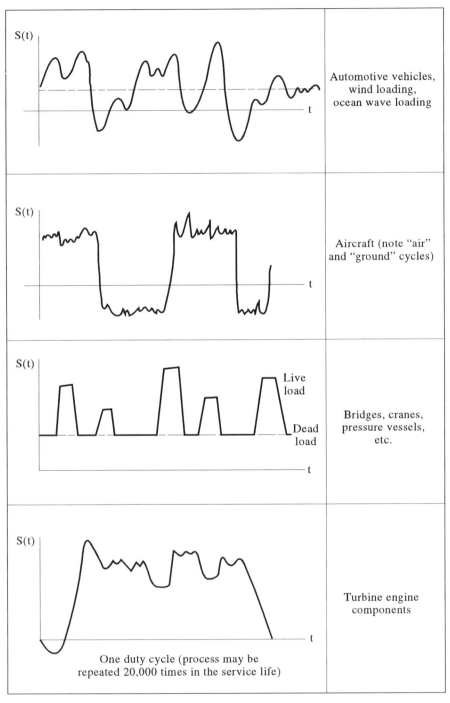

Figure 10.10 Examples of fatigue stress histories.

Block loading to simulate a random process

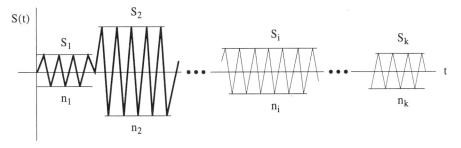

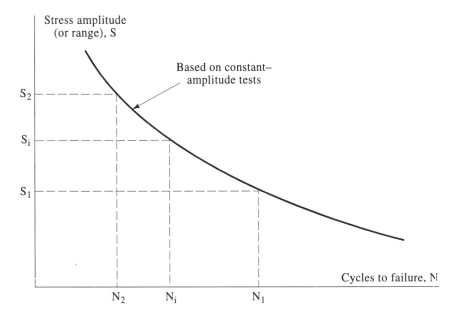

- Fractional damage at stress, $S_i = \dfrac{n_i}{N_i}$

- Total damage is a sum of fractional damages $\qquad D = \displaystyle\sum_{i=1}^{k} \dfrac{n_i}{N_i}$

- Failure $\longrightarrow$ $D \geq 1.0$

Figure 10.11 Illustration of Miner's rule.

Consider the first block having stress level S_1. From the S–N curve, we note that the number of cycles to failure at this level is N_1. But only $n_1 < N_1$ cycles are applied (assuming no failure). Therefore, we can define a fractional damage n_1/N_1. Clearly, failure would occur if this fraction exceeded unity.

This suggests that we can define a fractional damage at each stress level

n_i/N_i, define a total damage as the sum of all the fractional damages over a total of k blocks,

$$D = \sum_{i=1}^{k} \frac{n_i}{N_i} \qquad (10.31)$$

and define the event of failure as

$$D \geq 1.0 \qquad (10.32)$$

Comments with regard to the quality of this linear damage accumulation rule are presented later in this chapter.

10.3.4 Models of Fatigue Damage Under Narrow-Band Random Stress

In this section, practical models for random fatigue analysis are presented.

Discrete Stress Spectrum: Fundamental Case. Assume that the stress spectrum is presented as a discrete spectrum, that is, several blocks, as illustrated in Figure 10.11. A characteristic of this stress is that all stress cycles have the same mean value, for example, a narrow-band stress process. It is common practice to take a measured record as from a strain gauge or accelerometer and discretize it, as suggested by Figure 10.12. The number of peaks in the window ΔS_i are counted as n_i. Then f_i, which defines the fraction of stresses at level S_i, is

$$f_i = \frac{n_i}{n} \qquad (10.33)$$

where $n = \Sigma \, n_i$, the total number of cycles. Note that f_i is the probability that a single stress range will have magnitude S_i. Thus, f_i is the PDF of the random variable S, as shown in Figure 10.12. Total fatigue damage can be written as

$$D = \sum_{i=1}^{k} \frac{n_i}{N_i} = n \sum_{i=1}^{k} \frac{f_i}{N_i} \qquad (10.34)$$

Equation (10.34) is a general expression for D that can be used with any form of the S–N curve, for example, if the curve has an endurance limit. Because of this generality, this expression for D is widely used in practice.

Example 10.7. This example illustrates that direct application of Miner's rule can be applied to an S–N curve having any form. The example also illustrates how easy it is to use Miner's rule and, therefore, why it is so popular.

Analysis of a random stress (nominal) process in a structural component provides the stress spectrum, as defined in columns 1 and 2 of Table 10.2. The

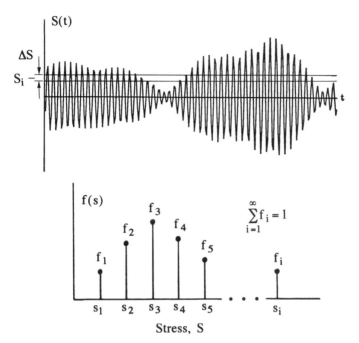

Figure 10.12 Discrete distribution of stress ranges.

cycles listed are the total that the component will experience over the service life. The fatigue strength of the aluminum alloy considered (including the effect of a stress concentration) is defined in Figure 10.13. Compute the total fatigue damage over the service life.

Solution. In column 3 of Table 10.2 the total cycles to failure of the material, N_i, is read off of the *S–N* curve at each stress level S_i. Fractional damages are computed in column 4. The sum of this column is the total computed damage,

$$D = \Sigma \frac{n_i}{N_i} = 0.51$$

Table 10.2   Computation of Total Fatigue Damage

Number of Cycles, n_i (1)	Stress amplitude, S_i (MPa) (2)	Cycles to Failure,[a] N_i (3)	$\dfrac{n_i}{N_i}$ (4)
2,750	200	20,000	0.1375
9,000	100	80,000	0.1125
27,000	80	0.5×10^5	0.1800
40,000	60	5×10^5	0.0800
		$D = \Sigma\, n_i/N_i =$	0.5100

[a]Read from Fig. 10.13.

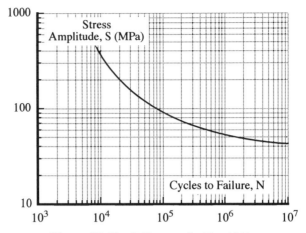

Figure 10.13 *S–N* curve for Ex. 10.7.

***Discrete Stress Spectrum: Linear Model for S–N* Curve.** There is a significant
payoff in simplicity of analysis if it can be assumed that fatigue strength can
be expressed by a linear function in log-log space. If so, the empirical form

$$NS^m = A \tag{10.35}$$

is implied. The empirical constants are A, the fatigue strength coefficient, and
m, the fatigue strength exponent. Note that either S or A can be based on range
or amplitude, but the distinction is important, as we will see later. While this
form of the *S–N* relationship is a special case, it will apply to the elastic strain
life curve (high cycle fatigue) in a strain-based approach, as well as to fracture
mechanics model of Eq. (10.30). Comparing Eqs. (10.30) and (10.35), the
fatigue strength coefficient based on a fracture mechanics formulation would
be

$$A = \frac{1}{C\pi^{m/2}} \int_{a_0}^{a_f} \frac{da}{Y^m(a)a^{m/2}} \tag{10.36}$$

Using the linear model, the expression for D becomes

$$D = \frac{n}{A} \sum_{i=1}^{k} f_i S_i^m \tag{10.37}$$

Note that S is a discrete random variable denoting stress range (or amplitude)
of a single cycle. The "expected value" of S^m is, by definition,

$$E(S^m) = \sum_{i=1}^{k} f_i S_i^m \tag{10.38}$$

Thus,

$$D = \frac{n}{A} E(S^m) \qquad (10.39)$$

Note that, in the constant-amplitude case, we would express D as

$$D = \frac{n}{A} S^m \qquad (10.40)$$

An equivalent constant-amplitude stress can be derived by comparing Eqs. (10.39) and (10.40),

$$S_e = [E(S^m)]^{1/m} \qquad (10.41)$$

Here, S_e is called *Miner's stress* or sometimes *equivalent stress*.

Example 10.8. A random axial load is applied to a butt-welded joint having a width $b = 18$ cm. Analysis of the random load spectrum over the service life is provided in Table 10.3, columns 1 and 2. It is required to design the joint, that is, determine the minimum plate thickness t. The design requirement is the total damage must be less than 0.50. (While in theory failure occurs at $D = 1.0$, this lower value is a safety factor of 2 to account for uncertainties in the life prediction process.)

The S–N curve for butt-welded joints is defined by the U.K. Department of Energy as $m = 3.5$ and $A = 1.08 \times 10^{14}$ (MPa units).

Solution. Because the S–N curve has a linear form, the simple fatigue models of Section 10.14 can be used.

Stress in the plate is $S = Q/A_r$, where $A_r = bt$ is the cross-sectional area. From the load spectrum, an equivalent constant-amplitude stress can be computed as

$$S_e = [E(S^m)]^{1/m} = \frac{[E(Q^m)]^{1/m}}{A_r}$$

Table 10.3 Analysis of Random Load Spectrum Over Service Life

Number of Cycles, n_i (1)	Load Range, Q_i (N) (2)	$f_i = \dfrac{n_i}{N_T}$ (3)	$f_i\, Q_i^{3.5}$ (4)
320	1400×10^3	0.000359	1.17×10^{18}
1,750	870×10^3	0.00196	1.20×10^{18}
28,300	540×10^3	0.0317	3.67×10^{18}
860,000	280×10^3	0.965	11.2×10^{18}
$N_T = 890{,}370$			$\Sigma f_i\, Q_i^{3.5} = 1.72 \times 10^{19}$

Here, Q is a discrete random variable. The expected or average value of Q^m is

$$E(Q^m) = \sum f_i Q_i^m$$

where $f_i = n_i/N_T$ and N_T is the total number of cycles in the service life. Thus, from Table 10.3,

$$E(Q^m) = 1.72 \times 10^{19}$$

and

$$S_e = \frac{(1.72 \times 10^{19})^{1/3.5}}{0.18t} = \frac{1.74 \times 10^6}{t} \text{ Pa} = \frac{1.74}{t} \text{ MPa}$$

The expression for the total damage can then be used to compute the requirement on plate thickness,

$$D = \frac{N_T S_e^m}{A}$$

$$0.5 = \frac{(8.9 \times 10^5)(1.74/t)^{3.5}}{1.08 \times 10^{14}}$$

and the minimum value of t,

$$t = 0.010 \text{ m} = 1.0 \text{ cm}$$

Continuous Stress Spectra. Frequently, a continuous model can be employed to describe a distribution of stress ranges, for example, the Rayleigh for a stationary narrow-band Gaussian process. Consider the PDF of stress range, S, as shown in Figure 10.14. The fraction of stresses in $(s, s + \Delta s)$ is given as

$$f_i \cong f_S(s) \, \Delta s \tag{10.42}$$

Total fatigue damage is the sum of incremental damage in each Δs,

$$D \cong n \sum_{i=1}^{k} \frac{f_S(s) \, \Delta s}{N(s)} \tag{10.43}$$

In the limit, as $\Delta s \to 0$,

$$D = n \int_0^\infty \frac{f_S(s) \, ds}{N(s)} \tag{10.44}$$

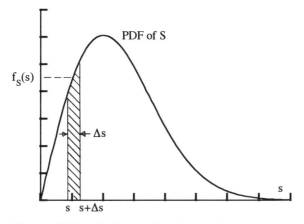

Figure 10.14 Continuous distribution of stess ranges.

For the linear $S-N$ curve, $NS^m = A$,

$$D = \frac{n}{A} \int_0^\infty s^m f_S(s)\, ds \tag{10.45}$$

The integral is, by definition, $E(S^m)$, and

$$D = \frac{n}{A} E(S^m) \tag{10.46}$$

Continuous Spectra: Special Cases.

1. Here, $S(t)$ is stationary narrow-band Gaussian and the stress amplitudes or ranges, S, are Rayleigh. It can be shown that

$$E(S^m) = S_e^m = \begin{cases} (\sqrt{2}\sigma)^m \Gamma\left(\frac{1}{2}m + 1\right) \cdots S, A \text{ based on amplitude} & (10.47) \\[2ex] (2\sqrt{2}\sigma)^m \Gamma\left(\frac{1}{2}m + 1\right) \cdots S, A \text{ based on range} & (10.48) \end{cases}$$

where $\Gamma(\cdot)$ is the gamma function. Fatigue damage at time τ can be written by combining Eqs. (10.47) and (10.48) with (10.46),

$$D = \begin{cases} \dfrac{v_0^+ \tau}{A} (\sqrt{2}\sigma)^m \Gamma\left(\frac{1}{2}m + 1\right) \cdots S, A \text{ based on amplitude} & (10.49) \\[3ex] \dfrac{v_0^+ \tau}{A} (2\sqrt{2}\sigma)^m \Gamma\left(\frac{1}{2}m + 1\right) \cdots S, A \text{ based on range} & (10.50) \end{cases}$$

where $n = \nu_0^+ \tau$, ν_0^+ being the rate of zero crossings with positive slope (an equivalent or average frequency of the narrow-band process).

2. Stress ranges S have a Weibull distribution. The CDF of S is

$$F_S(s) = 1 - \exp\left[-\left(\frac{s}{\delta}\right)^\xi \right]$$ (10.51)

The general expression for $E(S^m)$ is

$$E(S^m) = \delta^m \Gamma \left(\frac{m}{\xi} + 1\right)$$ (10.52)

For design purposes, redefine the parameters. Let N_S be total number of stress applications in the service life. Define "once-in-a-lifetime" stress range S_0,

$$P(S > S_0) = \frac{1}{N_S}$$ (10.53)

The term S_0 is used as a "design" stress for static failure modes. This concept is based on the geometric distribution. Here, S_0 is the value that will be exceeded by S, on the average, once every N_S times. The "return period" for S_0 is N_S.

Substitute Eq. (10.51) into Eq. (10.53),

$$S_0 = [\ln (N_S)]^{1/\xi} \delta$$ (10.54)

Substitute Eq. (10.54) into Eq. (10.52) to eliminate δ. The resulting expression for D is

$$D = \frac{N_S}{A} S_0^m [\ln (N_S)]^{-m/\xi} \Gamma \left(\frac{m}{\xi} + 1\right)$$ (10.55)

It has been observed that the Weibull is a reasonable model for long-term distributions of structural responses to nonstationary environmental loading, for example, stresses in marine and aircraft structures. This model may have wide application for structures that are exposed to random environmental loading that appear to be nonstationary.

Blocks of Continuous Spectra. The long-term distribution of fatigue stress ranges was modeled as a Weibull, for example, in the above section. Another approach is to separate the nonstationary process into k blocks of stationary processes. Assuming the process is Gaussian, the distribution of peaks in each block is known and damage can be computed.

For example, a long-term nonstationary stress process is decomposed into k blocks of stationary processes having assumed Rayleigh peaks. The RMS of each block is σ_i, the rate of zero crossings is ν_i^+, and the time of application is τ_i. The total damage then would be (for A based on amplitude)

$$D = \sum_{i=1}^{k} \frac{\nu_i^+ \tau_i}{A} (\sqrt{2}\sigma_i)^m \Gamma\left(\frac{1}{2} m + 1\right) \tag{10.56}$$

For A based on range, the additional factor of 2 should be included. [Eq. 10.50]

How to Treat Mean Stress in Model. In general, the stress process $S(t)$ will have a nonzero mean. The mean, denoted as μ_S, is a principal factor that influences fatigue life and it should be accounted for. There is no universal algorithm that has been fully verified, by testing, for general use. There are a number of models that have been presented throughout the literature (e.g., Dowling, 1993), and a comprehensive review here is out of the scope of this book.

Some comments, however, may be helpful. It is known that μ_S does not play an important role in fatigue in welded joints and is often ignored in the analysis. The Goodman model for mean stress (e.g., Dowling, 1993) can be used to adjust the fatigue strength coefficient A of the linear model of Eq. (10.35),

$$A = A_0 \left[1 - \frac{\mu_S}{S_u}\right]^m \quad \text{for} \quad \mu_S \geq 0 \tag{10.57}$$

where A_0 is the fatigue strength coefficient based on zero-mean tests and S_u is the ultimate strength of the material. This form was developed for constant-amplitude fatigue, and the analyst should be cautious in applying it to the case of random stresses. For the fracture mechanics model, [Eq. (10.36)], the Walker correction to C (e.g. Dowling, 1993), may provide a reasonable first approximation to life prediction in the presence of a mean stress.

Some constant-amplitude data and crack growth data are presented for specific values of mean stress (e.g., MIL-HDBK 5, 1987). Many S–N and crack growth curves are presented in terms of the stress ratio $R = S_{min}/S_{max}$, e.g., U.S. Air Force (1988)]. Here, it is not clear how to translate such data to a random process.

Using concepts of crack opening stress, fracture mechanics models have been developed by Veers, et al. (1989) and Perng (1989) that account for mean stress effects during crack growth under random loading. Perng has developed the concept of an equivalent R ratio.

Example 10.9. This example illustrates the use of a model to describe the long-term distribution of stress ranges. It also illustrates how the fracture mechanics model can be employed to predict fatigue under random stresses.

A component of a flight vehicle is made of a titanium alloy Ti–6V–4Al. Crack growth data on this alloy indicate a Paris exponent and coefficient of $m = 6.29$ and $C = 3.35 \times 10^{-22}$ (in units of newtons and millimeters, respectively). These are the parameters that are considered to be valid for the mean stress for this problem. Thus, mean stress is ignored in the analysis.

The component is part of a structure that responds to a random aerodynamic load environment. The response is narrow band and the environmental model is assumed to be stationary so that the stress will be assumed to have Rayleigh peaks. Structural analysis has indicated that the RMS of the stress process is $\sigma_S = 38$ MPa (in newtons per square millimeter).

The initial crack size is estimated as $a_0 = 0.40$ mm. Failure crack length is assumed to be $a_f = 2.50$ mm. The geometry factor is constant and equal to 1.10.

It is required to determine the total number of cycles to failure.

Solution. The expression for damage is

$$D = \frac{n S_e^m}{A}$$

At failure, $D = 1.0$ and $n = N_f$, the total number of cycles to failure. Thus,

$$N_f = \frac{A}{S_e^m}$$

The mth power of Miner's stress is

$$S_e^m = (2\sqrt{2}\sigma)^m \Gamma\left(\frac{1}{2} m + 1\right)$$

$$= (2\sqrt{2} \cdot 38)^{6.29} \, \Gamma\left(\frac{6.29}{2} + 1\right)$$

$$= 4.32 \times 10^{13}$$

Note that the expression for stress range is used. All fracture mechanics crack growth data is based on range. The fatigue strength coefficient is evaluated from integration of the Paris crack growth law [Eq. (10.36)],

$$A = \frac{1}{C\pi^{m/2}} \int_{a_0}^{a_f} \frac{da}{Y^m a^{m/2}}$$

$$= \frac{1}{(3.35 \times 10^{-22})\,\pi^{6.29/2}} \int_{0.40}^{2.50} \frac{da}{(1.1)^{6.29} a^{6.29/2}}$$

$$= 1.46 \times 10^{20}$$

Substitution into the above expression for fatigue life yields

$$N_f = \frac{1.46 \times 10^{20}}{4.32 \times 10^{13}} = 3.38 \times 10^6 \text{ cycles}$$

10.3.5 Models of Fatigue Damage Under Wide-Band Random Stresses

Equivalent Narrow-Band Process. Assume that $S(t)$ is a stationary zero-mean Gaussian process. Two realizations of $S(t)$ are shown in Figure 10.15. Stress cycles are easily identified for the narrow-band process having Rayleigh peaks. A closed-form expression for D has been derived for the case of a linear $S\text{--}N$ curve, $NS^m = A$, as Eqs. (10.49) or (10.50).

But upon examination of the realization of the wide-band process of Figure 10.15, it is not immediately obvious how to count stress cycles to be used with Miner's rule. One approximation is to assume that damage for a wide-band process having zero crossing rate v_0^+ and RMS σ will be the same as the narrow-band process. Thus, fatigue damage under a wide-band stress can be estimated using an equivalent narrow-band process; compute v_0^+ and σ and use Eqs. (10.49) or (10.50). Fortunately, this algorithm leads to a conservative damage estimate (see below).

Rainflow Algorithm. Over the years, a number of algorithms for identifying stress cycles in a given wide-band record have been proposed. A review of these methods is provided by Nelson (1978) and the *Fatigue Design Handbook* (SAE, 1988). But the method now widely accepted as providing the most accurate results is the rainflow method.

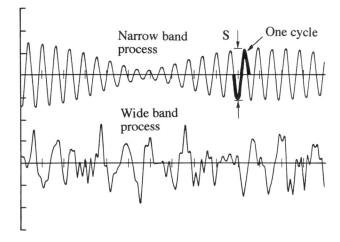

Figure 10.15 Samples of random processes. (Each process has the same RMS and expected rate of zero crossings.)

The rainflow method is used to identify stress cycles, that is, the stress range and mean stress for each cycle, from a realization of a wide-band stress process $S(t)$. The rainflow method was developed by Dowling (1972); a detailed description of the rainflow method is presented in the SAE *Fatigue Design Handbook* (1988); a FORTRAN program for performing rainflow analysis on a record is also provided. A simplified description of the method follows.

First, $S(t)$ is transformed to a process of peaks and troughs. Then rotate the time axis so that it points downward. Consider water sources at both the peaks and troughs. Water flows downward according to the following rules. An example is presented in Figure 10.16 that helps describe the process.

1. A rainflow path starting at a trough will continue down the "pagoda roofs" until it encounters a trough that is more negative than the origin. From Figure 10.16, the path that starts at 1 will end at 5.
2. A rainflow path is terminated when it encounters flow from a previous path. For example, the path that starts at 3 is terminated as shown.
3. A new path is not started until the path under consideration is stopped.
4. Trough-generated half-cycles are defined for the entire record. For each cycle, the stress range S_i is the vertical excursion of a path. The mean μ_{S_i} is the midpoint.
5. The process is repeated in reverse with peak-generated rainflow paths. For a sufficiently long record, each trough-generated half-cycle will match a peak-generated half-cycle to form a whole cycle. As a practical matter, one may choose only to analyze a record for peak- or trough-generated half-cycles and assume each is a full cycle.

Closed-Form Expression for Rainflow Damage. Attempts have been made to develop closed-form expressions for fatigue damage under wide-band stresses given a spectral density function, rather than having to execute a simulation, as suggested above. It is assumed in the following discussions on this topic that the S–N curve is linear [Eq. (10.35)] with no endurance limit.

Wirsching and Light (1980) proposed the following as an empirical form based on extensive simulation. The general expression for fatigue damage is

$$D = \lambda(\epsilon, m)D_{\mathrm{NB}} \tag{10.58}$$

where $\lambda(\epsilon, m)$ is the rainflow correction factor, and ϵ is the spectral width parameter. See Eq. (6.36). The term D_{NB} is damage as estimated using an equivalent narrow-band process,

$$D_{\mathrm{NB}} = \frac{\nu_0^+ \tau}{A} (\sqrt{2}\sigma_S)^m \Gamma\left(\frac{1}{2} m + 1\right) \tag{10.59}$$

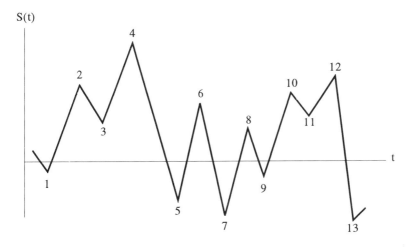

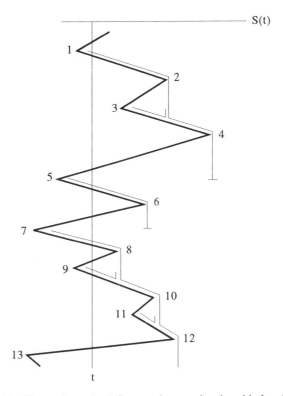

Figure 10.16 Illustration of rainflow cycle counting in wide-band process.

Note that A is based on *amplitude*; if A is based on *range*, the middle term should be $(2\sqrt{2}\sigma)^m$. Here, ν_0^+ is equivalent frequency (rate of zero crossings), τ is time of exposure, and σ_S is the RMS.

The empirical rainflow correction factor, derived by simulating processes having a variety of spectral shapes, is

$$\lambda(\epsilon, m) = a(m) + [1 - a(m)](1 - \epsilon)^{b(m)}$$

$$a(m) = 0.926 - 0.033 \, m$$

$$b(m) = 1.587 \, m - 2.323 \tag{10.60}$$

A more refined model was developed by Ortiz and Chen (1987) and Perng (1989). They have developed an approximate distribution for rainflow cycle counted effective stress ranges, H, as a Rayleigh distribution with parameter $2\beta_k \sigma_S$. The CDF of H is

$$F_H(h) = 1 - \exp\left[-\frac{1}{2}\left(\frac{h}{2\beta_k \sigma_S}\right)^2\right] \tag{10.61}$$

where β_k is the generalized spectral bandwidth

$$\beta_k = \sqrt{\frac{M_2 M_k}{M_0 M_{k+2}}} \tag{10.62}$$

and where

$$k = \frac{2.0}{m} \tag{10.63}$$

and M_j is the jth moment of the one-sided spectral density function

$$M_j = \int_0^\infty f^j W_S(f) \, df \tag{10.64}$$

Their general expression for wide-band damage is

$$D = \lambda_k D_{\text{NB}} \tag{10.65}$$

where

$$\lambda_k = \frac{\beta_k^m}{\alpha} \tag{10.66}$$

and α is the irregularity factor. See Eq. (6.34).

Another study by Lutes and Larsen (1990; see also Larsen and Lutes, 1991), produced the expression for damage under wide-band stresses as

$$D = \lambda_L D_{\text{NB}} \qquad (10.67)$$

where

$$\lambda_L = \frac{(M_{2/m})^{m/2}}{\nu_0^+} \qquad (10.68)$$

Using simulation, the authors of this work claim good agreement with the rainflow algorithm including cases where stress spectral density has a bimodal form.

Example 10.10. A wide-band stress process is defined by its spectral density function

$$W_S(f) = \begin{cases} 0.05 & 0 \le f \le 20 \text{ Hz} \\ 0 & \text{otherwise} \end{cases}$$

The slope of the S–N curve is given as $m = 3$, a typical value for the fracture mechanics model or for welded joints. Compare the rainflow correction factors of (a) Wirsching and Light, (b) Ortiz, Chen, and Perng, and (c) Lutes and Larsen.

Solution. For these models, the following parameters are calculated:

$$M_0 = \int_0^{20} W(f) \, df = 1.0$$

$$M_2 = \int_0^{20} f^2 W(f) \, df = 133.32$$

$$M_4 = \int_0^{20} f^4 W(f) \, df = 32{,}000$$

The rate of zero crossings and rate of peaks are given as

$$\nu_0^+ = \sqrt{M_2/M_0} = 11.55 \qquad \nu_p = \sqrt{M_4/M_2} = 15.49$$

The irregularity factor and spectral width parameter are

$$\alpha = \frac{\nu_0^+}{\nu_p} = 0.745$$

$$\epsilon = \sqrt{1 - \alpha^2} = 0.667$$

Other moments are defined as

$$k = \frac{2.0}{m} = 0.667$$

$$M_k = M_{2/m} = \int_0^{20} f^{0.667} W(f)\, df = 4.424$$

$$M_{k+2} = \int_0^{20} f^{2.667} W(f)\, df = 804.51$$

The three rainflow correction factors are:

(a) *Wirsching and Light.* From Eq. (10.60)

$$a = 0.827 \qquad b = 2.438 \qquad \lambda(\epsilon, m) = 0.839$$

(b) *Ortiz, Chen, and Perng.* From Eq. (10.62)

$$\beta_k = 0.856$$

and from Eq. (10.66)

$$\lambda_k = 0.843$$

(c) *Lutes and Larsen.* From Eq. (10.68)

$$\lambda_L = 0.806$$

10.3.6 Quality of Miner's Rule

Studies have shown that, in some cases, Miner's rule seems to provide reasonable approximations to fatigue life when the stress process is a stationary random process. Several random and variable-amplitude fatigue tests, as well as surveys of random fatigue tests, have confirmed this. Yet it is well known that Miner's rule can be "fooled" by certain choices of stress sequences, for example, applying several high stress cycles followed by several low stress cycles, and vice versa. And a summary of the performance of life prediction methods under variable-amplitude loading presented by Schütz (1979) showed evidence of the unpredictability of Miner's rule, citing results of studies of tests of crack initiation where strains were both elastic and plastic. Schütz (1979) also demonstrated unpredictability of crack propagation life prediction. Results of a study of Miner's rule under typical operating conditions in an automotive component, published by Schütz (1993), suggest that Miner's rule can be grossly unconservative. In summary, Schütz (1979) states that the main disadvantage of Miner's rule is it is impossible to predict if the rule will be conservative or nonconservative because it is not known which parameters influence the result.

A sampling of variable-amplitude fatigue test results is shown in Table 10.4. Damage at failure, the value of D at failure, is denoted as Δ. For a given test specimen, cycles to failure is observed and the value of Δ is computed. A random sample of Δ is collected as other specimens are tested. Probability plots of Δ have indicated that the lognormal distribution is generally a reasonable model. The sample median and coefficient of variation of Δ are the values in Table 10.4.

The numbers in Table 10.4 give evidence to the unpredictability of Miner's rule. It is important to note that any observation of Δ critically depends on a large number of factors such as the definition of the constant-amplitude S–N curve, the nature of the variable-amplitude stress, the type and material of specimen, and so on.

To account for uncertainty in the performance of Miner's rule (and other design factors), a design rule may require a target damage at failure, $\Delta_0 <$ 1.0. Typical values that have been used are $\Delta_0 = 0.50$ for routine structural components and $\Delta_0 = 0.10$ for those components where the consequences of failure are serious.

For those problems where sequence effects are considered to be important, there are several references in the literature that treat this topic, for example, Veers et al. (1989).

Table 10.4 Some Test Results on Damage at Failure

	Damage at Failure, Δ[a]	
	Median	Coefficient of Variation
Miner (1945); Miner's original work	0.95	0.26
Fatigue Under Complex Loading (SAE, 1977); a syntheses of results of the SAE Fatigue Design and Evaluation Committee	1.09	0.90
Schütz (1979), crack initiation		
(a) 29 random sequence test series	1.05	0.55
(b) Tests with large quasi-static mean load changes	0.60	0.60
(c) Significant plastic strains at notch	0.37	0.78
Schütz (1993), axle spindle of automobile, typical customer spectrum on normal roads	0.15	0.60
Shin and Lukens (1983); extensive survey of random test data	0.90	0.67
Gurney (1983); test data on welded joints	0.85	0.28
Default value used in reliability analysis for welded joints, Wirsching and Chen (1988)	1.0	0.30

[a]Statistics based on a lognormal distribution of Δ.

PROBLEMS

10.1 Shown in Figure P10.1 is a cantilever beam subjected to a stationary Gaussian narrow-band force $F(t)$. The spectral density of $F(t)$ is given in the figure. Assume that the central frequency of the force process is the midpoint of the frequency spectrum, i.e., 6 Hz. The loading environment defined by the force spectral density is to be applied for a service life of 24 min. It is required to design this element, i.e., determine the minimum dimension b. The material to be used is 6061-T6 aluminum having a mean yield strength of 40 ksi and a mean ultimate strength of 45 ksi. Assume that the coefficient of variation of each is 0.08. Also assume that the natural frequency is high enough so that there is no dynamic amplification.

a Design this beam using the three-sigma criterion applied to yield strength.

b Design the beam for a first-passage failure, with respect to ultimate. The reliability goal is 0.99 for the service life.

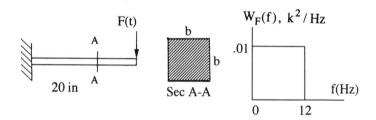

Figure P10.1

10.2 The plate shown in Figure P10.2 is subject to a stationary random force process $Q(t)$ whose mean is 17 kN and whose RMS is 22 kN. Brittle fracture is assumed to be the principal mode of failure. Fracture toughness is given as $K_C = 27$ MPa$\sqrt{m}$. The crack size is $a = 1.5$ mm. No subcritical crack propagation (fatigue) is assumed. The geometry factor is $Y = 1.2$.

Determine the minimum value required for t using the three-sigma criterion. Failure occurs when the stress intensity factor $K = YS\sqrt{\pi a} > K_C$, where S is stress.

10.3 Consider the plate of problem 10.2. Assume that $Q(t)$ is narrow band and the center frequency $f_0 = 2.0$ Hz and that $Q(t)$ is applied for 100 sec. Using a first-passage criterion, design the plate so that the probability of failure is less than 10^{-3}.

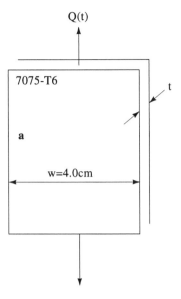

Figure P10.2

10.4 The beam shown Figure P10.4 is subjected to a white-noise base acceleration having a spectral density of $0.04g^2 \text{ Hz}^{-1}$. This base vibration is applied for 4 min and 25 sec in the service life. The beam is made of 7049-T73 aluminum having an ultimate strength of 483 MPa and a yield strength of 410 MPa. The modulus of elasticity is 69 GPa. The diameter $d = 1.0$ cm. Assume a damping factor $\zeta = 0.05$.

a Does the design satisfy the three-sigma criterion with respect to yield?

b Relative to ultimate strength, does the design satisfy a first-passage failure criterion? The maximum allowable first-passage failure is 10^{-3}.

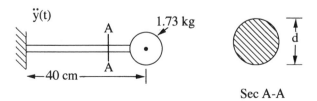

Figure P10.4

10.5   A force process $F(t)$ having a spectral density as shown in Figure P10.5
is applied as an axial load to a tie bar of Ti–6Al–4V. In addition to
the dynamic component, there is a mean load of 10 kN. The service
life of the bar under this load is 3.0 min. The titanium alloy has mean
yield and ultimate strengths of 1070 and 1180 MPa, respectively. The
coefficient of variation of both is 10%. The design values of yield and
ultimate strength are defined as the lower 1% point. Assume that both
are normal. Determine the minimum cross-sectional area A of the bar
using:

a The three-sigma criterion

b A first-passage criterion where the maximum allowable probability
of excursion is 10^{-3}

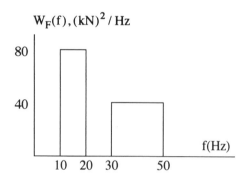

Figure P10.5

10.6   The long-term distribution of stress peaks on a structural component
has a Weibull distribution having a mean of 110 MPa and standard
deviation of 32 MPa. There are an estimated 27,000 cycles of this
stress applied in the service life. Use the extreme-value distribution
(EVD) approximation to estimate the probability of failure (first passage). The strength of the component is given as 380 MPa.

10.7   The long-term distribution of load peaks on a marine structure has a
Weibull distribution with a mean of 110 kN and a standard deviation
of 96 kN. During the service life, there will be 10^8 cycles applied.
Define the design load as that value having a return period equal to the
serivce life. Determine the design load.

10.8   It is required to perform a safety check, with respect to fatigue, on a
structural component. The design S–N curve for the material of the
component is given in Figure P10.8, as is the stress spectrum. The

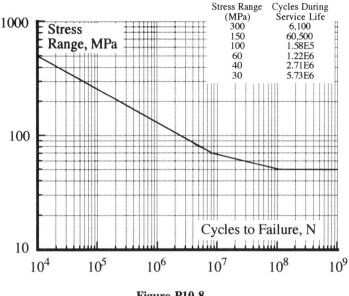

Figure P10.8

requirement is that the design life be twice the service life. This is equivalent to a safety check expression $D \leq 0.50$. Compute the total damage to check the design requirement.

10.9 Consider problem 10.8. It has been decided (because of operation of the component in a ''hostile'' environment) that a more prudent model of the fatigue strength would be a linear extrapolation of the first segment into the high-cycle range all the way down to $S = 0$.

a Compute the fatigue strength exponent and coefficient, m and A.

b Compute an equivalent constant-amplitude stress (Miner's stress).

c Use the information in parts a and b above to perform a safety check; i.e., will cycles to failure exceed twice the service life?

10.10 Consider the system shown in Fig. P10.4. The base acceleration is a Gaussian white noise having a spectral density of $0.04g^2$ Hz^{-1}. This base motion is applied for 100 hr. The bar has a solid circular cross section of diameter d. The material is 7049-T73 aluminum having a modulus of elasticity $E = 69$ GPa. The fatigue strength of this material is given by the fatigue strength coefficient and exponent $A = 2.00 \times 10^{18}$ (MPa units; based on amplitude) and $m = 5.60$, respectively. The damping factor is given as $\zeta = 0.08$. The basic design criterion is that at failure damage $D \leq 0.25$. Determine the minimum value of d to satisfy the fatigue criterion.

10.11 The long-term distribution of load Q on a butt-welded joint having width $w = 15$ cm and thickness t is Weibull. The once-in-a-lifetime load range is 750 kN; the Weibull exponent $\xi = 0.80$. The load is applied for 10^8 cycles over a service life of 20 years. Design the connection, i.e., determine the minimum plate thickness t. The fatigue design criterion is that at the service life, damage $D \leq 0.5$. The fatigue strength, based on range, is given by $A = 4.23 \times 10^{13}$ (MPa units) and $m = 3.5$.

10.12 This problem combines the fatigue- and time-dependent first-passage processes. This has the scope of a small project.

Consider the structural component subjected to a random stress process $S(t)$ such that fatigue is considered as a primary failure mode. As a fatigue crack grows, the residual strength $R(t)$ of the material becomes smaller. Fatigue failure occurs the first time that the peak fatigue stress exceeds the strength of the component.

Consider a component subjected to a stationary Gaussian fatigue stress process $S(t)$ having spectral density $W_S(f)$. It will be assumed that an initial crack of size a_0 is present in the component. The crack growth rate is defined by the Paris equation,

$$\frac{da}{dN} = C(\Delta K)^m$$

where

$$\Delta K = Y(a)S\sqrt{\pi a} \qquad Y(a) = \gamma a^{-\xi}$$

and C and m are the Paris coefficient and exponent, respectively, S is the equivalent (or Miner's) stress range, ΔK is the stress intensity factor, and γ and ξ are parameters of the geometry factor. Integration of the Paris law provides an equation that relates instantaneous crack length a to the number of cycles applied (or time).

Fracture in the part is assumed to be *brittle* fracture. From the basic fracture mechanics equation, the stress for fracture (strength) is

$$R = \frac{K_c}{Y(a)\sqrt{\pi a}}$$

where K_c is the fracture toughness. Note that as the crack grows, the strength R will decrease, and therefore, it will be a function of time or number of cycles.

a Derive the expression for crack length a as a function of cycles N.

b As an estimate of life, compute the value of N for which crack length becomes infinite.

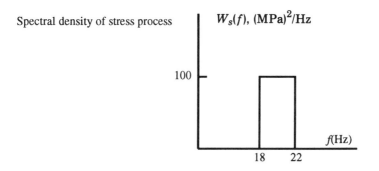

Figure P10.12

c As perhaps a more reasonable estimate of life, compute the value of N for which the instantaneous strength R equals Miner's stress.

d Determine the distribution function of cycles to failure. This can be done by computing the probability of failure p_f versus time (or cycles). Solution requires an integration of the expression for p_f in the time-dependent first passage problem.

Data are given in Fig. P10.12.

11

INTRODUCTION TO PARAMETER ESTIMATION

The practical study of random vibrations is based upon the theories of random processes and random signal analysis, and these, in turn, are based upon the theories of probability and statistics. As shown in previous chapters, the theory of probability is, in its most basic form, a collection of methods for using the axioms of probability and knowledge about the probabilities of some events to establish the probabilities of other events. In its more general form, the theory of probability introduces random variables and their probability distributions and parameters and seeks to find the probability distributions and parameters of functions of the original random variables.

The theory of statistics is complementary to the theory of probability. Whereas the theory of probability assumes prior knowledge of the probabilities of some events or probability distributions and parameters of some random variables, the theory of statistics estimates from measured data the probabilities of events and the parameters of random variable probability distributions and infers the probability distributions of random variables. Further, having established estimates for these quantities, the theory of statistics develops formulas useful for judging the quality of estimates. The part of the theory of statistics that deals with the estimation of probability distribution parameters is known as parameter estimation. The part of the theory of statistics that deals with the assessment of the quality of the estimates is known as interval estimation.

Some of the specific quantities that are estimated in statistical studies are the probabilities of events, parameters of the probability distributions of random variables, and moments of random variables like mean, variance, and standard deviation. Later chapters will show how the theory of statistics can be extended for the study of random signals that are functions of time. There, estimators for such quantities as means, variances, and spectral densities of random processes will be developed.

This chapter starts the development of a tie between the theory of random vibration and the practical world in which structural inputs and the responses they excite can only be known through their measured realizations.

This chapter covers only some of the most fundamental ideas in parameter estimation. For further information on statistics, see the texts by Ang and Tang (1975); Benjamin and Cornell (1970); Brownlee (1960); Hines and Montgomery (1990); Mood, Graybill, and Boes (1974); and Soong (1981).

11.1 ESTIMATION AND ANALYSIS OF MEAN

We now proceed with a fairly detailed statistical analysis of the estimation of the mean or expected value of a random variable. There are two reasons for this. First, the formulas developed in connection with estimation of the mean are typical of the formulas developed in estimation of other parameters of random variables and random processes. Second, techniques used to develop estimation formulas and judge the quality of estimates are typical of those used for other measures of random variables and random processes.

11.1.1 Maximum Likelihood Estimation

We first develop a formula for estimation of the mean using the method of maximum likelihood. However, before starting, note that we refer to *estimation* of the mean for the following reason. In classical statistics, it is assumed that the parameters of the probability distribution of random variables are unknown constants. Formulas that use finite collections of measured data can be developed to approximate these unknown constants, but the constants can never be known with exact precision.

It is well known that the average of a collection of numbers x_j, $j = 1$, . . . , n, is given by

$$\bar{x} = \frac{1}{n} \sum_{j=1}^{n} x_j \tag{11.1}$$

This formula is known as the sample mean of the data x_j, $j = 1, \ldots, n$. (In this text, quantities that refer to random variable realizations, like x_j, $j = 1, \ldots, n$, will be denoted with lowercase letters. Functions of random variable realizations will also be denoted with lowercase letters, as $\bar{x}$. Random variables and functions of random variables will be denoted with capital letters. We note this convention here because many statistics texts use capital X bar, $\bar{X}$, to denote both the mean estimator random variable and its realizations.)

Equation (11.1) is used to estimate the mean of a random variable, and it appears quite frequently, in various forms, in statistics. The formal development of this formula and its characteristics are given in the following.

Consider a random variable X with unknown probability distribution and moments. Denote its mean μ_X and its variance σ_X^2. In order to estimate its mean, we perform a sequence of n nominally identical and independent random experiments, drawing a realization x_j from a random variable source X_j, in each experiment. The realizations form a sequence $x_j, j = 1, \ldots, n$, and the corresponding random variables are $X_j, j = 1, \ldots, n$. Each X_j is a duplicate of the source random variable X and has the same probability distribution. A formula that uses the $x_j, j = 1, \ldots, n$, to approximate the mean is called an estimator formula.

One way to interpret the mean estimator formula [Eq. (11.1)] is in the framework of the discrete random variable (see Chapter 2). The mean of a discrete-valued random variable is the sum of the discrete realizations times their probabilities. If the $x_j, j = 1, \ldots, n$, were realizations of a discrete random variable, each with equal probability $1/n$, then Eq. (11.1) would be the mean of the random variable.

A broader and more widely applicable framework is presented by the method of maximum likelihood. Let the PDF of X be denoted $f_X(x, \{p_1\})$, and let the joint PDF of the $X_j, j = 1, \ldots, n$, be denoted $f_{X_1,\ldots,X_n}(x_1, \ldots, x_n, \{p_n\})$. The vector of parameters of the first PDF is explicitly denoted $\{p_1\}$, and the vector of parameters of the second PDF is explicitly denoted $\{p_n\}$. Because the random experiments that sample the $X_j, j = 1, \ldots, n$, are independent and the sampled random variables are identically distributed as X,

$$f_{X_1,\ldots,X_n}(x_1, \ldots, x_n, \{p_n\}) = \prod_{j=1}^{n} f_{X_j}(x_j, \{p_1\})$$

$$= \prod_{j=1}^{n} f_X(x_j, \{p_1\}) \tag{11.2}$$

where the $\prod$ denotes the product of the functions. Because the random variables $X_j, j = 1, \ldots, n$, are independent and identically distributed, the parameters $\{p_1\}$ that characterize an individual random variable are the only ones needed to characterize an entire sequence of random variables, and $\{p_1\} = \{p_n\}$. The methods of statistics seek, for one thing, to estimate $\{p_1\}$ given measured data $x_j, j = 1, \ldots, n$. When the parameters in $\{p_1\}$ are unknown in the above expressions and the $x_j, j = 1, \ldots, n$, are known, the expressions are known as likelihood functions. The method of maximum likelihood takes as "best" estimates of the parameters in $\{p_1\}$ those quantities that maximize the likelihood function. Let $L(\{p_1\})$ denote the likelihood function. Then, in this case,

$$L(\{p_1\}) = \prod_{j=1}^{n} f_X(x_j, \{p_1\}) \tag{11.3}$$

and the estimator formulas for the elements in $\{p_1\}$ are obtained by maximizing $L(\{p_1\})$ with respect to the elements in $\{p_1\}$. Sometimes this maximization is more easily achieved when the logarithm of $L(\{p_1\})$ is maximized. [The maximum of the logarithm of $L(\{p_1\})$ occurs at the same location as the maximum of $L(\{p_1\})$.] Therefore, the log-likelihood function is defined as

$$\mathcal{L}(\{p_1\}) = \ln[L(\{p_1\})] \tag{11.4}$$

and maximized to establish estimator formulas for the parameters in $\{p_1\}$.

Example 11.1. Consider the case where the random variable X is normally distributed with mean μ_X and variance σ_X^2. The log-likelihood function for the mean and variance is

$$\mathcal{L}(\mu_X, \sigma_X^2) = \ln\left[\prod_{j=1}^{n} \frac{1}{\sqrt{2\pi\sigma_x^2}} \exp\left(-\frac{(x_j - \mu_x)^2}{2\sigma_x^2}\right)\right]$$

$$= -\frac{n}{2}\ln(2\pi\sigma_x^2) - \frac{1}{2\sigma_x^2}\sum_{j=1}^{n}(x_j - \mu_x)^2 \tag{11.5}$$

Because this expression is continuous in μ_X and σ_X^2, its maximum can be established by taking the partial derivatives of $\mathcal{L}(\mu_X, \sigma_X^2)$ with respect to μ_X and σ_X^2, then equating the resulting expressions to zero and solving for μ_X and σ_X^2. The results are

$$\bar{x} = \frac{1}{n}\sum_{j=1}^{n}x_j \tag{11.1}$$

$$\hat{\sigma}_X^2 = \frac{1}{n}\sum_{j=1}^{n}(x_j - \mu_X)^2 \tag{11.6}$$

where, in the first equation, μ_X has been replaced by $\bar{x}$, the estimator for the mean. In the second equation, the caret indicates that $\hat{\sigma}_X^2$ is the estimator for the variance. Note that the right side of the second equation involves the unknown mean of X. This will be elaborated upon later. The development presented here is an example of how the method of maximum likelihood can be applied in the solution of a basic, though important, problem.

We will not go into much more detail here, but the reader should note that the method of maximum likelihood has been very widely studied, and it has been shown that estimators developed using the method of maximum likelihood have several important characteristics. Some of these are enumerated here and discussed later.

1. Maximum likelihood estimators (MLEs) are asymptotically unbiased. (Bias is explained in the following section.)

2. MLEs are always consistent (i.e., along with characteristic 1 this implies that as $n \to \infty$, the estimator approaches the parameter to be estimated).
3. MLEs are efficient (i.e., they have minimum variance among all unbiased estimators).
4. MLEs are sufficient (i.e., they use all available data in the parameter estimate).
5. MLEs are invariant [i.e., they provide an estimate of a parameter p that satisfies $g(\hat{p}) = \hat{g}(p)$]. For this reason, an estimator for the standard deviation is simply the square root of Eq. (11.6).
6. MLEs are asymptotically normally distributed. This will be seen in following sections.

For a more detailed description of the method of maximum likelihood estimation and specific maximum likelihood estimators, see Brownlee (1960) and Benjamin and Cornell (1970).

It is important to note that although the method of maximum likelihood is widely used to establish estimator formulas, it is not the only technique for doing so. Other techniques for estimating probability distribution parameters are the method of mean likelihood estimation, the method of moments, and the use of probability paper. The latter two methods are especially widely used in engineering applications. The former method is discussed in Jenkins and Watts (1969). The latter two methods are discussed in Ang and Tang (1975) and Benjamin and Cornell (1970).

Example 11.2. Use of the formulas in Eqs. (11.1) and (11.6) is easily demonstrated. Consider the data listed in Table 11.1. These are 40 realizations of a normal random variable generated during independent random experiments. The mean and variance estimates were calculated using Eqs. (11.1) and (11.6) and are listed at the bottom of the table. Figure 11.1 is a plot of the log likelihood function [Eq. (11.5)] for $\sigma_x^2 = (1.5)^2$ and variable μ_x. It is clear that the curve is maximum for μ_x near 2.1.

Table 11.1 40 Random Numbers

3.747	2.940	2.113	2.527	0.955
4.544	2.088	4.696	2.396	3.308
−0.169	0.948	3.869	1.042	2.866
1.460	1.797	−0.024	0.094	3.477
1.933	0.802	0.852	3.293	1.916
2.770	2.595	3.134	2.601	−0.012
2.563	3.688	3.093	−1.566	1.589
1.516	2.477	1.233	1.997	4.410

Estimated mean = 2.139
Estimated standard deviation = 1.401

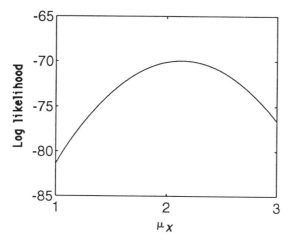

Figure 11.1 Log likelihood function of data in Table 11.1 for $\sigma_X^2 = (1.5)^2$, variable μ_X, and X normal.

11.1.2 Bias and Consistency of Mean Estimator

The previous section described some of the reasons why Eq. (11.1) is taken as an estimator for the mean of a random variable. Recall that because $\bar{x}$ is based upon measured samples that come from a random source, it will not equal the (unknown) mean μ_X that it is meant to estimate. An important question is: How close is $\bar{x}$ to μ_X? Of course, because we do not know the value of μ_X, we cannot answer the question exactly. However, we can draw some conclusions about how close $\bar{x}$ is to μ_X on the average. We commence this process in this section and continue it in later sections.

In order to evaluate the quality of the mean estimator $\bar{x}$, we recall that each realization $x_j, j = 1, \ldots, n$, was drawn from a random variable source X_j, $j = 1, \ldots, n$. When we use the X_j in Eq. (11.1) in place of the x_j, we obtain

$$\bar{X} = \frac{1}{n} \sum_{j=1}^{n} X_j \tag{11.7}$$

This formula indicates that the random variable $\bar{X}$ is formed by summing the random variables $X_j, j = 1, \ldots, n$, then multiplying the result by $1/n$. Here, $\bar{X}$ is the source of realizations $\bar{x}$ that are specific mean estimators. In other words, if we performed sequences of n experiments several times over, where we sampled the X_j to obtain x_j, then averaged the x_j to obtain $\bar{x}$, we would draw different realizations x_j from the X_j, and these would yield different $\bar{x}$. These $\bar{x}$ would be realizations of the random variable $\bar{X}$ in Eq. (11.7). The important point is that $\bar{X}$ is a random variable and it reflects the characteristics of the $X_j, j = 1, \ldots, n$, which, in turn, reflect the characteristics of X. Because $\bar{X}$ is a random variable, it has a probability distribution with mean, variance, and so on. The following section analyzes the probability distribution of $\overline{X}$. Here, we analyze its mean and variance.

The mean of $\overline{X}$ can be written using the rules developed in Chapter 2:

$$E[\overline{X}] = E\left[\frac{1}{n}\sum_{j=1}^{n} X_j\right] = \frac{1}{n} E\left[\sum_{j=1}^{n} X_j\right]$$

$$= \frac{1}{n}\sum_{j=1}^{n} E[X_j] = \frac{1}{n}\sum_{j=1}^{n} \mu_X = \mu_X \tag{11.8}$$

That is, the expected value of the mean estimator $\overline{X}$ is the mean of the random variable X. This is good and indicates that $\overline{X}$ is an unbiased estimate of the mean of X. When the mean of an estimator formula, like Eq. (11.7), does not equal the quantity the formula is meant to estimate, then the estimator is said to be biased.

In addition to unbiasedness in an estimator, we would also hope to find a quality that enables us to somehow improve the estimate of the parameter we seek, for example, by increasing the amount of data used to estimate the parameter. Consistency is such a quality, and it is analyzed as follows. Again, using the rules developed in Chapter 2, evaluate the variance of $\overline{X}$. It is

$$V[\overline{X}] = V\left[\frac{1}{n}\sum_{j=1}^{n} X_j\right] = \frac{1}{n^2} V\left[\sum_{j=1}^{n} X_j\right]$$

$$= \frac{1}{n^2}\left(\sum_{j=1}^{n} V[X_j] + \sum_{j \neq k}\sum \text{Cov}\,[X_j, X_k]\right)$$

$$= \frac{1}{n^2}\sum_{j=1}^{n} \sigma_X^2 = \frac{\sigma_X^2}{n} \tag{11.9}$$

That is, the variance of the mean estimator $\overline{X}$ is the variance of the random variable X divided by n. (The sum of covariance terms equals zero because the X_j, $j = 1, \ldots, n$, are independent.) The variance of the mean estimator decreases with increasing n; therefore $\overline{X}$ is a consistent estimator of the mean. This characteristic of $\overline{X}$, coupled with the unbiasedness of the estimator, indicates that, with increasing n, the estimates of the mean $\bar{x}$ based on measured data will tend toward the true mean of X.

Example 11.3. In order to demonstrate the ideas behind bias and consistency, consider the following example. We are able to generate independent realizations x from a theoretical random variable X. The random variable X is normally distributed with mean zero and unit variance. We perform two numerical studies on the mean estimator of X. In the first study, we draw 10 realizations from the random variable X and use these to estimate the mean of X. We repeat these operations 999 more times to obtain a total of 1000 mean estimates, based on 10 realizations each. The histogram of the 1000 mean estimates is shown in Figure 11.2a. (The histogram of a sequence of numbers is a bar chart where the height of each bar indicates the number of samples that fall in a particular abscissa range.) The theoretical result in Eq. (11.8) indicates that the mean of the PDF whose shape is approximated by Figure 11.2a is zero; this appears plausible from the plotted results. The theoretical result in Eq. (11.9) indicates that the standard deviation of the PDF whose shape is approximated by Figure 11.2a is $1/\sqrt{10}$; this appears plausible from the plotted results. The (nonnormalized) theoretical PDF of $\overline{X}$ is also shown in Figure 11.2a.

In the second study, we draw 100 realizations from the random variable X and use these to estimate the mean of X. We repeat these operations 999 more times to obtain a total of 1000 mean estimates, based on 100 realizations each. The histogram of the 1000 mean estimates is shown in Figure 11.2b. The theoretical result in Eq. (11.8) indicates that the mean of the PDF whose shape is approximated by Figure 11.2b is zero; this appears plausible from the plotted results. The theoretical result in Eq. (11.9) indicates that the standard deviation of the PDF whose shape is approximated by Figure 11.2b is $1/\sqrt{100}$; this appears plausible from the plotted results. The (nonnormalized) theoretical PDF of $\overline{X}$ is also shown in Figure 11.2b.

The results shown here indicate that the random variable that estimates the mean [Eq. (11.7)] is indeed unbiased and consistent and that the quality of mean estimates improves as $1/\sqrt{n}$. This general result is seen throughout statistics.

11.1.3 Sampling Distribution of Mean Estimator and Confidence Intervals on Mean

It is interesting and important to know the mean and variance of an estimator formula, but beyond this, there is a great deal of value in knowing the probability distribution of an estimator. Consider the probability distribution of $\overline{X}$. The random variable $\overline{X}$ equals $(1/n)$ times a sum of random variables, $X_j, j = 1, \ldots, n$. Based on the developments of Chapter 2, it is clear that if the underlying random variable X is normally distributed, then the $X_j, j = 1, \ldots, n$, are normally distributed and $\overline{X}$ is normally distributed. If X is not normally distributed, then $\overline{X}$ is $1/n$ times the sum of n independent, identically distributed nonnormal random variables. The central limit theorem (Appendix A) indicates that, in this case, the distribution of $\overline{X}$ approaches the normal dstribution as n increases. The rate at which the distribution of $\overline{X}$ approaches the normal depends upon the smoothness and symmetry of the probability density function of X. When the PDF of X is smooth, the PDF of $\overline{X}$ approaches the normal distribution rapidly, and otherwise, it does not. The probability distribution of an estimator for a parameter of a random variable is known as the sampling distribution of the parameter estimate.

Assume that the distribution of $\overline{X}$ is normal and that the variance σ_X^2 of X is known. Then the probability distribution of the quantity $(\overline{X} - \mu_X)/(\sigma_X/\sqrt{n})$ is

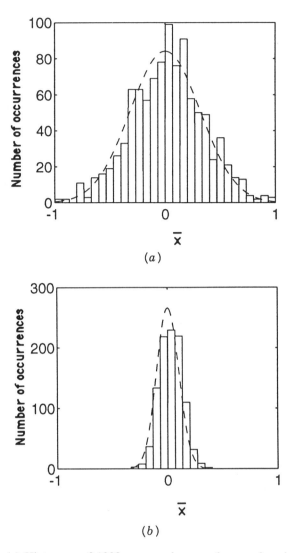

Figure 11.2 (*a*) Histogram of 1000 mean estimates where each estimate is average of 10 realizations. (Underlying random variable is normal with zero mean and unit variance.) Smooth curve is theoretical shape histogram should assume if mean estimator is normal with zero mean and variance equal to 0.1. (*b*) Histogram of 1000 mean estimates where each estimate is average of 100 realizations. (Underlying random variable is normal with zero mean and unit variance.) Smooth curve is theoretical shape histogram should assume if mean estimator is normal with zero mean and variance equal to 0.01.

standard normal, and we can write

$$P\left(-\infty < \frac{\overline{X} - \mu_X}{\sigma_X/\sqrt{n}} \leq z_{1-\alpha}\right) = \Phi(z_{1-\alpha}) = 1 - \alpha \qquad (11.10)$$

The expression on the left is a probability statement on the random variable $\overline{X}$. The quantities μ_X and σ_X are probability distribution parameters. The σ_X is assumed known; μ_X is unknown, yet assumed constant. The n is an experiment parameter. The $z_{1-\alpha}$ is an inequality limit. The value of $z_{1-\alpha}$ is related to the magnitude of α and the relation between these quantities is defined in Section E.1 in the table of the standard normal cumulative distribution function. When we are interested in establishing bounds on the unknown mean of X, μ_X, we can solve the double inequality in the argument of the left-hand expression in Eq. (11.10) for μ_X and replace the random variable $\overline{X}$ with the realized mean estimator $\overline{x}$. The bounds on the mean μ_X so established define the interval

$$\left(\overline{x} - \frac{z_{1-\alpha}\sigma_X}{\sqrt{n}}, \infty\right) \qquad (11.11)$$

This is known as the $(1 - \alpha) \times 100\%$ one-sided confidence interval on the mean of X, μ_X. It is called a confidence interval because μ_X is assumed to be a constant and does not have probabilities associated with its values. The expression reflects our relative confidence that the true mean falls in the interval identified in Eq. (11.11). Note that as $1 - \alpha$ approaches 1, $z_{1-\alpha}$ increases and the confidence interval widens. In other words, we have greatest confidence that the mean falls in the widest intervals.

In a manner analogous to that in which we wrote Eq. (11.10), we can also write

$$P\left(-z_{1-\alpha} < \frac{\overline{X} - \mu_X}{\sigma_X/\sqrt{n}} < \infty\right) = 1 - \Phi(-z_{1-\alpha}) = 1 - \alpha \qquad (11.12)$$

This expression leads to the $(1 - \alpha) \times 100\%$ confidence interval on the mean,

$$\left(-\infty, \overline{x} + \frac{z_{1-\alpha}\sigma_X}{\sqrt{n}}\right) \qquad (11.13)$$

Its interpretation and behavior are analogous to that described in the previous paragraph.

Finally, a two-sided confidence interval can be written starting with the probability statement

$$P\left(-z_{1-\alpha/2} < \frac{\overline{X} - \mu_X}{\sigma_X/\sqrt{n}} \le z_{1-\alpha/2}\right) = \Phi(z_{1-\alpha/2}) - \Phi(-z_{1-\alpha/2}) = 1 - \alpha$$

(11.14)

Solve the double inequality in the argument on the left to establish the $(1 - \alpha) \times 100\%$ two-sided confidence interval on the mean,

$$\left(\overline{x} - \frac{z_{1-\alpha/2}\,\sigma_X}{\sqrt{n}},\ \overline{x} + \frac{z_{1-\alpha/2}\,\sigma_X}{\sqrt{n}}\right)$$

(11.15)

This reflects our confidence that the mean falls within the specified interval. As before, when $1 - \alpha$ approaches 1, the magnitude of $z_{1-\alpha/2}$ increases, and the confidence interval widens. Further, increasing n causes the width of the confidence interval to decrease. This reflects the consistency of the mean estimator.

Figure 11.3 plots the normalized confidence intervals for the mean of a random variable with known variance over a range of averages. The normalized ordinates of the plot are the limit values minus the sample mean divided by σ_X. For example, the fact that the curve marked 10% has an ordinate of -0.2 where $n = 40$ indicates that we are 10% confident that the quantity (true mean minus the estimated mean) divided by the quantity (true standard deviation) is lower than -0.2 when 40 samples are used to estimate the mean. Note also that the curve marked 90% has an ordinate of 0.2 where $n = 40$. This can be interpreted in a similar manner as for the 10% curve. Together, these two facts indicate that we are 80% confident that the normalized mean falls in the interval $(-0.2, 0.2)$ when the estimate of the mean is based on 40 samples.

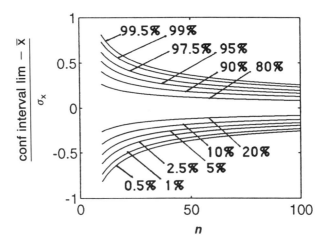

Figure 11.3 Normalized confidence intervals for mean with known variance as function of number of measured values used to form mean estimate.

Under certain circumstances, it is quite desirable to establish confidence intervals on most of the random variable parameters we estimate.

11.1.4 Bias in Variance Estimator

We conclude this section with an analysis, the results of which will prove useful in the following sections. Specifically, we analyze the bias of the variance estimator [Eq. (11.6)]. To do this, we replace the x_j, $j = 1, \ldots, n$, on the right-hand side of Eq. (11.6) with the corresponding random variable X_j, $j = 1, \ldots, n$. This substitution yields the estimator formula

$$\hat{\Sigma}_X^2 = \frac{1}{n} \sum_{j=1}^{n} (X_j - \mu_X)^2 \tag{11.16}$$

where $\hat{\Sigma}_X^2$ is the random variable that is sampled to obtain variance estimates, $\hat{\sigma}_X^2$. When we evaluate the expected value of $\hat{\Sigma}_X^2$, we find that its mean is $\hat{\sigma}_X^2$. Therefore, Eq. (11.16) provides an unbiased estimate of the variance where the mean μ_X is known.

Usually, though, the mean μ_X is not known, and this affects the bias in estimation of the variance. In fact, it is a direct matter to show that when $\overline{X}$ is used in place of μ_X, the unbiased estimate of the variance is

$$S_X^2 = \frac{1}{n-1} \sum_{j=1}^{n} (X_j - \overline{X})^2 \tag{11.17}$$

The mean of S_X^2 is evaluated as follows:

$$E[S_X^2] = E\left[\frac{1}{n-1} \sum_{j=1}^{n} (X_j - \overline{X})^2\right] = E\left[\frac{1}{n-1} \sum_{j=1}^{n} [(X_j - \mu_X) - (\overline{X} - \mu_X)]^2\right]$$

$$= \frac{1}{n-1} \sum_{j=1}^{n} \{E[(X_j - \mu_X)^2] - 2E[(X_j - \mu_X)(\overline{X} - \mu_X)]$$

$$+ E[(\overline{X} - \mu_X)^2]\}$$

$$= \frac{1}{n-1} \sum_{j=1}^{n} \left(\sigma_X^2 - \frac{2}{n} \sigma_X^2 + \frac{1}{n} \sigma_X^2\right) = \sigma_X^2 \tag{11.18}$$

In view of this, the variance of random variable X is normally estimated using the formula

$$s_X^2 = \frac{1}{n-1} \sum_{j=1}^{n} (x_j - \overline{x})^2 \tag{11.19}$$

with the realizations x_j, $j = 1, \ldots, n$. This is the unbiased estimator for the variance of X when the mean of X is unknown. Clearly, when n is large, there is not much difference between the formulas in Eqs. (11.6) and (11.19).

11.2 OTHER IMPORTANT PROBLEMS IN PARAMETER ESTIMATION

Section 11.1 introduced the ideas of bias, consistency, and sampling distribution for the mean estimator, and confidence intervals for the mean of a random variable with known variance. In general, though, the moments of a random variable are not known; therefore, distributions that involve probability distribution parameters and estimates of parameters must be developed. This is done in the following, and the results are used to develop confidence intervals for the mean with unknown variance and the variance itself.

11.2.1 Sampling Distribution for Variance

Because our development of confidence intervals on the mean value of the random variable X, when the variance is unknown, relies on knowledge of the distribution of the variance estimator S_X^2, we now proceed to develop the sampling distribution for the variance of X. We start by assuming that the source random variables, $X_j, j = 1, \ldots, n$, leading to an estimate for the variance are normally distributed. Consider Eq. (11.17); multiply it by $n - 1$ and divide by σ_X^2. The result is

$$(n - 1) \frac{S_X^2}{\sigma_X^2} = \frac{1}{\sigma_X^2} \sum_{j=1}^{n} (X_j - \overline{X})^2$$

$$= \sum_{j=1}^{n} \left(\frac{X_j - \mu_X}{\sigma_X} \right)^2 - \left(\frac{\overline{X} - \mu_X}{\sigma_X/\sqrt{n}} \right)^2 \qquad (11.20)$$

To establish the final expression on the right, μ_X was subtracted from each term on the right side of the first equation, the square was executed, and the three resulting terms were summed separately. Recognize that the first term in the final right-hand expression above is the sum of the squares of n independent standard normal random variables. Therefore, the first term is chi-square distributed with n degrees of freedom. (See Chapter 2 for a complete discussion of the chi-square distribution.) The second term in the final right-hand expression is the square of a single standard normal variable; therefore, its distribution is chi square with one degree of freedom. Because of the regenerative character of chi-square random variables, we conclude that the quantity on the left-hand side is chi square distributed with $n - 1$ degrees of freedom.

The result developed here can be used to establish confidence intervals on the variance following the procedure presented in Section 11.1.3. For example, the probability statement involving the random variable S_X^2,

$$P\left((n - 1) \frac{S_X^2}{\sigma_X^2} > x_{1-\alpha} \right) = 1 - F_{\chi_{n-1}^2}(x_{1-\alpha}) = 1 - \alpha \qquad (11.21)$$

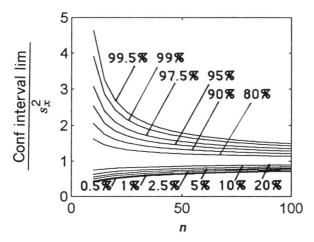

Figure 11.4 Normalized confidence intervals for variance.

leads to the following $(1 - \alpha) \times 100\%$ confidence interval for the variance:

$$\left(0, \ \frac{(n - 1)s_X^2}{x_{1 - \alpha}}\right) \tag{11.22}$$

Because of the form of its definition, s_X^2 is guaranteed nonnegative. Confidence intervals for the variance exhibit the same general behavior as confidence intervals for the mean. Figure 11.4 shows the normalized confidence intervals for the variance. The ordinates are the bound of Eq. (11.22) divided by s_x^2. Figure 11.4 can be interpreted in the same general way as Figure 11.3.

It is important to note that whenever we deal with sums of squares of normally distributed random variables, their distribution is related to the chi-square distribution. In particular, later when we consider spectral density estimates, the chi-square distribution will be important.

11.2.2 Confidence Intervals for Mean with Unknown Variance

In Section 11.1.3, we developed confidence intervals for the mean based on our knowledge of the probability distribution of the quantity $(\overline{X} - \mu_X)/(\sigma_X/\sqrt{n})$. We assumed that the variance σ_X^2 is known, but this is typically not so. Rather, the variance estimator s_X^2 is used to approximate σ_X^2. In view of this, we seek to establish the PDF of $(\overline{X} - \mu_X)/(s_X/\sqrt{n})$. To do this, we note first that the random variable defined $(\overline{X} - \mu_X)/(\sigma_X/\sqrt{n})$ is a standard normal random variable, and second that the random variable defined $\sqrt{n - 1}\,S_X/\sigma_X$ is a chi-distributed random variable with $n - 1$ degrees of freedom. The former fact is the basis of the development in Section 11.1.3 and was discussed there. The latter arises from the discussion following Eq. (11.20) that $(n - 1)S_X^2/\sigma_X^2$ is a chi-square distributed random variable with $n - 1$ degrees of freedom. (See Chapter 2 for a complete discussion of the chi-distribution.)

In view of these facts, the ratio of the first random variable to the second divided by $\sqrt{n - 1}$ forms a t-distributed random variable with $n - 1$ degrees of freedom. (See Chapter 2 for a complete discussion of the t distribution.) That is,

$$t_{n-1} = \frac{\overline{X} - \mu_X}{S_X/\sqrt{n}} \tag{11.23}$$

is t distributed with $n - 1$ degrees of freedom. (In addition, independence of the random variables is also required, and this can be demonstrated.)

We can use this fact to develop confidence intervals for the mean of the random variable X when the variance of X is unknown. This is done following the general approach of Section 11.1.3. For example, the two-sided confidence interval for the mean is based on the probability statement

$$P\left(-b_{1-\alpha/2} < \frac{\overline{X} - \mu_X}{s_X/\sqrt{n}} \le b_{1-\alpha/2}\right)$$

$$= F_{t_{n-1}}(b_{1-\alpha/2}) - F_{t_{n-1}}(-b_{1-\alpha/2})$$

$$= 1 - \alpha \tag{11.24}$$

We solve the double inequality in the argument on the left for μ_X to establish the $(1 - \alpha) \times 100\%$ confidence interval for the mean. It is

$$\left(\overline{x} - \frac{b_{1-\alpha/2}s_X}{\sqrt{n}}, \quad \overline{x} + \frac{b_{1-\alpha/2}s_X}{\sqrt{n}}\right) \tag{11.25}$$

Note that the realization $\overline{x}$ has been used in place of $\overline{X}$.

Application of the procedure for specifying confidence intervals shown in the previous paragraph leads to the conclusion that the confidence intervals at a particular confidence level generated using the t distribution are wider than confidence intervals generated using the normal distribution when s_X^2 is used to estimate σ_X^2. The reason is that the t distribution accounts for variability in the variance estimator s_X^2, and this leads to a less precise knowledge of the distribution of $(\overline{X} - \mu_X)/(s_X/\sqrt{n})$. Figure 11.5 shows the normalized confidence intervals for the mean of a random variable with unknown variance over a range of numbers of degrees of freedom in the t distribution. The normalized ordinates of the plot are the limit values minus the sample mean $\overline{x}$ divided by s_X. The curves in Figure 11.5 can be interpreted in the same way as those in Figure 11.3.

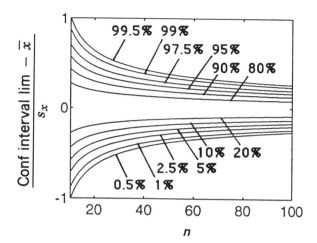

Figure 11.5 Normalized confidence intervals for mean with unknown variance.

In general, confidence intervals like the one in Eq. (11.25) exhibit the same behavior as the confidence interval in Eq. (11.15).

11.3 CLOSURE

We reiterate, in closing, that only a few specialized topics in statistics have been covered in this chapter. There are many other important topics, including some relevant to the study of random vibrations, that might be covered. Important among these is the subject of hypothesis testing. Hypothesis testing permits us to pose hypotheses about collections of random data and confirm or deny the hypotheses based on the data. For example, under certain circumstances, it may be useful to compare the magnitudes of measures of structural excitation or response for their severity.

The field of hypothesis testing also includes methods to confirm the probability distribution of random data. These methods are especially important in random vibration applications where the response of a system may not be normally distributed (as is usually the case when we consider the response of a nonlinear system). The probability distribution of random data is often tested using the chi-square test or the Kolmogorov–Smirnov test (see Benjamin and Cornell, 1970; Miller and Freund, 1985; and Soong, 1981). Probability paper is also frequently used to establish, approximately, whether or not a particular probability distribution is a plausible choice for a collection of random data (see Ang and Tang, 1975; and Benjamin and Cornell, 1970).

We also often need to develop a sense for the probability distribution of a collection of data without any prior idea about the parametric model from which the data might arise. In these cases, we seek to empirically approximate the data PDF. Frameworks for accomplishing this are the histogram or the kernel density estimator. For more information on these subjects, see Parzen (1962a), Cacoullos (1966), or Tapia and Thompson (1978).

PROBLEMS

11.1 Use the method of maximum likelihood to write formulas for estimation of the parameters of
  a An exponentially distributed random variable
  b A Rayleigh distributed random variable

11.2 Estimate the mean, variance, and standard deviation of the following data:

3.42	5.68	1.32	4.22	3.65	4.96	2.33	1.07	6.58	4.75
2.92	5.49	3.77	7.01	4.55	3.79	4.91	5.19	4.82	4.44

11.3 Compute the 80 and 98% two-sided confidence intervals on the mean estimated using the data of problem 11.2. State your assumptions.

11.4 Compute the 90 and 99% upper confidence intervals for the variance estimated using the data of problem 11.2. State your assumptions.

11.5 Write the formula for the variance of the mean estimator when the samples used to obtain the estimate are correlated. [Use Eq. (11.9), but do not assume that the covariance terms are zero.]

11.6 Write a computer subroutine or function to generate random variates from a uniform distribution on the interval $[a, b]$. To accomplish this, use the call available on most computers to generate a uniform $[0, 1]$ variate, then transform the variate linearly to an interval $[a, b]$. [If such a call is not available on your computer, then use the code in Press et al. (1988)]. Use the subroutine or function to generate 20 variates in the interval $[-10, 20]$, and print them out.

11.7 Write a computer subroutine or function to generate random variates from a normal distribution with mean μ and variance σ^2. To accomplish this, first write a subroutine or function to generate a random number from a standard normal source. This can be accomplished using one of three possible approaches:

  a *Inversion* (based on equations in Abramowitz and Stegun, 1964). Generate a random number u, from a uniform $[0, 1]$ random source (as in problem 11.6). Define p as follows:

  $$p = \begin{cases} u & \text{if } u \leq 0.5 \\ 1 - u & \text{if } u > 0.5 \end{cases}$$

Define

$$t = \sqrt{\ln\left(\frac{1}{p^2}\right)}$$

Compute

$$x_p = t - \frac{a_0 + a_1 t}{1 + b_1 t + b_2 t^2}$$

where

$$a_0 = 2.30753 \qquad b_1 = 0.99229$$

$$a_1 = 0.27061 \qquad b_2 = 0.04481$$

The random number from the standard normal source is

$$x = \begin{cases} x_p & \text{if } u \le 0.5 \\ -x_p & \text{if } u > 0.5 \end{cases}$$

b *Sum of uniform random numbers* (based on the central limit theorem). Generate 12 independent random numbers u_j, $j = 1, \ldots ,$ 12, from a uniform [0, 1] random source. The random number from the standard normal source is

$$x = \left(\sum_{j=1}^{12} x_j\right) - 6$$

c *Change of variables* (based on bivariate transformation example in Chapter 2). Generate two independent random numbers u_1 and u_2 from a uniform [0, 1] random source. Two independent random numbers from a standard normal source are

$$x_1 = \sqrt{-2 \ln u_1} \, \cos 2\pi u_2$$

$$x_2 = \sqrt{-2 \ln u_1} \, \sin 2\pi u_2$$

Transform the random number from the standard normal source to obtain the random number from a normal source with mean μ and variance σ^2:

$$y = \sigma x + \mu$$

11.8   Write a computer subroutine or function to estimate the mean, variance, and standard deviation of a data sequence. Generate the signal

$$x_j = \sin \frac{2\pi j}{32} \quad j = 0, \ldots , 511$$

Compute its mean, variance, and standard deviation. Demonstrate, using an analytical approach, that the results obtained are to be expected.

11.9   Generate 100 random numbers from a uniform $[-\sqrt{3}, \sqrt{3}]$ source using the tool developed in problem 11.6. Estimate the mean, variance, and standard deviation using the tool developed in problem 11.8.

11.10   Generate 100 random numbers from a standard normal random source using the tool developed in problem 11.7. Estimate the mean, variance, and standard deviation using the tool developed in problem 11.8.

11.11   Compute the 98% two-sided confidence intervals for the means and the 90% one-sided upper confidence intervals for the variances estimated in problems 11.9 and 11.10. Assume first for the confidence intervals on the mean that the variance is known to be 1; next, assume that the variance in unknown.

12

TIME-DOMAIN ESTIMATION OF RANDOM PROCESS PARAMETERS

To characterize the random processes that generate realizations in practical applications, we require techniques that use realizations measured from random process sources to estimate the random process parameters. Individual random processes have many characterizing features, including mean function, variance and standard deviation functions, first and higher order PDFs, higher order moments, autocorrelation functions, and so on. Random process pairs have all the above characterizing features for the individual random processes, plus the cross-correlation function, cross-covariance function, higher order cross-moment functions, and so on. When the individual or multiple random processes under consideration are weakly stationary, they also possess auto- and cross-spectral densities. Because we deal with normal random processes in so many practical applications, the most important quantities characterizing random processes are their first- and second-order moments. Therefore, we devote the majority of our attention in this chapter and the following chapters to estimation of these moments.

It will become apparent in the following sections that there is a great difference in the statistical analysis of stationary and nonstationary random processes. Techniques for the statistical estimation of stationary random processes have been thoroughly developed and can be excuted in a very practical framework. Techniques for the statistical estimation of nonstationary random processes require amounts of data that are sometimes unrealistic, restrictive assumptions about the behavior of the random process or they are quite ad hoc in nature. Nevertheless, nonstationary random processes occur widely in practice and will be considered in the following.

This chapter starts with a presentation of the general method for analysis of random processes, either nonstationary or stationary, via the use of ensemble averages. Then, we develop methods for the estimation of the parameters of stationary random processes with time averages. Finally, we comment on some models and techniques used for nonstationary random process analysis.

It is assumed that the random processes considered in this chapter are continuous valued, with a continuous parameter, time. The measurements of their realizations used for statistical analysis are sampled at a discrete time interval Δt, and the measurements are not aliased. All the measured signals used to characterize a random process come from the same measurement source. Each measured signal used in the analysis is the sum of a random process realization plus noise. The moment estimators developed in this and the following chapters minimize the effects of noise. For analyses and derivations that explicitly consider the effects of noise, see Bendat and Piersol (1986).

12.1 RANDOM PROCESS PARAMETER ESTIMATION VIA ENSEMBLE AVERAGE

All random processes generated from real physical sources are nonstationary because no real random process can have the moment or PDF characteristics required of a stationary random process over all time or all space. Even so, when a random process has realizations that appear stationary over a "substantial period of time," then it is analyzed as stationary. Some methods for stationary signal analysis are developed in the following section.

There is a general method for analyzing the parameters of an arbitrary random process that assumes several measurements from the ensemble of the random process are available. Let $X(t)$, $t \in T$, be an arbitrary random process, and let x_{mj}, $m = 1, \ldots , M$, $j = 0, \ldots , n - 1$, denote independent realizations sampled in discrete time, $t_j = j \, \Delta t$, $j = 0, 1, \ldots , n - 1$, from the complete ensemble of the random process. The index m numbers the realization; the index j numbers the time in the mth measurement. The time interval $[0, T]$ is represented in the measurements, and $n \, \Delta t = T$. Figure 12.1 is an example that shows three members of the ensemble of a nonstationary random process that is defined on the time interval $[0, 1]$.

12.1.1 Mean, Variance, and Standard Deviation Function

The moments of the random process that characterize the individual behavior of the random variables in the random process can be easily estimated using the formulas of Chapter 11. For example, the estimator for the mean function of the random process is

$$\bar{x}_j = \frac{1}{M} \sum_{m=1}^{M} x_{mj} \qquad j = 0, \ldots , n - 1 \tag{12.1}$$

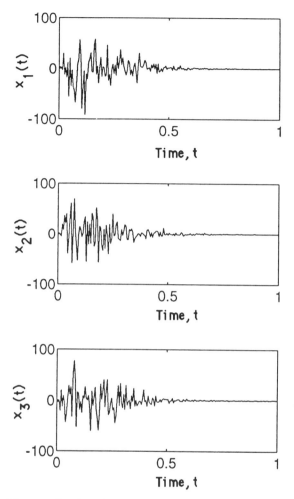

Figure 12.1   Three realizations from ensemble of nonstationary random process.

This estimator for the value of the mean function at time t_j has all the characteristics developed in Chapter 11. Most important, it is unbiased and consistent.

The estimator for the variance function is

$$s_X^2(t_j) = \frac{1}{M-1} \sum_{m=1}^{M} (x_{mj} - \bar{x}_j)^2 \qquad j = 0, \ldots, n-1 \qquad (12.2)$$

and the corresponding estimator for the standard deviation function is

$$s_X(t_j) = \sqrt{s_X^2(t_j)} \qquad j = 0, \ldots, n-1 \qquad (12.3)$$

These are unbiased and consistent estimators of the random process variance and standard deviation functions.

12.1.2 Correlation

The next level of characterization of a random process is the second-order moments of different random variables in the random process, namely the autocorrelation, autocovariance, and autocorrelation coefficient functions. Estimator formulas for these moment functions can be established using the method of maximum likelihood. For example, consider the estimation of correlation defined $R = E[XY]$ between two normally distributed random variables X and Y with zero means and arbitrary variances σ_X^2 and σ_Y^2. The joint PDF of the two random variables is

$$f_{XY}(x, y) = \frac{1}{2\pi\sqrt{\sigma_X^2\sigma_Y^2 - R^2}} \exp\left[-\frac{1}{2(\sigma_X^2\sigma_Y^2 - R^2)}(\sigma_Y^2 x^2 - 2Rxy + \sigma_X^2 y^2)\right]$$

$$-\infty < x, y < \infty \qquad\qquad (12.4)$$

It is assumed that estimates of the variances and correlation are sought and will be based on joint measurements of realizations of the random variables x_j, y_j, $j = 1, \ldots, n$. The separate measurements are independent. The likelihood function based on the n independent measured sample pairs x_j, y_j, $j = 1, \ldots, n$, is

$$L(\sigma_X^2, \sigma_Y^2, R) = \frac{1}{(2\pi)^n(\sigma_X^2\sigma_Y^2 - R^2)^{n/2}}$$

$$\times \exp\left[-\frac{1}{2(\sigma_X^2\sigma_Y^2 - R^2)}\sum_{j=1}^{n}(\sigma_Y^2 x_j^2 - 2Rx_jy_j + \sigma_X^2 y_j^2)\right]$$

$$(12.5)$$

The log-likelihood function is the natural logarithm of this function and is given by

$$\mathcal{L}(\sigma_X^2, \sigma_Y^2, R) = -n\ln(2\pi) - \frac{n}{2}\ln(\sigma_X^2\sigma_Y^2 - R^2)$$

$$-\frac{1}{2(\sigma_X^2\sigma_Y^2 - R^2)}\sum_{j=1}^{n}(\sigma_Y^2 x_j^2 - 2Rx_jy_j + \sigma_X^2 2y_j^2) \quad (12.6)$$

It is through maximimization of the log-likelihood function that estimates for the variances and correlation are established. To accomplish the maximization, take the partial derivative of $\mathcal{L}$ with respect to σ_X^2, σ_Y^2, and R, then equate the results to zero. These computations yield

$$n + \frac{1}{\hat{\sigma}_Y^2} \sum_{j=1}^{n} y_j^2 = \frac{1}{\hat{\sigma}_X^2 \hat{\sigma}_Y^2 - \hat{r}^2} \sum_{j=1}^{n} (\hat{\sigma}_Y^2 x_j^2 - 2\hat{r}x_j y_j + \hat{\sigma}_X^2 y_j^2) \qquad (12.7)$$

$$n + \frac{1}{\hat{\sigma}_X^2} \sum_{j=1}^{n} x_j^2 = \frac{1}{\hat{\sigma}_X^2 \hat{\sigma}_Y^2 - \hat{r}^2} \sum_{j=1}^{n} (\hat{\sigma}_Y^2 x_j^2 - 2\hat{r}x_j y_j + \hat{\sigma}_X^2 y_j^2) \qquad (12.8)$$

$$n\hat{r} + \sum_{j=1}^{n} x_j y_j = \frac{\hat{r}}{\hat{\sigma}_X^2 \hat{\sigma}_Y^2 - \hat{r}^2} \sum_{j=1}^{n} (\hat{\sigma}_Y^2 x_j^2 - 2\hat{r}x_j y_j + \hat{\sigma}_X^2 y_j^2) \qquad (12.9)$$

where we have used $\hat{\sigma}_X^2$, $\hat{\sigma}_Y^2$, and $\hat{r}$ in place of σ_X^2, σ_Y^2, and R, respectively, to show that the solution to these equations governs estimates for the parameters σ_X^2, σ_Y^2, and R. Now add Eqs. (12.7) and (12.8) and divide the result by 2 to obtain

$$n + \frac{1}{2}\left[\frac{1}{\hat{\sigma}_Y^2} \sum_{j=1}^{n} y_j^2 + \frac{1}{\hat{\sigma}_X^2} \sum_{j=1}^{n} x_j^2\right] = \frac{1}{\hat{\sigma}_X^2 \hat{\sigma}_Y^2 - \hat{r}^2} \sum_{j=1}^{n} (\hat{\sigma}_Y^2 x_j^2 - 2\hat{r}x_j y_j + \hat{\sigma}_X^2 y_j^2)$$

$$(12.10)$$

Use Eq. (12.10) in Eq. (12.9) to obtain

$$\sum_{j=1}^{n} x_j y_j = \frac{\hat{r}}{2}\left(\frac{1}{\hat{\sigma}_X^2} \sum_{j=1}^{n} x_j^2 + \frac{1}{\hat{\sigma}_Y^2} \sum_{j=1}^{n} y_j^2\right) \qquad (12.11)$$

The estimate of the variance $\hat{\sigma}_X^2$ is established by substituting Eq. (12.11) into (12.7). It is

$$\hat{\sigma}_X^2 = \frac{1}{n} \sum_{j=1}^{n} x_j^2 \qquad (12.12)$$

Likewise, the variance estimator $\hat{\sigma}_Y^2$ is given by

$$\hat{\sigma}_Y^2 = \frac{1}{n} \sum_{j=1}^{n} y_j^2 \qquad (12.13)$$

Substitution of the variance estimators [Eqs. (12.12) and (12.13)] into Eq. (12.11) produces the correlation estimator

$$\hat{r} = \frac{1}{n} \sum_{j=1}^{n} x_j y_j \qquad (12.14)$$

This formula is effective for estimating the correlation between normally distributed random variables when data are available. In fact, it is used to estimate the correlation between any two random variables, regardless of their probability distribution.

In order to investigate the characteristics of the correlation estimator formula [Eq. (12.14)], we substitute the random variables X_j and Y_j, $j = 1, \ldots, n$, which are the sources of the realizations x_j and y_j, $j = 1, \ldots, n$, into Eq. (12.14). This yields

$$\hat{R} = \frac{1}{n} \sum_{j=1}^{n} X_j Y_j \tag{12.15}$$

We assume that the random variables X_j, $j = 1, \ldots, n$, are independent and identical in distribution to a random variable X and that the Y_j, $j = 1, \ldots, n$, are independent and identical in distribution to a random variable Y. The random variables X and Y are jointly normally distributed with means zero, variances σ_X^2 and σ_Y^2, respectively, and correlation, R.

Any bias in the estimator is established by evaluating the mean of $\hat{R}$. This is

$$E[\hat{R}] = E\left[\frac{1}{n} \sum_{j=1}^{n} X_j Y_j\right] = \frac{1}{n} \sum_{j=1}^{n} E[X_j Y_j] = R \tag{12.16}$$

This result indicates that the correlation estimate is unbiased. The variance of the correlation estimator is

$$V[\hat{R}] = V\left[\frac{1}{n} \sum_{j=1}^{n} X_j Y_j\right] = \frac{1}{n^2} \sum_{j=1}^{n} V[X_j Y_j] = \frac{1}{n} (R^2 + \sigma_X^2 \sigma_Y^2) \tag{12.17}$$

The middle equation is established due to the fact that the samples are independent. The final equation on the right is based on the fact that $E[X^2 Y^2] = \sigma_X^2 \sigma_Y^2 + 2R^2$ when X and Y are zero-mean, normal random variables (see Papoulis, 1965). This result makes it clear that the correlation estimator is consistent.

12.1.3 Autocorrelation Function

In order to develop an estimate for the autocorrelation function of a random process, we simply estimate the correlation between every pair of random variables in the random process. Assume that $X(t)$, $t \in T$, is a normal random process with autocorrelation function $R_X(t, s)$, $t, s \in T$. When M discrete-time sampled measurements from the ensemble of the random process are available at times $t_j = j \, \Delta t$, $j = 0, \ldots, n - 1$, and are denoted x_{mj}, $m = 1, \ldots,$ M, the autocorrelation function estimate is

$$\hat{r}_X(t_i, t_j) = \frac{1}{M} \sum_{m=1}^{M} x_{mi} x_{mj} \qquad t_i, t_j \in T \tag{12.18}$$

Because the autocorrelation functions of mechanical system excitation and response random processes frequently alternates sign, it is not typical that we attempt to conservatively bound the functions using confidence intervals. Moreover, the sampling distribution of the autocorrelation function estimator is quite complicated when the autocorrelation function is nonzero. If confidence intervals are required for the autocorrelation function estimator, some approximations are available. Because the summands in Eq. (12.18) are independent and identically distributed, the central limit theorem applies when M is large, and the distribution of $\hat{R}_X(t_i, t_j)$ approaches the normal distribution with mean and variance derived from Eqs. (12.16) and (12.17), respectively. The mean of $\hat{R}_X(t_i, t_j)$ is $R_X(t_i, t_j)$, and its variance is $(1/n) [R_X^2(t_i, t_j) - \sigma_X^2(t_i)\sigma_X^2(t_j)]$.

Example 12.1. As an illustration of the statistical analysis of a random process, consider the following example. A zero-mean, nonstationary random process is denoted $X(t)$, $0 \le t \le 1$, and 20 realizations of the random process were measured. One realization of the random process is shown in Figure 12.2, and this is typical of all the others. The mean of the random process was estimated using Eq. (12.1), and the results are plotted in Figure 12.3 using the same scale as Figure 12.2. Though the estimated mean is not zero, it may plausibly be assumed zero since it is small compared to the realization values. Recall, based on Eq. (11.9), that the standard deviation of the mean estimator equals the standard deviation of the underlying random variable divided by the square root of the number of realizations used to generate the estimate (in this case, 20). If the standard deviation of the random process between 0.2 and 0.4 sec is on the order of 100, then the standard deviation of the mean estimator should be on the order of 22 $(=100/\sqrt{20})$, and the oscillations of the mean estimator may be ascribed to random estimation error.

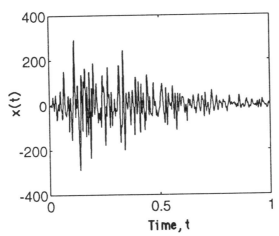

Figure 12.2 Realization of nonstationary random process.

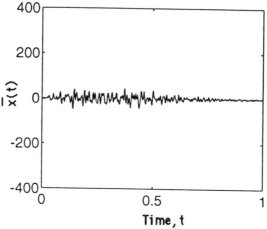

Figure 12.3   Estimated mean function of random process that is source of signal shown in Figure 12.2. Estimate based on 20 realizations.

The variance and standard deviation functions of the random process were estimated using Eqs. (12.2) and (12.3), and the latter is plotted in Figure 12.4. The plot indicates that the random process has RMS signal content that increases from time zero until about 0.2 sec; then it decays. For the same general reasons listed in the previous paragraph, the oscillations in the standard deviation estimator may reflect random estimation error in the estimate.

The autocovariance function of the random process is a surface defined above the (t, s) plane in the region $0 < t < 1, 0 < s < 1$. It could be estimated for every pair of points in the (t, s) plane using Eq. (12.18); however, if it were, the surface would be so complicated that it could not be practically

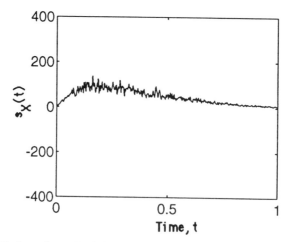

Figure 12.4   Estimated standard deviation function of random process that is source of signal shown in Figure 12.2. Estimate based on 20 realizations.

interpreted. A function that can be interpreted is the autocorrelation function estimate at a constant lag value, plotted over the entire time domain (0, 1). The autocovariance function was estimated at three lag values: 0, 0.002, and 0.004 sec. The estimator plots are shown in Figures 12.5a, 12.5b, and 12.5c, respectively, all using the scale of Figure 12.5a. (Note that the first plot is simply the estimated variance of the random process.) There is substantial

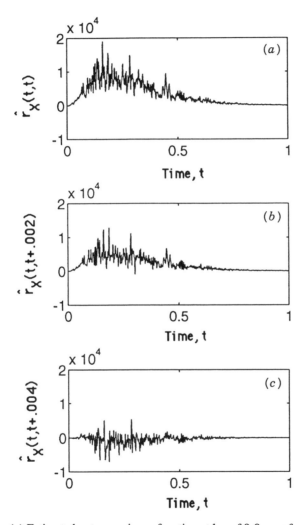

Figure 12.5 (a) Estimated autocovariance function at lag of 0.0 sec of random process that is source of signal shown in Figure 12.2 Estimate based on 20 realizations. (This equals variance estimate for random process.) (b) Estimated autocovariance function at lag of 0.002 sec of random process that is source of signal shown in Figure 12.2. Estimate based on 20 realizations. (c) Estimated autocovariance function at lag of 0.004 sec of random process that is source of signal shown in Figure 12.2. Estimate based on 20 realizations.

variability in all three estimates, due, perhaps, to random error. Figure 12.5a shows that the variance varies as the RMS in Figure 12.4 indicates it should (i.e., the curve in Figure 12.5a is the square of the curve in Figure 12.4). Figure 12.5b shows that the autocovariance is small, but predominantly positive at a lag of 2 msec. Figure 12.5c shows that the autocovariance is near zero at a time lag of 4 msec. A sequence of plots like these three could be used to completely characterize the mean-square behavior of the nonstationary random process.

We note, finally, that although there is no minimum number of realizations required (as long as $M > 1$) to execute the averages defined in this chapter, the 20 used in the above example is usually considered fairly minimal. When fewer realizations are available, they can obviously be used, but the analyst must be very careful to note that estimation errors may be quite substantial. The greater the amount of data used in these formulas, the better, and it is not at all unreasonable to execute averages with 100 or more measured realizations.

12.1.4 Cross-Correlation Function

Because Eq. (12.14) provides an estimate for the correlation between any two random variables, it can be used to develop an estimate for the cross-correlation function between two random processes. Assume that $X(t)$, $t \in T$, and $Y(t)$, $t \in T$, are jointly distributed, normal random processes with cross-correlation function $R_{XY}(t, s)$, $t, s \in T$. Further, assume that M discrete-time sampled measurements from the ensemble of realizations of the random process are available at times $t_j = j \, \Delta t, j = 0, \ldots , n - 1$, and are denoted x_{mj}, y_{mj}, $m = 1, \ldots , M$. (Since we are attempting to establish a relation between two random processes, it is clear that we must sample their realizations simultaneously and in phase. That is, there can be no time lag between the sampling of x_{mj} and y_{mj}.) Then the cross-correlation function estimate is

$$\hat{r}_{XY}(t_i, t_j) = \frac{1}{M} \sum_{m=1}^{M} x_{mi} y_{mj} \qquad t_i, t_j \in T \qquad (12.19)$$

Each element in the estimate is unbiased and consistent, as in the case of the autocorrelation function estimate. The asymptotic distribution of the estimator $\hat{R}_{XY}(t_i, t_j)$ is normal with mean $R_{XY}(t_i, t_j)$ and variance $(1/n) \, [R_{XY}^2(t_i, t_j) - \sigma_X^2(t_i) \, \sigma_Y^2(t_j)]$.

12.1.5 Covariance and Correlation Coefficient Function

We chose to estimate and analyze the autocorrelation and cross-correlation functions in this section because they are so frequently used in practice. When a random process has nonzero mean, the mean function is frequently estimated and treated separately. However, auto- and cross-covariance functions and/or auto- and cross-correlation coefficient functions could as well have been estimated. For purposes of generality, we note that the maximum likelihood estimator for the cross-covariance function between random processes $X(t)$, $t \in T$, and $Y(t)$, $t \in T$, is

$$\hat{C}_{XY}(t_i, t_j) = \frac{1}{M} \sum_{m=1}^{M} (x_{mi} - \bar{x}_i)(y_{mj} - \bar{y}_j) \qquad t_i, t_j \in T \qquad (12.20)$$

where the same notation as in Eq. (12.19) is used for the measured realizations and $\bar{x}_i$ and $\bar{y}_j$ denote the sample means of the random processes $X(t)$, $t \in T$, and $Y(t)$, $t \in T$, at times t_i and t_j, respectively. This formula yields a biased estimate for the cross-correlation function, but it can be adjusted as in the case of the variance estimator.

The cross-correlation coefficient estimator for the random processes $X(t)$, $t \in T$, and $Y(t)$, $t \in T$, is

$$\hat{\rho}_{XY}(t_i, t_j) = \frac{\hat{C}_{XY}(t_i, t_j)}{s_X(t_i)s_Y(t_j)} \qquad t_i, t_j \in T \qquad (12.21)$$

where the notation on the right-hand side comes from Eqs. (12.20) and (12.3). This is also a biased estimator of the cross-correlation coefficient function, but it cannot be adjusted easily because the bias depends on the level of correlation.

In order to develop the autocovariance and autocorrelation coefficient functions, we simply substitute the random process $X(t)$, $t \in T$, wherever the random process $Y(t)$, $t \in T$, is used above, and we use x_{mj} in place of y_{mj}, $m = 1, \ldots, M, j = 0, \ldots, n - 1$.

12.2 STATIONARY RANDOM PROCESS PARAMETER ESTIMATION VIA TEMPORAL AVERAGING

The formulas for random process parameter estimation developed in the previous section apply to any random process. However, it is typical in practical applications that we do not have multiple realization measurements from the ensemble of a random process with which to work. Some reasons are that it may be quite costly to obtain measurements of random process realizations or nature may simply not produce them.

When the random processes under consideration are weakly stationary, alternatives to the techniques presented in the previous section exist. To commence our presentation, consider the following facts. The mean, variance, and standard deviations of a weakly stationary random process are constants. Yet, for example, when we estimate the mean function of a stationary random process using Eq. (12.1), the values $\bar{x}_j$, $j = 0, \ldots, n - 1$, will typically not be equal. The reason is that estimation error affects each value, $\bar{x}_j$. When $X(t)$, $-\infty < t < \infty$, is a weakly stationary random process and when we have a measured ensemble from a time period $(0, T)$ of this random process, denoted

x_{mj}, $m = 1, \ldots, M$, $j = 0, \ldots, n - 1$, with $n \, \Delta t = T$, we can estimate the mean using Eq. (12.1). However, we can improve on the estimate of the mean by further averaging all the values $\bar{x}_j$, $j = 0, \ldots, n - 1$. This is appropriate because they are all estimates of the same quantity, $E[X(t)] = \mu_X(t) = \mu_X$. The formula is

$$\bar{x} = \frac{1}{n} \sum_{j=0}^{n-1} \bar{x}_j \qquad (12.22)$$

Likewise, the variance estimate for the weakly stationary random process is

$$s_X^2 = \frac{1}{n} \sum_{j=0}^{n-1} s_X^2(t_j) \qquad (12.23)$$

and the standard deviation estimate is

$$s_X = \sqrt{s_X^2} \qquad (12.24)$$

These estimates of random process moments normally provide improvements over the estimates $\bar{x}_j$ and $\hat{\sigma}_x^2(t_j)$, $j = 0, \ldots, n - 1$, because the averaging operation tends to diminish the variance of the estimator (see Section 11.1.2). The amount of improvement obtained through the use of Eqs. (12.22) and (12.23) is not usually on the order of $1/n$, though, because that sort of improvement is only realized when the samples are independent, and sequential moment averages are typically not independent. Rather, they have a dependence that is related to the autocorrelation function of the random process itself. The process employed above is called temporal averaging because averages are taken across time.

In the same way that the mean, variance, and standard deviation estimates of stationary random processes can be improved through temporal averaging, the autocorrelation function estimate can also be improved. Recall that the autocorrelation function of a stationary random process is a function only of the time lag between random variables in the random process. In view of this, $\hat{r}_X(t_i, t_j)$ estimates only one quantity as long as t_i and t_j are separated by a constant amount of time. Averaging all such estimates will improve the estimate of the underlying quantity. Note that Eq. (12.18) estimates the autocorrelation function only on the time interval $(0, T)$; therefore, only the autocorrelation functions with a time lag $|t_j - t_i| < T$ can be temporally averaged to improve their characteristics, and only autocorrelation function estimates, where the time lag $|t_j - t_i|$ is much smaller than T, can be substantially improved. The reason is that a large number of averages is required to improve an estimate. When the autocorrelation function of Eq. (12.18) is available and the random process under consideration is weakly stationary, the autocorrelation function estimate can be improved using

$$\hat{r}_X(\tau_j) = \frac{1}{n-j} \sum_{i=0}^{n-1-j} \hat{r}_X(t_i, t_i + \tau_j) \qquad 0 \le t_i, \tau_j \le T \qquad (12.25)$$

where $\tau_j = j\,\Delta t$, $j = 0, \ldots, n - 1$, is a time lag, and the expression is a function of τ_j only because the underlying random process is weakly stationary. Clearly, the more ensemble elements that go into the estimate of $\hat{r}_X(t_i, t_i + \tau_j)$ and the more elements that go into the average of Eq. (12.25), the better. The estimates in Eqs. (12.22)–(12.25) are mixed ensemble and temporal averages because both types of averaging are used in their development.

A question that cannot be answered, in general, is: How many averages are sufficient? We simply note that increasing the number of averages that lead to the mean, variance, or autocorrelation function estimate improves the estimate. This question has very special meaning, though, in the important case where only one measured signal from the random process ensemble is available. Equation (12.1) can be used in Eq. (12.22), Eq. (12.2) can be used in Eq. (12.23), and Eq. (12.18) can be used in Eq. (12.25), and when $M = 1$, the estimates of the mean, variance, and autocorrelation function become purely temporal averages. The mean estimate is

$$\bar{x} = \frac{1}{n} \sum_{j=0}^{n-1} x_j \qquad (12.26)$$

where x_j, $j = 0, \ldots, n - 1$, is the single measured member of the ensemble of the stationary random process $X(t)$, $-\infty < t < \infty$. The estimate is unbiased and consistent. The variance estimate is

$$s_X^2 = \frac{1}{n} \sum_{j=0}^{n-1} (x_j - \bar{x})^2 \qquad (12.27)$$

and the autocorrelation function estimate is

$$\hat{r}_X(\tau_j) = \frac{1}{n-j} \sum_{i=0}^{n-1-j} x_i x_{i+j} \qquad 0 \le \tau_j \le T \qquad (12.28)$$

where $n\,\Delta t = T$. The standard deviation estimate is obtained by using Eq. (12.27) in Eq. (12.24).

The issue that is raised in writing Eqs. (12.26)–(12.28) is that estimates for the first- and second-order moments are based entirely on a single realization of the stationary random process $X(t)$, $-\infty < t < \infty$. It is clear that this can only be reasonable when the realization used is representative of practically all the others in the random process. A random process for which this is true is called ergodic. It is easy to construct random processes that are not ergodic. The quality of ergodicity in a random process has specific mathematical requirements that relate directly to how fast the autocorrelation function magnitude diminishes with increasing lag. We will not pursue these mathematical requirements here because most of the oscillatory excitation and response random processes that we need to consider and that can be approximated as weakly stationary can also be considered ergodic. We simply note that the analyst needs to be careful regarding this issue of "representativeness" [see Papoulis (1965) and Lin (1967), on ergodicity]. In fact, a more important consideration is that the experimentalist needs to be careful that different measured random process realizations truly come from a single source, when it is assumed they are members of the ensemble in a single random process.

It is important to mention, at this point, that although the random process under consideration is assumed weakly stationary and possesses a spectral density, it is not usually prudent to estimate the spectral density by Fourier transforming the estimated autocorrelation function [Eq. (12.28)]. Some reasons are that (1) the number of computations required to establish the spectral density estimate via this sequence of operations is much greater than necessary (see Chapter 14) and (2) the spectral density thus estimated is not guaranteed nonnegative. Indeed, it is usually considered more efficient to estimate the spectral density via the methods of Chapter 14 and then inverse Fourier transform the result to obtain an estimate of the autocorrelation function.

The estimate for the cross-correlation function of a pair of jointly distributed, weakly stationary, ergodic random processes $X(t)$, $-\infty < t < \infty$, and $Y(t)$, $-\infty < t < \infty$, can be written as

$$\hat{r}_{XY}(\tau_j) = \frac{1}{n-j} \sum_{i=0}^{n-1-j} x_i y_{i+j} \qquad 0 \le \tau_j \le T \qquad (12.29)$$

where x_j, y_j, $j = 0, \ldots, n-1$, are the lone measured members of the ensembles of the random processes. This estimate is unbiased and consistent. The cross-covariance function estimate for the same pair of random processes is

$$\hat{C}_{XY}(\tau_j) = \frac{1}{n-j} \sum_{i=0}^{n-1-j} (x_i - \bar{x})(y_{i+j} - \bar{y}) \qquad 0 \le \tau_j \le T \qquad (12.30)$$

and the estimate for the cross-correlation coefficient function is

$$\hat{\rho}_{XY}(\tau_j) = \frac{\hat{C}_{XY}(\tau_j)}{s_X s_Y} \qquad 0 \le \tau_j \le T \qquad (12.31)$$

The estimates in Eqs. (12.30) and (12.31) becomes autocovariance function and autocorrelation coefficient function estimates when y is replaced with x wherever it occurs.

Example 12.2. As an illustration of the estimation of the autocorrelation function of a stationary, ergodic random process, based on one measured segment

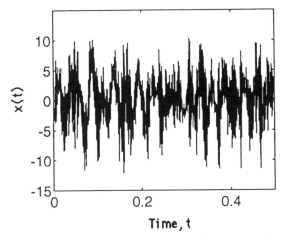

Figure 12.6 Segment of measured realization from stationary, ergodic, random source.

of one element in the ensemble of the random process, consider the following example. A zero-mean, stationary random process is denoted $X(t)$, $-\infty < t < \infty$, and one finite-duration realization is measured from the random process. The duration of the measurement is 8 sec. A 0.5-sec portion of the measured signal is plotted in Figure 12.6. It is assumed that the random process under consideration is ergodic. The autocorrelation function of the random process was estimated using Eq. (12.28) and the results are plotted in Figure 12.7. It is apparent from the plot that the estimated mean square of the random process is about 15 units. (This is the ordinate at a lag of zero sec.) Because the mean of the random process is zero, the estimated mean square equals the variance

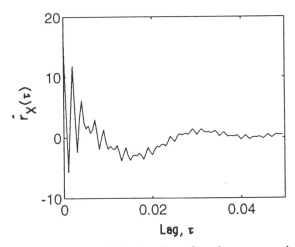

Figure 12.7 Estimated autocorrelation function of random process that is source of signal shown in Figure 12.6. Estimate based on one measured signal.

estimate. The autocorrelation function is clearly composed of two main components: one with frequency of about 30 Hz and period 0.033 sec, and the other with frequency of about 500 Hz and period 0.002 sec. When a random process is not composed of only a few distinct components, it is not so easily interpreted. In such cases, the spectral density must be used to interpret the frequency content of the random process. The formulas for estimating spectral density are developed in Chapter 14.

12.3 OTHER METHODS FOR NONSTATIONARY RANDOM PROCESS ANALYSIS

It is interesting to note that in spite of its omnipresence and consequent importance, the nonstationary random process has not been as widely investigated as might be anticipated. Of course, some reasons for this are that (1) nonstationary random processes are more complicated to investigate, in a rigorous sense, than stationary random processes and (2) they are much more difficult to characterize, as noted at the beginning of this chapter. There are, however, several methods for characterizing random sources, and we will briefly discuss some of them in this section. The fall into two categories: direct methods that seek to characterize, and perhaps simulate, nonstationary random processes and indirect methods that characterize nonstationary random processes in terms of their severity, damage-causing potential, or effects on structures.

12.3.1 Direct Analysis of Nonstationary Random Processes

There are two general approaches to the direct analysis of nonstationary random processes. One approach treats nonstationary random processes as outputs of filters whose characteristics vary with time and that are excited by broadband random noise. This approach is described in many texts and some papers by Priestly (1965, 1967). A simpler approach is to treat a nonstationary random process as the product between two time-varying functions, one deterministic and the other random. Let $X(t)$, $-\infty < t < \infty$, be a zero-mean nonstationary random process with autocorrelation function $R_X(t, s)$, $-\infty < t, s < \infty$. The random process can be expressed as

$$X(t) = a(t)U(t) \qquad -\infty < t < \infty \qquad (12.32)$$

where $a(t)$ is a deterministic function and $U(t)$, $-\infty < t < \infty$, is a zero-mean (potentially nonstationary) random process with a constant mean square of 1. In general, $a(t)$ and $U(t)$ are completely arbitrary: that is, they can have any characteristics we wisk to assign them. But for practical purposes, we might require, for example, that $a(t)$ be finite valued and Fourier transformable. We might require that $U(t)$ be a normally distributed random process, and of course, since its mean square is constant, it manifests its nonstationarity through changes in its spectral distribution of signal content (see Chapters 13 and 14).

The autocorrelation of the random process $X(t)$ can be expressed in terms of the function $a(t)$ and the autocorrelation function of the random process $U(t)$, $R_U(t, s)$, $-\infty < t, s < \infty$. It is

$$R_X(t, s) = a(t)a(s)R_U(t, s) \qquad -\infty < t, s < \infty \qquad (12.33)$$

The mean square of the random process is simply the autocorrelation function evaluated at $s = t$. This is

$$\sigma_X^2(t) = a^2(t) \qquad (12.34)$$

since the mean square of $U(t)$ equals 1.

A more specialized form for the representation of the nonstationary random process requires that the zero-mean random process $U(t)$, $-\infty < t < \infty$, be weakly stationary and that the function $a(t)$ be nonzero only for $t \geq 0$. These requirements translate into a simplification in the autocorrelation function of the random process $X(t)$. It can now be expressed

$$R_X(t, s) = a(t)a(s)R_U(s - t) \qquad -\infty < t, s < \infty \qquad (12.35)$$

The mean square of the random process is the same here as in the previous case, Eq. (12.34).

An even greater specialization of the representation for the nonstationary random process requires that the highest frequency where the Fourier transform of $a(t)$ displays a substantial signal component be lower than the lowest frequency where the spectral density of $U(t)$ displays substantial signal components. These requirements do not change the autocorrelation function or mean-square representations of Eqs. (12.35) and (12.34), but they do considerably simplify the equivalent frequency-domain representations.

There are methods for simplifying the models described above even further. For example, the function $a(t)$ might be assumed to have a specific parametric form involving only a few parameters. The reason for making such an assumption would be to minimize the number of parameters that need to be estimated with the given data; this typically improves the quality of the parameter estimates.

As one might expect, there are approaches to the estimation of the nonstationary random process moments and parameters connected with any model. For example, the method of maximum likelihood can usually be employed to establish formulas for nonstationary random process parameters. We will not go into detail here on methods for estimating nonstationary random process moments, except to show that some characteristics, like the modulating function $a(t)$ in Eq. (12.32), can be very easily estimated. Let us take $a(t)$ to be a nonnegative function. Then it is clear, from Eq. (12.34), that if we can accurately estimate the mean square of the nonstationary random process $X(t)$, then we have $a(t)$. Assume that a single sampled realization of the nonstationary random process is available and is given by x_j, $j = 0, \ldots, n - 1$. This is one element of the ensemble of the random process $X(t)$ measured at times $t_j = j \, \Delta t$, $j = 0, \ldots, n - 1$. An estimate for the mean square of the nonstationary random process at time t_j can be written as

$$\hat{\sigma}_X^2(t_j) = \sum_{k=-N}^{N} w_k x_{j+k}^2 \qquad j = N, \ldots, n - 1 - N \qquad (12.36)$$

This is a weighted estimate of the random process mean square because it involves the weighting function w_k, $k = -N, \ldots, N$. The weighting function used here is taken to be an even nonnegative function about $k = 0$ that satisfies the requirement

$$\sum_{k=-N}^{N} w_k = 1 \qquad (12.37)$$

and permits the analyst to emphasize certain values in the estimate. Specifically, it is normally used to emphasize values near $k = 0$; the reason for this is to diminish the bias in the mean-square estimate. It can be shown that the expected value of the mean-square estimate is

$$E[\hat{\sigma}_X^2(t_j)] = \sum_{k=-N}^{N} w_k \sigma^2(t_{j+k}) \qquad j = N, \ldots, n - 1 - N \qquad (12.38)$$

When σ_{j+k}^2 varies linearly for $-N \leq k \leq N$, the estimate is clearly unbiased. When σ_{j+k}^2 has a negative second derivative for $-N \leq k \leq N$, the estimate must clearly have a negative bias, and when its derivative is positive, it has a positive bias. The above equation makes it clear that the greater w_0 is [within the constraint imposed by Eq. (12.37)], the lower the bias.

It can be shown that the variance of the mean-square estimator is given by

$$V[\hat{\sigma}_X^2(t_j)] = 2 \sum_{k=-N}^{N} \sum_{m=-N}^{N} w_k w_m R_X^2(j + k, j + m)$$

$$j = N, \ldots, n - 1 - N \qquad (12.39)$$

where $R_X(j, k)$ represents the autocorrelation function of the random process $X(t)$ at $t = j \, \Delta t$ and $s = k \, \Delta t$. Because the autocorrelation function $R_X(t, s)$ of a nonstationary random process tends to be (but is not strictly) maximal at $t = s$, it is clear that the variance of the mean-square estimate is diminished by reducing the values of w_k near $k = 0$. This indicates that the requirements of making the estimator for the mean square unbiased and minimizing its variance are in conflict with one another. In order to optimize the estimate of the random process mean square, the analyst should try different weighting functions for estimation of the nonstationary random process mean square and the corresponding modulating function $a(t)$.

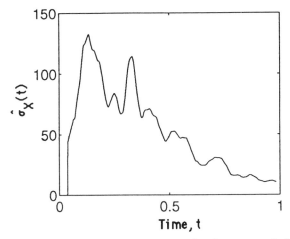

Figure 12.8 Estimated RMS of random process that is source of signal shown in Figure 12.2. Estimate uses window defined in Eq. (12.40) and is based on signal realization shown in Figure 12.2.

Example 12.3. Consider the nonstationary random process realization shown in Figure 12.2. We estimate the random process RMS function using Eq. (12.36). The window function we use is given by

$$w_k = \frac{1}{20}\left(1 - \frac{|k|}{20}\right) \qquad |k| \leq 20 \qquad (12.40)$$

The result is shown in Figure 12.8. It is clearly similar to the result shown in Figure 12.4, but smoother. It should be pointed out that although the process described above can be used to estimate the random process mean square at each time t_j, $j = 0, \ldots, n - 1$, the mean-square estimates so defined are correlated because of the commonality between data used to estimate $\hat{\sigma}_X^2(t_j)$ and $\hat{\sigma}_X^2(t_{j+1})$. In view of this, the estimates for the mean square developed using the present technique are simply interpolated estimates of the mean-square estimates at intervals where the correlation is greatly diminished.

Much more information on the types of estimates discussed here is given in the references by Himelblau and Piersol (1989), Bendat and Piersol (1986), Paez and Baca (1987), and Paez (1985).

12.3.2 Indirect Analysis of Nonstationary Random Processes: Method of Shock Response Spectra

During the past several decades, several methods for assessing the severity of nonstationary random processes and accounting for their effects in tests have been developed. The oldest of these methods is the method of shock response spectra proposed by Biot (1941) and described by Thomson (1972) and Jacobsen and Ayre (1958). This technique is widely used and has been investigated by hundreds of researchers. Alternate methods, such as the method of least favorable response (Drenick, 1970; Shinozuka, 1970), Fourier methods (Baca, 1982, 1984), temporal moment methods (Smallwood, 1989, 1992, 1993), and other methods have sought to do the same thing, usually with substantial advantages over the method of shock response spectra. Nevertheless, the method of shock response spectra has become entrenched in mechanical and structural engineering practice, particularly in aerospace testing laboratories and in earthquake engineering design, and it is unlikely that it will soon be dislodged. In view of this, we discuss the method and some of its probabilistic implications in the following.

The concept of the shock response spectrum was first introduced by Biot in the 1930s. Though not explicitly stated, its purpose was to develop a means for the representation of nonstationary random processes. This representation would allow the user to specify a test on available shock test hardware, and the test would be conservative, in some sense.

We note in passing that the shock response spectrum is often confused with the spectral density. The shock response spectrum does not have the same, or even similar, properties as the spectral density, particularly in connection with the input–output relations of spectral densities and their relation to the frequency response function. (See Chapter 14 for a development of this subject.) The shock response spectrum does approximately evaluate the severity of shock signals, and it does present a method for specification of shock tests.

The shock response spectrum of a single shock is defined as follows. Let $\ddot{z}_j, j = 0, \ldots, n - 1$, represent a shock input excitation, absolute acceleration sampled at a Δt time interval. The acceleration has corresponding velocity and displacement signals $\dot{z}_j, z_j, j = 0, \ldots, n - 1$. To establish the shock spectral ordinate, we write the differential equation governing absolute response of a single-degree-of-freedom system to the base motion defined by $\ddot{z}_j, j = 0, \ldots, n - 1$. It is (see section 7.2.2)

$$\ddot{x} + 2\zeta\omega\dot{x} + \omega^2 x = 2\zeta\omega\dot{z} + \omega^2 z \qquad (12.41)$$

We compute the response, usually assuming zero initial conditions, and for example, the acceleration is $\ddot{x}_j, j = 0, 1, 2, \ldots$. [The method of computation used here must be efficient because this computation will be repeated many times. An optimal technique for response computation is given by Smallwood (1981).] The absolute acceleration shock spectral ordinate at frequency $f = \omega/2\pi$ (corresponding to the single-degree-of-freedom system natural frequency) is

$$B(f) = \max_t |\ddot{x}(t)| \qquad (12.42)$$

This quantity clearly depends on damping. We compute this value at a sequence of frequencies in a set $\mathscr{F}$, and these form the shock response spectrum of the acceleration, $\ddot{z}_j, j = 0, \ldots, n - 1$.

Many types of shock response spectra are defined. These depend on when the maximum response is taken (during the input, after the input ends, over all time), the value measured (maximum positive, maximum negative, maximum absolute value), and the measure of response considered (absolute acceleration, velocity, or displacement; relative acceleration, velocity, or displacement). The one we have developed in Eq. (12.42) is known as the absolute-acceleration, maximax shock response spectrum.

Figures 12.9a and 12.9b show an example of a shock signal (a nonstationary random process realization) and its absolute-acceleration, maximax shock response spectrum.

When a number of shock signal realizations are measured from the ensemble

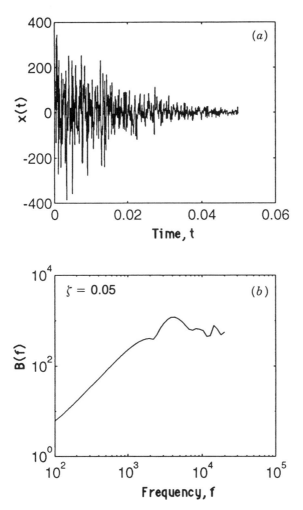

Figure 12.9 (a) Shock signal that is realization of nonstationary random process. (b) Shock response spectrum of signal shown in part (a).

of a single nonstationary random process source, they may be denoted $\ddot{z}_{mj}$, $m = 1, \ldots, M, j = 0, \ldots, n - 1$. The subscript m indexes the number of the measurement from the ensemble; the subscript j indexes time. Each shock signal has a corresponding shock response spectrum, and the collection of shock spectra can be denoted $B_m(f)$, $m = 1, \ldots, M$.

In order to develop a shock test that is representative of the measured shocks, a shock spectrum that is a representative of the shock spectra of the measured shocks is defined. Denote this latter shock spectrum $B_c(f)$. (Means for defining a shock whose shock spectrum represents many others will be discussed later). Let $\ddot{z}_T(t_j)$, $j = 0, \ldots, n - 1$, denote the test shock that will be used to represent the measured shocks. The test shock has a corresponding shock spectrum $B_T(f)$. The principle of shock response spectra holds that if $B_T(f)$ matches $B_c(f)$ "as nearly as possible," then $\ddot{z}_T(t_j)$ is a representation of the measured shocks $\ddot{z}_{mj}$, $m = 1, \ldots, M, j = 0, \ldots, n - 1$. It is important to note that the form of the shock $\ddot{z}_T(t_j)$ does not (in its practical implementation) need to match the form of the underlying shocks $\ddot{z}_{mj}$. Indeed, this is the original reason why the method was invented: to permit the representation of complex shocks with a simple shock that could be generated on a simple (impact-type) shock machine.

In typical shock spectral analyses, the representative shock spectrum $B_c(f)$ is taken as a simple envelope of the shock spectra $B_m(f)$, $m = 1, \ldots, M$. When M is large, this may yield a conservative test $z_T(t_j)$. However, when M is sufficiently large, a more controlled approach to test specification involves the computation of the moments of the $B_m(f)$, $m = 1, \ldots, M$, and the use of these to define $B_c(f)$. For example, the mean and standard deviation estimates of the shock response spectra are

$$\overline{B}(f) = \frac{1}{M} \sum_{m=1}^{M} B_m(f)$$

$$s_B(f) = \left(\frac{1}{M-1} \sum_{m=1}^{M} [B_m(f) - \overline{B}(f)]^2 \right)^{1/2} \tag{12.43}$$

A representative shock spectrum with a controlled level of conservatism is defined as

$$B_c(f) = \overline{B}(f) + K s_B(f) \tag{12.44}$$

where K is a constant greater than zero. Clearly, as K is increased, the level of conservatism in the test $\ddot{z}_T(t_j)$ is increased. More specifically, as K is increased, the probability that a random process realization, drawn from the same source as the $\ddot{z}_{mj}$, will have a shock response spectrum enveloped by the shock spectrum $B_c(f)$ increases.

Example 12.4. Figures 12.10a, b, and c present an example that uses the above approach. The shock response spectra of five shock signals (nonstation-

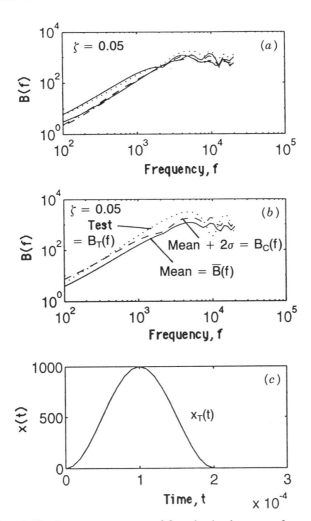

Figure 12.10 (a) Shock response spectra of five shocks that come from same source as that shown in Figure 12.9a. (b) Estimated mean and mean plus two standard deviations of shock response spectra for shocks shown in part (a) and shock response spectrum of test shock based on latter curve. (c) Time history of test shock corresponding to test shock response spectrum shown in part (b).

ary random process realizations) that come from the same source that produced the shock in Figure 12.9a are shown in Figure 12.10a. The mean and standard deviation functions of the shock spectra were computed, and the mean and mean plus two standard deviations are plotted in Figure 12.10b. The shock response spectrum of a haversine pulse, shown in Figure 12.10c, is also plotted in Figure 12.10b. Because the shock response spectrum of the Haversine pulse "matches" the mean plus two standard deviation curve, the pulse in Figure 12.10c is "representative" of the underlying shocks.

We note that, in order to use the approach presented above, all the shocks to be analyzed must come from the same random source. Obviously, the shock test derived according to this method may be unrepresentative of the random process source in some very important aspects. For example, the test need not have the same frequency content as the represented shocks. Further, the effect that the test shock has on a complex structure may be substantially different from the effects that actual shocks from the random source have on the complex structure. However, as mentioned earlier, the method of shock response spectra is so widely used that at least some small consideration of its characteristics is required.

PROBLEMS

12.1 Generalize the results obtained in Secton 12.1.2 by using the method of maximum likelihood to establish formulas for $\bar{x}$, $\bar{y}$, $\hat{\sigma}_X^2$, $\hat{\sigma}_Y^2$, and $\hat{r}$ when the random variables X and Y are jointly normally distributed with non-zero means (the derivation in the text is for the case where $\mu_X = \mu_Y = 0$) and realizations of the random variables x_j, y_j, $j = 1, \ldots, n$, are available.

12.2 Derive Eqs. (12.38) and (12.39).

12.3 Use the data listed in Table P12.3 to estimate the mean, variance, standard deviation, and mean square of the random process [Eqs. (12.1)–(12.3)]. Plot the results when the time increment between samples is $\Delta t = 0.1$ sec. Does the random process appear to be stationary or nonstationary?

12.4 Write a computer program to generate random signals using the results of the Chapter 11 problem set. A realization of a band-limited white-noise random signal is a sequence of independent random variable realizations. Write a computer program that generates and stores sequences of n independent normal random variates, each with mean zero and unit variance. Next, in the program, multiply the white-noise random process realization by a modulating function

$$a(t) = a_0(e^{-\alpha t} - e^{-\beta t}) \qquad t \geq 0$$

where the variates in the white-noise realization are interpreted as occurring at Δt time intervals. The random process source has the theoretical form

$$X(t) = a(t)U(t) \qquad t \geq 0$$

where $U(t)$ is a normally distributed band-limited white-noise random process with mean zero and unit variance. This random process reali-

Table P12.3 Measured Random Process Realizations

Time Index	1	2	3	4	5
			Signal Number		
1	17.289157	−0.66607636	12.580013	−7.5348173	5.8562911
2	13.852585	−17.655961	5.9247888	19.564328	14.130294
3	1.8591155	−18.947003	−22.867199	−6.1432287	21.647035
4	8.6998956	21.322117	−1.7443836	−17.969757	43.360105
5	−16.196992	−1.3074824	3.4391270	−10.349152	−7.4425950
6	35.699772	10.807496	−11.725499	−12.900324	−2.8922344
7	1.0972533	7.3698109	−6.2555508	−3.8856231	11.440193
8	28.956993	12.185283	6.6907392	9.0581306	15.757229
9	3.6434899	5.5257125	21.493958	−14.679494	−15.389029
10	10.200290	−15.696790	−28.603099	4.1142814	−6.4363314
11	−14.248523	3.6951224	−10.820060	11.162975	0.39296878
12	−5.7837872	9.2812580	9.2605208	1.2372756	−20.480523
13	8.5670010	5.0099220	3.9993704	4.8346085	7.9665638
14	−3.6593381	−13.578164	−1.5497641	−0.29933409	−5.8613077
15	2.7313467	−1.2952185	1.9594693	9.5491653	5.4569527
16	−1.4077391	−1.2627160	−3.8233169	3.6135230	−3.0750962
17	−0.43736228	1.0258130	−3.2951983	−5.8522424	2.0478604
18	−3.5861553	1.3589494	0.84432368	9.2949071×10^{-2}	2.1804142
19	−2.7785146	$−4.4644877 \times 10^{-3}$	3.3157763	−3.9538523	−0.38497562
20	1.7702747	2.888535	1.3474928	1.8487083	1.0113379

zation has signal content up to a frequency of $1/2 \; \Delta t$. (See Chapter 13 for the reason why.) For $n = 200$, $\Delta t = 0.1$ sec, $\alpha = 0.2$, $\beta = 0.4$, and $a_0 = 100$, generate a random process realization and plot it. Compute the theoretical mean, variance, and standard deviation functions of $X(t)$.

12.5 Generate 20 nonstationary random process realizations using the parameters in problem 12.4. Write a computer program to execute the computations in Eqs. (12.1)–(12.3) and use it to estimate the moments of the random process based on the 20 realizations. Plot the results.

12.6 Write a computer program to execute the computations in Eq. (12.18). Use it with the 20 realizations generated in problem 12.5 to estimate the autocorrelation function of the source random process at lag values of τ equal to 0, 0.1, 0.2 sec. Plot the results.

12.7 Generate one realization of a stationary, white-noise random signal as described in problem 12.4. Let $n = 16{,}000$ and assume $\Delta t = 0.001$ sec. Write a computer program to execute the computations of Eq. (12.28). Use it to estimate the autocorrelation function of the generated signal for $\tau = 0, 0.001, \ldots, 0.010$. Plot the results and explain their meaning.

13

DISCRETE FOURIER TRANSFORM

We showed, starting in Chapter 5, that some of the clearest measures of stationary random processes are defined in the frequency domain. Establishing these definitions involves use of the Fourier transform. The discrete Fourier transform (DFT) can be used to approximate the continuous Fourier transform, and the DFT is developed in this chapter. The following chapter will show that the practical estimation of frequency-domain measures of stationary random processes can be obtained using the DFT. Chapter 4 presents the theory of continuous Fourier transforms.

13.1 DEFINITION OF DFT AND FUNDAMENTAL CHARACTERISTICS

The DFT provides a finite series representation for finite-duration, sampled (discrete-time) signals, sometimes called time series. Let the time-domain signal of interest be denoted $x_j = x(j \, \Delta t)$, $j = 0, \ldots, n - 1$, where Δt is a nonnegative time increment and n is an even number. For simplicity, we assume that Δt is measured in seconds, so that the frequencies discussed later will be in units of hertz. In general, the signal can be complex valued. The need for the DFT arises when we seek a harmonic series representation for x_j, $j = 0, \ldots, n - 1$. Such a representation is given by

$$x_j = \frac{1}{n} \sum_{k=0}^{n-1} \xi_k e^{+i2\pi jk/n} \qquad j = 0, \ldots, n - 1 \qquad (13.1)$$

where the ξ_k, $k = 0, \ldots, n - 1$, are complex constants, i is the imaginary unit, and k is the arbitrary index of summation. The function $e^{i\alpha}$ is known as the complex exponential, or Euler's function, and it is related to the sine and cosine functions through

$$e^{i\alpha} = \cos(\alpha) + i \sin(\alpha) \qquad e^{-i\alpha} = \cos(\alpha) - i \sin(\alpha) \qquad (13.2)$$

There are n components in the representation (13.1). The DFT representation expresses the signal x_j, $j = 0, \ldots, n - 1$, in terms of a constant component ($k = 0$), a harmonic component whose period is $n \, \Delta t$ ($k = 1$), a component whose period is $n \, \Delta t/2$ ($k = 2$), and so on.

To represent a signal using Eq. (13.1), we must choose the coefficients of the harmonic components correctly, and our intuition indicates that this should be possible because we are using n components in a series, and their corresponding coefficients, ξ_k, $k = 0, \ldots, n - 1$, to represent n points, x_j, $j = 0, \ldots, n - 1$. The problem here is to find the appropriate coefficients, ξ_k, $k = 0, \ldots, n - 1$, to use in the series that represents x_j, $j = 0, \ldots, n - 1$. It is relatively easy to solve because the harmonic functions are orthogonal.

To establish a formula for the ξ_k, $k = 0, \ldots, n - 1$, multiply both sides of Eq. (13.1) by $\exp(-i2\pi jm/n)$ and sum the results on both sides over the index, $j = 0, \ldots, n - 1$. The result of these operations is expressed as

$$\sum_{j=0}^{n-1} x_j \, e^{-i2\pi jm/n} = \sum_{j=0}^{n-1} e^{-i2\pi jm/n} \frac{1}{n} \sum_{k=0}^{n-1} \xi_k \, e^{+i2\pi jk/n} \qquad (13.3)$$

The right-hand side can be simplified by doing three things. First, take the constant $1/n$ outside the sum; this is allowable since it is not a function of either summation index j or k. Second, combine the two exponentials into one expression. Third, exchange the order of summations; this is allowable since neither sum has a limit that is a function of the other index. The result is

$$\sum_{j=0}^{n-1} x_j \, e^{-i2\pi jm/n} = \frac{1}{n} \sum_{k=0}^{n-1} \xi_k \sum_{j=0}^{n-1} e^{+i2\pi j(k - m)/n} \qquad (13.4)$$

Now consider the second sum on the right:

$$S(k - m) = \sum_{j=0}^{n-1} e^{i2\pi j(k - m)/n} \qquad k - m = 0, \ldots, n - 1 \qquad (13.5)$$

For now, we consider only index differences $k - m$, equal to $0, 1, \ldots, n - 1$. The sum is a function of $k - m$ and not j, because dependence on j is eliminated when we sum over j. When $k = m$, it is clear that $\exp[i2\pi j(k - m)/n] = 1$, and the value of the sum is n, because the value "one" is added n times. When $k - m$ is not zero, the sum $S(k - m)$ equals zero. Here is the reason why. When $k - m = r \neq 0$, the sum $S(k - m) = S(r)$ adds together the values of

$$e^{i2\pi jr/n} = \cos\left(\frac{2\pi jr}{n}\right) + i \sin\left(\frac{2\pi jr}{n}\right) \qquad j = 0, \ldots, n - 1 \quad (13.6)$$

where $r \neq 0$. Each of the harmonic functions on the right side executes r complete cycles as the index j varies from zero through $n - 1$. The sum of ordinate values in each such function is zero because there are as many positive ordinates in a sine or cosine function as negative ordinates, and the positive and negative ordinates match one to one as long as n is even (which we assumed). In view of these facts, we can write

$$S(k - m) = \begin{cases} n & k = m \\ 0 & k \neq m \end{cases} \qquad k - m = 0, \ldots, n - 1 \quad (13.7)$$

This behavior can also be expressed using the Kronecker delta δ_r. The Kronecker delta function is a discrete function defined as follows:

$$\delta_r = \begin{cases} 1 & r = 0 \\ 0 & r \neq 0 \end{cases} \qquad (13.8)$$

Therefore, we can write

$$S(k - m) = n\delta_{k-m} \qquad k - m = 0, \ldots, n - 1 \quad (13.9)$$

We are now interested in returning to Eq. (13.4) to use this result. Substitution of Eq. (13.9) into Eq. (13.4) yields

$$\sum_{j=0}^{n-1} x_j\, e^{-i2\pi mj/n} = \frac{1}{n} \sum_{k=0}^{n-1} \xi_k n\delta_{k-m} \qquad (13.10)$$

The product in the sum on the right, $\xi_k \delta_{k-m}$, is zero for all values of the index k except $k = m$. When $k = m$, the product equals ξ_m; therefore, Eq. (13.10) reduces to

$$\xi_m = \sum_{j=0}^{n-1} x_j\, e^{-i2\pi mj/n} \qquad m = 0, \ldots, n - 1 \quad (13.11)$$

This formula is called the forward discrete Fourier transform, or simply, the DFT. Equation (13.1) is called the inverse DFT. Together, the formulas define the DFT pair x_j, $j = 0, \ldots, n - 1$, and ξ_k, $k = 0, \ldots, n - 1$. There is a symmetry in the definitions of Eqs. (13.1) and (13.11). The right-hand side of Eq. (13.1), the inverse DFT, has a $1/n$ factor and a plus sign in the exponential. The right-hand side of Eq. (13.11), the DFT, simply has a negative sign in the exponential.

Because Eq. (13.1) uniquely establishes x_j, $j = 0, \ldots, n - 1$, in terms of the sequence ξ_k, $k = 0, \ldots, n - 1$, and Eq. (13.11) uniquely establishes ξ_k, $k = 0, \ldots, n - 1$, in terms of x_j, $j = 0, \ldots, n - 1$, knowing one of the sequences is equivalent to knowing the other. The sequence x_j, $j = 0, \ldots, n - 1$, is a function defined in time—a time history. The sequence ξ_k, $k = 0, \ldots, n - 1$, is a function of frequency.

It will prove useful in the following chapter to have a relation between the sums of squares of the signal x_j, $j = 0, \ldots, n - 1$, and its DFT ξ_k, $k = 0, \ldots, n - 1$. We can obtain this as a byproduct of the development that led to Eq. (13.9). Because, in general, the x_j and their corresponding DFT coefficients ξ_k can be complex valued, we develop the relation for a complex-valued time series. To start, multiply each side of Eq. (13.1) by its own complex conjugate. The result is

$$|x_j|^2 = \frac{1}{n^2} \sum_{k=0}^{n-1} \sum_{m=0}^{n-1} \xi_k \xi_m^* \, e^{i2\pi j(k-m)/n} \qquad j = 0, \ldots, n - 1 \qquad (13.12)$$

where the star superscript indicates the operation of complex conjugation. Now sum each side over the index values $j = 0, \ldots, n - 1$, and change the order of summation on the right. This yields

$$\sum_{j=0}^{n-1} |x_j|^2 = \frac{1}{n^2} \sum_{k=0}^{n-1} \sum_{m=0}^{n-1} \xi_k \xi_m^* \sum_{j=0}^{n-1} e^{i2\pi j(k-m)/n} \qquad (13.13)$$

Recognize the j-indexed sum, on the right, as the series denoted $S(k - m)$ in Eqs. (13.5) and (13.9). Use Eq. (13.9) in Eq. (13.13) to obtain

$$\sum_{j=0}^{n-1} |x_j|^2 = \frac{1}{n} \sum_{k=0}^{n-1} |\xi_k|^2 \qquad (13.14)$$

This is known as Parseval's relation for DFTs.

13.2 PERIODICITY OF DFT; DFT OF REAL SIGNAL

Though no prior assumption was made about the behavior of x_j, $j = 0, \ldots,$ $n - 1$, for index values of j less than zero or j greater than $n - 1$, the representation established in Eq. (13.1) does imply values for x_j when j is outside the interval $[0, n - 1]$. Specifically, when we evaluate x_j at an index value equal to $j + mn$, where m is an integer, we obtain

$$x_{j+mn} = \frac{1}{n} \sum_{k=0}^{n-1} \xi_k \, e^{i2\pi(j+mn)k/n}$$

$$= \frac{1}{n} \sum_{k=0}^{n-1} \xi_k \, e^{i2\pi jk/n} e^{i2\pi km}$$

$$= x_j \qquad j = 0, \ldots, n-1 \tag{13.15}$$

In the middle step, use was made of the fact that $\exp(i2\pi km) = 1$, because k and m are integers, and therefore, km is an integer. The DFT representation, therefore, presents x_j as a periodic function with period n. In a similar manner, it is easy to show that the DFT of x_j, $j = 0, \ldots, n-1$, is periodic. That is,

$$\xi_{k+mn} = \xi_k \qquad k = 0, \ldots, n-1 \tag{13.16}$$

where m is an integer. The periodicity of x_j, $j = 0, \ldots, n-1$, and its DFT have important implications in certain numerical applications. For example, when evaluating structural system response using the DFT of the convolution integral, we must use specific precautions to avoid a phenomenon known as circular convolution and ensure accuracy in the computation. For a description of this phenomenon and the reasons why it occurs, see Stearns (1975).

We are usually interested in evaluating the DFT of signals that are real valued in the time domain. For example, measured excitation and response signals are real valued. It is important to know the characteristics of the DFT of a real signal for several reasons. Among these are the following:

1. When we evaluate the DFTs of real-valued functions in order to perform computations in the frequency domain and then inverse DFT the result to observe it in the time domain, we must be sure that the frequency-domain result is appropriate to yield a real-valued function of time.

2. When we perform computations in the frequency domain on functions that are DFTs of real-valued signals, the fact that we started in the time domain with a real-valued signal usually diminishes our frequency-domain calculations by a factor of 2.

Consider the real-valued function x_j, $j = 0, \ldots, n-1$, and its DFT, Eq. (13.11). When the exponential in Eq. (13.11) is written in terms of its real and imaginary parts, Eq. (13.11) can be written as two sums,

$$\xi_k = \sum_{j=0}^{n-1} x_j \cos\left(\frac{2\pi jk}{n}\right) - i \sum_{j=0}^{n-1} x_j \sin\left(\frac{2\pi jk}{n}\right) \qquad k = 0, \ldots, n-1$$

$$\tag{13.17}$$

Based on this expression, it is clear that the DFT of x_j, $j = 0, \ldots, n - 1$, has real and imaginary parts that are

$$\text{Re}(\xi_k) = \sum_{j=0}^{n-1} x_j \cos\left(\frac{2\pi jk}{n}\right) \qquad k = 0, \ldots, n - 1 \qquad (13.18\text{a})$$

$$\text{Im}(\xi_k) = -\sum_{j=0}^{n-1} x_j \sin\left(\frac{2\pi jk}{n}\right) \qquad k = 0, \ldots, n - 1 \qquad (13.18\text{b})$$

If we evaluate the real and imaginary parts at the index value $-k$ (we can do this, as seen in the previous discussion on periodicity) and compare the results to the real and imaginary parts of the DFT at the index value k, then it is clear that

$$\text{Re}(\xi_{-k}) = \text{Re}(\xi_k) \qquad k = 0, \ldots, n - 1 \qquad (13.19\text{a})$$

$$\text{Im}(\xi_{-k}) = -\text{Im}(\xi_k) \qquad k = 0, \ldots, n - 1 \qquad (13.19\text{b})$$

because the cosine is an even function of its argument and the sine is an odd function of its argument. Further, the DFT is a periodic function with period n. Therefore, $\xi_{-k} = \xi_{n-k}$. Because of this,

$$\text{Re}(\xi_{n-k}) = \text{Re}(\xi_k) \qquad k = 0, \ldots, n - 1 \qquad (13.20\text{a})$$

$$\text{Im}(\xi_{n-k}) = -\text{Im}(\xi_k) \qquad k = 0, \ldots, n - 1 \qquad (13.20\text{b})$$

The first expression states that the real part of the DFT of a real signal is an even function about the index $k = \frac{1}{2}n$. The second expresion states that the imaginary part of the DFT of a real signal is an odd function about the index $k = \frac{1}{2}n$.

These facts imply that half the information contained in the DFT of a real signal is redundant. This is a logical condition if we take into account the following facts. Based on Eq. (13.18b),

$$\text{Im}(\xi_0) = \text{Im}(\xi_{n/2}) = 0 \qquad (13.21)$$

because $\sin(m\pi) = 0 = \sin(0)$ when m is an integer. The $\frac{1}{2}n + 1$ values $\text{Re}(\xi_k)$, $k = 0, \ldots, \frac{1}{2}n$, and the $\frac{1}{2}n - 1$ values $\text{Im}(\xi_k)$, $k = 1, \ldots, \frac{1}{2}n - 1$, a total of n numbers, completely and uniquely specify in the frequency domain the n real values x_j, $j = 0, \ldots, n - 1$, from the time domain.

Another way of expressing the above facts is to write

$$\xi_{n-k} = \xi_k^* \qquad k = 0, \ldots, \frac{1}{2}n \qquad (13.22)$$

This relation is said to demonstrate a conjugate symmetry.

Example 13.1. For example, consider the DFT of the real-valued signal x_j, $j = 0, \ldots , 63$, shown in Figure 13.1a. The time increment Δt is 0.1 sec. (Straight lines have been drawn between the known, discrete-time values.) The signal has a duration of 6.4 sec. The DFT of the signal has real and imaginary parts, and these can alternately be expressed in terms of modulus and phase. The real and imaginary parts of the DFT are shown in Figures 13.1b and c. The DFT is evaluated at the frequencies $f_k = k/(n\, \Delta t)$, $k = 0, \ldots , n - 1$. (Straight lines are drawn between the evaluated DFT values.) To see why the DFT is evaluated at these frequencies, note that the kernel in Eq. (13.1) can

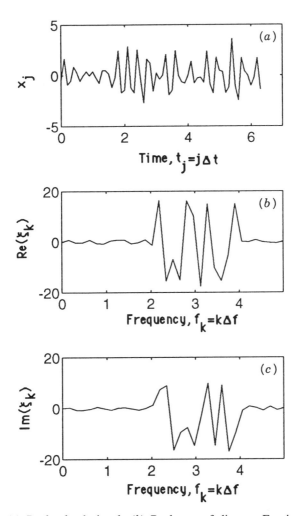

Figure 13.1 (a) Real-valued signal. (b) Real part of discrete Fourier transform of signal shown in part (a). (c) Imaginary part of discrete Fourier transform of signal shown in part (a).

be witten $\exp(i2\pi jk/n) = \exp\{i2\pi(j \ \Delta t)[k/(n \ \Delta t)]\}$; the $j \ \Delta t$ corresponds to a time t_j, and $k/(n \ \Delta t)$ corresponds to a frequency f_k. Because of the symmetry relations described in Eqs. (13.20), only the parts of the DFT with indices $k = 0, \ldots, \frac{1}{2}n$ are shown in these graphs and most of those that follow.

Finally, we note some important implications of the results derived in this section involving the DFTs of real signals. It was shown above that (1) the DFT ξ_k, $k = 0, \ldots, n - 1$, of a signal x_j, $j = 0, \ldots, n - 1$, is periodic with period n and (2) when x_j, $j = 0, \ldots, n - 1$, is a real-valued signal, the real part of ξ_k, $k = 0, \ldots, n - 1$, is an even function about $k = \frac{1}{2}n$, and the imaginary part of ξ_k, $k = 0, \ldots, n - 1$, is an odd function about $k = \frac{1}{2}n$. These facts imply that (1) the values of ξ_k, $k = -\frac{1}{2}n + 1, \ldots, -1$, equal the values of ξ_k, $k = \frac{1}{2}n + 1, \ldots, n - 1$, and (2) the real part of ξ_k, $k = -\frac{1}{2}n + 1, \ldots, \frac{1}{2}n - 1$, is an even function about the index $k = 0$ and the imaginary part of ξ_k, $k = -\frac{1}{2}n + 1, \ldots, \frac{1}{2}n - 1$, is an odd function about the index $k = 0$. In writing formulas for the DFT, the index values $k = 0, \ldots, n - 1$, are almost always used. The reason is that the index values can be used directly (or translated simply) as index values of a subscripted variable in a computer program. However, to understand the DFT, the index values $k = -\frac{1}{2}n + 1, \ldots, \frac{1}{2}n$ are frequently more useful.

One situation where it is more useful to consider the DFT indices $k = -\frac{1}{2}n + 1, \ldots, \frac{1}{2}n$ occurs when we wish to observe the frequency content in x_j, $j = 0, \ldots, n - 1$, that can be represented with the DFT. When we write Eq. (13.1) with the indices $k = -\frac{1}{2}n + 1, \ldots, \frac{1}{2}n$, it becomes

$$x_j = \frac{1}{n} \sum_{k=-n/2+1}^{n/2} \xi_k \ e^{+i2\pi jk/n} \qquad j = -\tfrac{1}{2}n + 1, \ldots, \tfrac{1}{2}n \qquad (13.23a)$$

and correspondingly

$$\xi_k = \sum_{j=-n/2+1}^{n/2} x_j \ e^{-i2\pi jk/n} \qquad k = -\tfrac{1}{2}n + 1, \ldots, \tfrac{1}{2}n \qquad (13.23b)$$

Equation (13.23a) makes it clear that the signal x_j, $j = -\frac{1}{2}n + 1, \ldots, \frac{1}{2}n$, is being represented with components whose highest frequency is $1/(2 \ \Delta t)$ hertz, where the signal x_j, $j = -\frac{1}{2}n + 1, \ldots, \frac{1}{2}n$, is taken over a time period of $n \ \Delta t$ seconds. If the signal x_j, $j = -\frac{1}{2}n + 1, \ldots, \frac{1}{2}n$, has content in the frequency range beyond $1/(2 \ \Delta t)$ hertz, then that portion of the signal content cannot be accurately represented. The frequency $f_{\max} = 1/(2 \ \Delta t)$ is called the Nyquist frequency.

The previous statements are clarified when we consider the graphs in Figures 13.2a and 13.2b. A segment of a harmonic function with constant amplitude and frequency $1/(2 \ \Delta t)$ hertz is shown. Figure 13.2a shows the representation we obtain (straight lines) when we sample the signal at a time interval Δt. It is clear that the harmonic function is accurately represented by the samples taken Δt seconds apart. However, two other facts are also apparent. First, if

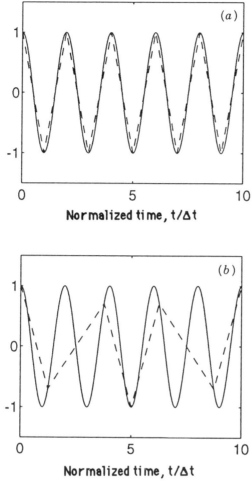

Figure 13.2 (*a*) Cosine wave with period 2 Δt and sampled version of signal with straight lines drawn between sample points. (*b*) Cosine wave with period 2 Δt and version of signal sampled at times $\Delta t' = (\frac{5}{4}) \Delta t$ with straight lines drawn between sample points.

the sampling times were slightly shifted, then the representation of the harmonic signal at the sample points would be less accurate, and if the sample times were shifted by $\frac{1}{2} \Delta t$ seconds, then the representation of the harmonic signal would be completely inaccurate. (The signal would be sampled where it equals zero.) Therefore, if the signal x_j, $j = 0, \ldots, n - 1$, contains components with frequencies as high as $1/(2 \Delta t)$ hertz, we must be very fortunate to obtain an accurate representation of the high-frequency components. Second, if the signal contains components whose frequencies are higher than $1/(2 \Delta t)$ hertz, then it is impossible for the DFT to accurately represent the signal at these frequencies because the sampling rate is too slow. Figure 13.2*b* demonstates this fact for a specific signal. The signal is the same as that shown in Figure 13.2a, but this signal is sampled at a time interval equal to $(\frac{5}{4})$ Δt. The signal sampled as shown is clearly not an accurate reflection of the underlying, continuous signal.

The problem described here—sampling a signal with frequency content beyond $1/(2 \Delta t)$ using a sampling time increment Δt—is called aliasing and will be discussed further in the following section along with the precautions that must be taken to avoid it. Another problem, known as the leakage problem, will also be discussed.

13.3 DFT OF CONTINUOUS-TIME SIGNALS: ALIASING AND LEAKAGE

Most of the signals we encounter in practical random vibration and random signal analysis applications come from continuous sources. These signals are measured, perhaps filtered during measurement, perhaps modulated, then usually sampled temporally, to generate a digitized signal ready for analysis. When the analysis to be performed on the signal is the DFT and the DFT is meant to serve as an approximation to the Fourier transform of the underlying signal, special care must be taken to ensure accuracy in the approximation. Two important things are demonstrated in this section. First, when a measured signal is a relatively short segment drawn from a long-duration, continuous-time signal, the Fourier transform approximation generated using the measured signal can only be an approximation to the Fourier transform of the underlying source, and the accuracy of the approximation is influenced by the manner in which the measured signal is drawn from the underlying source. Second, the rate at which a continuous-time signal is digitized is related to the capacity to analyze high frequencies in the original source, and precautions must always be taken in practical situations to avoid obtaining an inaccurate representation of the original source Fourier transform.

For purposes of the following development, denote the continuous-time source signal $s(t)$, $-\infty < t < \infty$, and let its Fourier transform be denoted $\psi(f)$, $-\infty < f < \infty$. Chapter 4 shows that $s(t)$ and $\psi(f)$ are related by

$$s(t) = \int_{-\infty}^{\infty} \psi(f)e^{i2\pi ft}\, df \quad -\infty < t < \infty$$

$$\psi(f) = \int_{-\infty}^{\infty} s(t)e^{-i2\pi ft}\, dt \quad -\infty < f < \infty$$

(13.24)

(Here, cyclic frequency $f = \omega/2\pi$ is used in place of circular frequency ω.) When we measure a signal, we record a portion of $s(t)$, $-\infty < t < \infty$, and a convenient approach to describing this operation uses the terminology of windowing. Let $x(t)$, $-\infty, < t < \infty$, denote the measured signal; then

$$x(t) = w(t)s(t) \quad -\infty < t < \infty$$

(13.25)

where $w(t)$, $-\infty < t < \infty$, is known as the temporal weighting function and the operation described here is known as windowing the signal $s(t)$. A family of useful temporal weighting functions will be described in the following section, but for now, consider that a very obvious form for $w(t)$ is

$$w(t) = \begin{cases} 1 & |t| \leq \frac{1}{2}T \\ 0 & |t| > \frac{1}{2}T \end{cases} \tag{13.26}$$

The multiplication of $s(t)$ by $w(t)$ will obviously affect the relation between the Fourier transforms of $s(t)$ and $x(t)$. To explore this relation, recall (as shown in Chapter 4) that the Fourier transform of the product of two functions can be expressed as the convolution integral involving the Fourier transforms of the two functions. In view of this, the Fourier transform of $x(t)$ is

$$\xi(f) = \int_{-\infty}^{\infty} \psi(\alpha)W(f - \alpha)\, d\alpha \quad -\infty < f < \infty \tag{13.27}$$

where $W(f)$ is the Fourier transform of $w(t)$. The Fourier transform of a temporal weighting is known as its associated spectral window. In this particular case, the Fourier transform of $w(t)$ is given by (see Example 4.2)

$$W(f) = \frac{\sin(2\pi fT)}{2\pi f} \quad -\infty < f < \infty \tag{13.28}$$

A normalized form of this function is plotted in Figure 13.3 as a function of fT. The ordinate is plotted in terms of decibels, referenced to the maximum value of $W(f)$, which always occurs at $f = 0$. The number of decibles (denoted db in the figure) that the spectral window $W(f)$ is above (positive decibles) or below (negative decibels) the reference is given by $10 \log[W(f)/W(0)]^2$, where

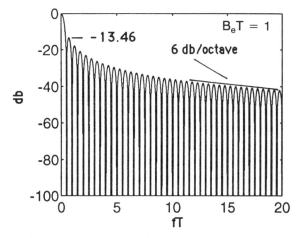

Figure 13.3 Spectral window for rectangular spectral weighting.

the logarithm is taken as base 10. Spectral windows are traditionally plotted in this way. Further, slopes on the spectral window curve are traditionally described in terms of decibels per octave. The octave of a particular frequency f is double that frequency, $2f$. Decay rates for spectral windows are traditionally specified in terms of their decibel-per-octave decay at large values of $|f|$. Note, as shown in the figure, that the highest side lobe is 13.46 dB down from the peak and the decay rate is 6 dB/octave.

Now consider what Eq. (13.27) and Figure 13.3 are saying. When we seek to establish an approximation to the Fourier transform of $s(t)$ by measuring a segment of $s(t)$, then Fourier transforming the measured segment, what we actually obtain is the convolution between $\psi(f)$ [the Fourier transform of $s(t)$] and a spectral window function $W(f)$. This convolution weighs $\psi(f)$ with a function $W(f)$ to obtain $\xi(f)$. Clearly, the form of $W(f)$ influences how well $\xi(f)$ approximates $\psi(f)$. The effect shown in Figure 13.3 that allows values of $\psi(f)$ away from $f = f_0$ to influence the value of $\xi(f_0)$ is known as leakage.

If $W(f)$ were a delta function, then Eq. (13.27) would yield $\xi(f) = \psi(f)$, but of course, this is not possible in a practical implementation. Therefore, we seek in $w(t)$, the temporal weighting, and $W(f)$, its corresponding spectral window, a good compromise. This compromise will be discussed at greater length in the following section.

Note that the reason this whole issue is important is that we are frequently interested in characterizing stationary random processes using measured realizations. We obviously cannot measure an entire stationary random process (they have infinite duration), so we must settle for finite-duration measurements. Indeed, because of economic limitations, we are frequently constrained to characterize random processes with very short measured signals.

After windowing a measured signal, we usually sample it to generate a discrete-time representation of the original signal. This discretization also has an effect on the representation, and we now explore that effect. Recall first, from Eq. (13.26), that the signal to be analyzed is nonzero in $[-\frac{1}{2}T, \frac{1}{2}T]$ and zero elsewhere. Therefore, we assume that the signal can be exactly represented with a Fourier series (see Chapter 4). The signal to be analyzed is $x(t)$ from Eq. (13.25) and has the representation

$$x(t) = \frac{1}{T} \sum_{k=-\infty}^{\infty} \xi\left(\frac{k}{T}\right) e^{i2\pi kt/T} \qquad |t| \le \frac{1}{2}T \qquad (13.29)$$

where $\xi(f)$ is the Fourier transform of the signal $x(t)$ and $\xi(k/T)$ is a Fourier series coefficient equal to $\xi(f)$ at $f = k/T$. The sampling process evaluates $x(t)$ at times $t_j = j \, \Delta t, j = -\frac{1}{2}n + 1, \ldots, \frac{1}{2}n$. Based on the above representation, $x_j = x(j \, \Delta t)$ can be expressed as

$$x_j = \frac{1}{T} \sum_{k=-\infty}^{\infty} \xi\left(\frac{k}{T}\right) e^{i2\pi kj\Delta t/T} = \frac{1}{T} \sum_{k=-\infty}^{\infty} \xi\left(\frac{k}{T}\right) e^{i2\pi kj/n}$$

$$k = -\frac{1}{2}n + 1, \ldots, \frac{1}{2}n \qquad (13.30)$$

because $T = n \, \Delta t$. We use this expression in Eq. (13.23b) to establish the DFT of $x_j, \; j = -\frac{1}{2}n + 1, \; \ldots \; , \frac{1}{2}n$:

$$
\begin{aligned}
\xi_m &= \sum_{j=-n/2+1}^{n/2} \left[\frac{1}{T} \sum_{k=-\infty}^{\infty} \xi\!\left(\frac{k}{T}\right) e^{i2\pi jk/n} \right] e^{-i2\pi mj/n} \\
&= \frac{1}{T} \sum_{k=-\infty}^{\infty} \xi\!\left(\frac{k}{T}\right) \left[\sum_{j=-n/2+1}^{n/2} e^{i2\pi j(k-m)/n} \right] \qquad m = -\tfrac{1}{2}n + 1, \; \ldots \; , \tfrac{1}{2}n
\end{aligned}
$$

(13.31)

A term similar to the one in brackets in the final expression was analyzed in Eqs. (13.5)–(13.9), and it was shown to have a Kronecker delta function behavior for $k - m = 0, \; \ldots \; , n - 1$. In fact, a similar analysis can be used to show that the present series has periodic behavior, with period n, given by

$$
\sum_{j=-n/2+1}^{n/2} e^{i2\pi j(k-m)/n} = n \sum_{p=-\infty}^{\infty} \delta_{k-m-pn}
$$

$$
k - m = \; \ldots \; , -1, 0, 1, \; \ldots \quad (13.32)
$$

When we use this in Eq. (13.31) we obtain

$$
\xi_m = \frac{n}{T} \sum_{k=-\infty}^{\infty} \xi\!\left(\frac{k}{T}\right) \sum_{p=-\infty}^{\infty} \delta_{k-m-pn} = \frac{1}{\Delta t} \sum_{p=-\infty}^{\infty} \xi\!\left(\frac{m+pn}{T}\right)
$$

$$
m = -\tfrac{1}{2}n + 1, \; \ldots \; , \tfrac{1}{2}n \quad (13.33)
$$

Now use Eq. (13.27) in Eq. (13.33) to obtain

$$
\begin{aligned}
\xi_m &= \frac{1}{\Delta t} \sum_{p=-\infty}^{\infty} \int_{-\infty}^{\infty} \psi(\alpha) W\!\left(\frac{m+pn}{T} - \alpha\right) d\alpha \\
&= \frac{1}{\Delta t} \int_{-\infty}^{\infty} \psi(\alpha) \left[\sum_{p=-\infty}^{\infty} W\!\left(\frac{m+pn}{T} - \alpha\right) \right] d\alpha
\end{aligned}
$$

$$
m = -\tfrac{1}{2}n + 1, \; \ldots \; , \tfrac{1}{2}n \quad (13.34)
$$

That is, when we sample the signal $x(t)$, $|t| \le \frac{1}{2}T$, to obtain $x_j, \; j = -\frac{1}{2}n + 1$, $\ldots$, $\frac{1}{2}n$ and then compute the DFT of x_j, the result is a function of the Fourier transform of $s(t)$, $\psi(f)$, which we wish to approximate, given by Eq. (13.34). This includes an effect caused by the windowing, described previously, and another effect caused by the sampling. The series in brackets in the final expression on the right is a periodic series of spectral windows, each centered at the frequency $\alpha = (m + pn)/T, \; p = \; \ldots \; , -1, 0, 1, \; \ldots$. The expression implies that ξ_m, the DFT at $f = m/T$, reflects the weighted Fourier transform $\psi(f)$ at all these frequencies. In other words, ξ_m not only reflects signal frequency content $\psi(f)$ near $f = m/T$, the quantity we wish to approximate, but it also reflects signal frequency content $\psi(f)$ near $f = (m \pm n)/T, f(m \pm 2n)/T, \ldots$, if there is any signal content at these other frequencies. If $x(t)$ does have signal content outside the frequency interval $(-1/(2\ \Delta t), 1/(2\ \Delta t))$, then the effect is known as aliasing, and it occurs for the practical reason discussed at the end of the previous section. Its effects cannot be completely eliminated in practical signal analysis, but they can be greatly mitigated.

Because of the periodicity of the quantity in brackets in the final expression of Eq. (13.34), we can only hope to uniquely approximate the Fourier transform of the original signal $s(t)$ in the frequency range $(-1/(2\ \Delta t), 1/(2\ \Delta t))$. (This identifies with the frequencies $f = m/t$ for $-\frac{1}{2}n \le m \le \frac{1}{2}n$.) In view of this, in practical applications, it is usually considered prudent to low-pass filter the original signal $s(t)$ at or below the frequency $1/(2\ \Delta t)$, with an analog filter, prior to windowing and sampling the signal. In doing this, we seek to zero $\psi(f)$, the Fourier transform of $s(t)$, for $|f| > 1/(2\ \Delta t)$, thereby eliminating the effects of all the terms in the series in brackets in Eq. (13.34), except for the $p = 0$ term.

In fact, the suggestion in the previous paragraph does not completely eliminate the possibility of aliasing. A real low-pass filter cannot eliminate all the signal information beyond the cutoff frequency $f = 1/(2\ \Delta t)$ while leaving intact the signal information below $f = 1/(2\ \Delta t)$, and a compromise must be made. Specifically, real analog and digital filters decrease (roll off) at a finite rate. Regardless of the frequency at which we wish to low-pass filter a signal, there will always be some signal content beyond that frequency, and the signal content will always be affected below the cutoff frequency. Frequently, in practical applications, the signal $s(t)$ is low-pass filtered with an analog filter at some frequency between $1/(4\ \Delta t)$ and $1/(2.5\ \Delta t)$ when it is to be sampled at a time interval of Δt seconds. The filters used to perform the low-pass filtering operations described here are known as antialiasing filters because their purpose is to prevent (mitigate the effects of) aliasing. A lower cutoff frequency is used when the antialiasing filters have a slow rolloff; a higher cutoff frequency is used when the antialiasing filters have a fast rolloff. The antialiasing filters must be analog because once a signal is sampled, it can never be known if it was aliased.

We conclude this section with a word of warning to the analyst. It is practically always a good idea to filter measured data with an antialiasing filter, even when one is "certain" that there is no signal content beyond the frequency $1/(2\ \Delta t)$. The reason is that mechanical systems can generate signal components at tens of thousands and even hundreds of thousands of hertz as a result of component impact, or rattling. Any system that has joints, is joined to another, or has the potential for internal component impacts can rattle and, with the appropriate excitation, will rattle.

Example 13.2. This example shows how aliasing appears in a time signal and how it is manifested in a DFT. Figure 13.4a shows a time history that is simply the sum of three harmonic components with frequencies 250, 350, and 600 Hz

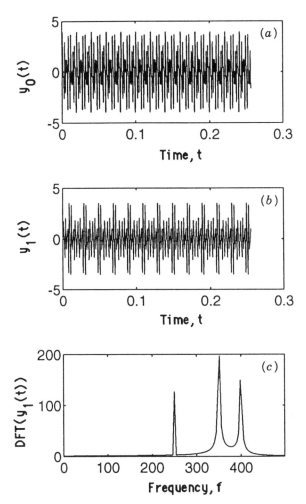

Figure 13.4 (a) Sum of three sinusoidal components. (b) Sampled version of signal shown in part (a) sampled at 0.001-sec interval. (c) Discrete Fourier transform of signal in part (b).

and amplitudes 1.0, 2.0, and 1.5, respectively. The DFT of the signal shown should appear as three spikes, centered at the frequencies of the sinusoidal components. Because the maximum frequency in the signal is 600 Hz, the sampling time interval should be, at greatest, $1/(2 \times 600) = 0.000833$ sec. To demonstate the effects of aliasing, we choose to sample the signal at a time interval of 0.001 sec and take 256 samples. The sampled signal appears as shown in Figure 13.4b. If the analyst had no foreknowledge that the sampled signal shown in the figure was aliased, he or she certainly could not develop that information through observation of the time history alone. The DFT of the signal shown in Figure 13.4b was taken, and the result is plotted in Figure 13.4c. The first two components appear at the frequencies where they should, but the third component appears aliased at 400 Hz. Note that the frequency where the third component appears is $400 = 500 - (600 - 500)$ Hz; the component is folded around the Nyquist frequency. For this reason, the Nyquist frequency is sometimes called the Nyquist folding frequency. This folding behavior is a feature of the aliasing process.

13.4 DATA WINDOWS

The concept of temporal weighting and data windows was introduced in the previous section where the temporal weighting in Eq. (13.26) was shown to have the corresponding spectral window (Fourier transform) shown in Figure 13.3. Many other temporal weightings are available, and all of them display superior performance to the rectangular temporal weighting of Eq. (13.26) that can be taken advantage of in practically all situations. Two important papers on windows in general and the characteristics of numerous specific windows are the ones by Harris (1978) and Nuttall (1981). In order to gain an appreciation for the full variety of data windows and their characteristics, the analyst should consult these references.

There is, however, one collection of data windows that can be described as a family, and these are presented in the paper by Nuttall (1981). In order to describe these data windows, let us point out some important features of windows, in general, using the window in Figure 13.3. That spectral window, and all others, are characterized by (1) the width of their central lobe, (2) the height of the highest side lobe, and (3) the decay rate of the side lobes for large $|fT|$.

The family of temporal weightings to be developed in this section has the form

$$
w(t) = \begin{cases} \dfrac{1}{T} \displaystyle\sum_{k=0}^{K} a_k \cos\left(\dfrac{2\pi k t}{T}\right) & |t| \le \tfrac{1}{2}T \\ 0 & |t| > \tfrac{1}{2}T \end{cases} \tag{13.35}
$$

where the a_k, $k = 0, \ldots, K$, are real constants. All the temporal weightings are symmetric functions about $t = 0$ for nonnegative K; therefore, their Fourier transforms will be real and symmetric. The weightings possess all orders of derivatives for $|t| < \tfrac{1}{2}T$, but $w(t)$ and/or its derivatives may be discontinuous at $|t| = \tfrac{1}{2}T$, depending on our choice for the values of the a_k, $k = 0, \ldots, K$. The discontinuities in $w(t)$ and its derivatives at $|t| = \tfrac{1}{2}T$ establish the asymptotic behavior of the spectral windows associated with the temporal weightings in Eq. (13.35) at large $|fT|$. All the temporal weightings considered in the following are normalized so that

$$
\sum_{k=0}^{K} a_k = 1 \tag{13.36}
$$

The spectral window corresponding to the temporal weighting in Eq. (13.35) is its Fourier transform, and this is

$$W(f) = \frac{Tf}{\pi} \sin(\pi f T) \sum_{k=0}^{K} \frac{(-1)^k a_k}{f^2 T^2 - k^2} \qquad -\infty < f < \infty \qquad (13.37)$$

At the discrete frequencies $f = m/T$, the spectral window equals

$$W\left(\frac{m}{T}\right) = \begin{cases} a_0 & m = 0 \\ \frac{1}{2} a_{|m|} & m \neq 0 \end{cases} \qquad (13.38)$$

We can develop an asymptotic expression for the decay of the spectral window at large values of $|fT|$ by noting that

$$\frac{1}{f^2 T^2 - k^2} = \frac{1}{f^2 T^2} \frac{1}{1 - (k/fT)^2} = \frac{1}{f^2 T^2} \sum_{m=0}^{\infty} \left(\frac{k}{fT}\right)^{2m} \qquad |fT| > k \quad (13.39)$$

This follows from the expression for the geometric series

$$\sum_{m=0}^{\infty} b^m = \frac{1}{1 - b} \qquad |b| < 1$$

Use of Eq. (13.39) in Eq. (13.37) results in

$$W(f) = \frac{\sin(\pi f T)}{\pi f T} \sum_{m=0}^{\infty} \frac{1}{(fT)^{2m}} \sum_{k=0}^{K} (-1)^k k^{2m} a_k \qquad |fT| > K \quad (13.40)$$

This expression establishes the asymptotic behavior of $W(f)$ for large $|fT|$ because the term

$$\sum_{k=0}^{K} (-1)^k k^{2m} a_k \qquad (13.41)$$

can be zero for some values of the index m. For example, if the expression (13.41) is zero for $m = 0, \ldots, M - 1$, then the first nonzero term in the series indexed with m has a coefficient that varies as $|fT|^{-(2M+1)}$. The overall series varies as its largest nonzero coefficient.

Consider some specific cases. First, consider the $m = 0$ case. When

$$\sum_{k=0}^{K} (-1)^k a_k \neq 0 \qquad (13.42)$$

the temporal weighting is discontinuous at $|t| = \frac{1}{2}T$ since

$$w(|\tfrac{1}{2}T|) = \lim_{t \to T/2^-} w(t) = \frac{1}{T} \sum_{k=0}^{K} (-1)^k a_k \qquad (13.43)$$

where the limit is taken from below $\frac{1}{2}T$. In this case, based on Eq. (13.40), $W(f)$ diminishes as $|fT|^{-1}$ for large $|fT|$. (This is a decay of 6 Db/octave.) However, if expression (13.41) is zero for $m = 0$ and nonzero for $m = 1$, then

$$\sum_{k=0}^{K} (-1)^k a_k = 0 \qquad \sum_{k=0}^{K} (-1)^k k^2 a_k \neq 0 \qquad (13.44)$$

and the second derivative of the temporal weighting is discontinuous at $|t| = \frac{1}{2}T$ since

$$w''(|\tfrac{1}{2}T|) = \lim_{t \to T/2^-} w''(t) = -\frac{4\pi^2}{T^3} \sum_{k=0}^{K} (-1)^k k^2 a_k \qquad (13.45)$$

(Note that the odd-numbered derivatives of the spectral window are all equal to zero and thus are continuous at $|t| = \frac{1}{2}T$.) In this case, based on Eq (13.40), $W(f)$ diminishes as $|fT|^{-3}$ for large $|fT|$. (This is a decay of 18 dB/octave.) Following the approach shown here and using Eq. (13.40), it can be demonstrated that the decay rates of the spectral windows, $W(f)$, of Eq. (13.37) at large $|fT|$ depend on the continuity of the even derivatives of the temporal weighting [Eq. (13.35)] at $|t| = \frac{1}{2}T$.

One of the features that characterizes a spectral window, mentioned at the beginning of this section, is the width of its central lobe. There are many measures of window bandwidth, useful for different purposes, but one that will be essential for the normalization of spectral density estimates is developed in the following. First, note that the bandwidth associated with the DFT is $B = 1/T$, where T is the time span represented by the data in the DFT. Here, B is the frequency spacing between elements in the DFT. The effective bandwidth of a spectral window is some multiple of this frequency increment and is defined as

$$B_e = \frac{\displaystyle\int_{-\infty}^{\infty} w^2(t)\, dt}{\left[\displaystyle\int_{-\infty}^{\infty} w(t)\, dt\right]^2} = \frac{\displaystyle\int_{-\infty}^{\infty} W^2(f)\, df}{[W(0)]^2} \qquad (13.46)$$

The effective bandwidth, as defined here, is given in Figure 13.3 for the spectral window of the rectangular temporal weighting and is $B_e = 1/T$.

In the particular case where the temporal weighting is defined as in Eq.

(13.35), the effective bandwidth is

$$B_e = \frac{1}{T}\left[1 + \frac{1}{2}\sum_{k=1}^{K}\left(\frac{a_k}{a_0}\right)^2\right]$$ (13.47)

The development presented here should make it clear that the windows described by Eq. (13.35) and other windows discussed in the references can be adjusted to optimize window behavior. Typically, the qualities we seek in a temporal weighting and its corresponding spectral window are simplicity and favorable leakage characteristics. The favorable leakage characteristics we seek are a low peak side-lobe height and a rapid decay rate for side lobes far from the center lobe. Using the present temporal weighting, simplicity is achieved by choosing K small, say 1 or 2. Favorable side-lobe characteristics are achieved by optimizing the coefficients a_k, $k = 0, \ldots, K$, to yield the desired behavior. These things are done in the referenced papers, and some of the windows whose temporal weightings have the form of Eq. (13.35) are summarized here.

The first window we consider is the frequently used Hanning window. For this window $K = 1$, and the coefficients are

$$a_0 = 0.5 \qquad a_1 = 0.5$$ (13.48)

These coefficients satisfy Eq. (13.36) and set expression (13.41) to zero for $m = 0$. The spectral window is plotted in Figure 13.5. Comparison to Figure 13.3 shows that the Hanning window is superior to the rectangular window in both peak side lobe and side-lobe decay rate.

An even better window can be obtained by using $K = 2$. We can use the window coefficients to satisfy Eq. (13.36) and set expression (13.41) to zero for $m = 0$ and $m = 1$. The coefficients are

$$a_0 = 0.375 \qquad a_1 = 0.5 \qquad a_2 = 0.125$$ (13.49)

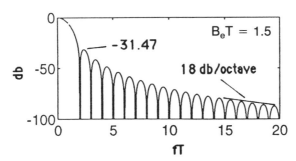

Figure 13.5 Hanning spectral window. Nuttall two-term temporal weighting with coefficients (0.5, 0.5).

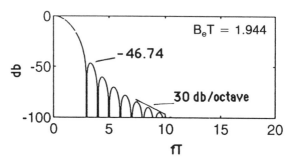

Figure 13.6 Spectral window of Nuttall three-term temporal weighting with coefficients (0.375, 0.5, 0.125).

The spectral window is plotted in Figure 13.6. It is superior to the Hanning window in both peak side-lobe and side-lobe decay characteristics.

Another window that also uses $K = 2$ can be developed. With this window, we satisfy Eq. (13.36), force expression (13.41) to be zero for $m = 0$, and optimize the coefficients so that the peak side lobe is minimal. The coefficients are

$$a_0 = 0.40897 \qquad a_1 = 0.5 \qquad a_2 = 0.09103 \qquad (13.50)$$

The spectral window is plotted in Figure 13.7. Its peak side-lobe behavior is superior to the previous window, whose coefficients are given in Eqs. (13.49), but its decay rate is inferior.

The temporal weightings for the windows we have analyzed in this section are shown in Figure 13.8 plotted on $[-0.5, 0.5]$. They are obviously all quite similar and differ mainly in their widths and continuity conditions at $|t| = \frac{1}{2}T$.

Many other windows can be defined, but the ones summarized here provide a sample of the characteristics available in a relatively simple class of windows.

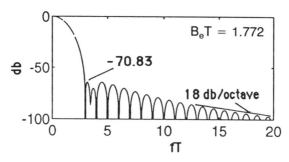

Figure 13.7 Spectral window of Nuttall three-term temporal weighting with coefficients (0.40897, 0.5, 0.09103).

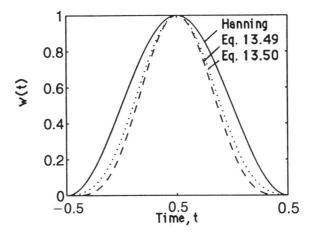

Figure 13.8 Temporal weightings of data windows analyzed.

13.5 FAST FOURIER TRANSFORM

It will be seen in the following chapter that the most important operations in random vibration signal analysis involve computation of the DFT. Our ability to execute DFT computations rapidly is especially important for two reasons. First, when we perform off-line analysis (i.e., recording data, storing it, then analyzing it later), we can analyze more data in a given amount of time when computations are fast. Second, when we perform on-line or real-time analyses, we can analyze a signal source without losing data when computations are fast. Modern computers can certainly execute DFTs rapidly, but there is an approach to computation of the DFT known as fast Fourier transform (FFT) that optimizes the speed of execution of the DFT by eliminating redundant operations.

There are several methods for computation of the FFT. Here, we shall only consider one approach to give the reader a general feel for the steps in execution of the FFT. Numerous texts cover the FFT, and some of the clearest descriptions are given in Stearns (1975) and Otnes and Enochson (1978).

The first step in demonstrating the efficiency that can be achieved with the FFT is to show how one step in the improvement can be developed. Consider the definition of the DFT [Eq. (13.11)]. Following standard FFT notation, we let W_n represent the invariant part of the exponential in the series $e^{i2\pi/n}$. Then, Eq. (13.11) can be written as

$$\xi_k = \sum_{j=0}^{n-1} x_j W_n^{jk} \qquad k = 0, \ldots, n-1 \qquad (13.51)$$

When the sequence x_j, $j = 0, \ldots, n-1$, is complex valued (as, in general, it may be), evaluation of the n values of ξ_k, $k = 0, \ldots, n-1$, takes n^2 complex multiplications. There are n products like $x_j W_n^{jk}$ in the evaluation of each ξ_k, and there are n ξ_k values.

When n is even (as assumed in Section 13.1), consider how the sum on the right side of Eq. (13.51) can be divided. Let $2N = n$. The series can be divided into one sum that involves terms with even index and another that involves terms with odd index:

$$\xi_k = \sum_{j=0}^{N-1} x_{2j} W_n^{2jk} + W_n^k \sum_{j=0}^{N-1} x_{2j+1} W_n^{2jk} \qquad k = 0, \dots, n-1 \qquad (13.52)$$

Each of the sums on the right side is, itself, a DFT. We recognize this by defining the sequences of numbers

$$y_j = x_{2j} \qquad z_j = x_{2j+1} \qquad j = 0, \dots, N-1 \qquad (13.53)$$

and noting that $W_n^2 = W_N$. Based on these facts, we can write

$$\eta_k = \sum_{j=0}^{N-1} y_j W_N^{jk} \qquad \zeta_k = \sum_{j=0}^{N-1} z_j W_N^{jk} \qquad k = 0, \dots, N-1 \qquad (13.54)$$

and finally,

$$\xi_k = \eta_k + W_n^k \zeta_k \qquad k = 0, \dots, N-1 \qquad (13.55)$$

Consideration of Eqs. (13.54) indicates that it takes $N^2 = (\frac{1}{2}n)^2$ complex multiplications to evaluate all the η_k, $k = 0, \dots, N-1$, and the same number of operations to evaluate all the ζ_k, $k = 0, \dots, N-1$. The total complex multiplications required to evaluate ξ_k, $k = 0, \dots, N-1$, is $(\frac{1}{2}n)(n+1)$. (The extra term accounts for the multiplications of W_n^k times ζ_k.) Of course, this only generates half the desired DFT values ξ_k, $k = 0, \dots, n-1$. Note, however, that the remaining DFT values can be evaluated using η_k and ζ_k, $k = 0, \dots, N-1$. Equation (13.16) indicates that the DFT is periodic; therefore,

$$\eta_{k+N} = \eta_k \qquad \zeta_{k+N} = \zeta_k \qquad k = 0, \dots, N-1 \qquad (13.56)$$

In view of this, the remaining DFT values ξ_k, $k = N, \dots, n-1$, not evaluated in Eq. (13.55), can be evaluated using Eqs. (13.55) and (13.56), with $k = N, \dots, n-1$, in Eq. (13.55).

A substantial efficiency is introduced into the computation of the DFT by dividing the DFT formula into two parts. Consider now that when the number of values in the original time series is divisible by 4, each of the series in Eqs. (13.54) can, in turn, be divided into two parts, introducing further efficiency. In fact, such divisions can occur until the time series has been divided into the set of sequences whose lengths are prime factors of n. Then the DFT of each of these sequences is computed. Optimal efficiency is achieved when n is a power of 2: $n = 2^p$. Many FFT computer programs are written to accommodate this requirement, and when n is not a power of 2, we often zero pad the time series x_j, $j = 0, \ldots, n - 1$, to make its length a power of 2. Zero padding involves the addition of zeros to the time series to make its length a power of 2.

When $n = 2^p$, the time series can be successively subdivided in p stages until each subgroup contains one element. Consider, for example, the division of a signal of eight numbers in a time series, shown in Figure 13.9. In $3 = \log_2 8$ stages, the original signal is subdivided until, in the final stage, there are eight sequences, each containing one number. Recalling the subdivisions performed above in Eqs. (13.52)–(13.55), we note that the DFTs of the subsets are required. Further, we note that the DFT of a single number, x, is the number, x, itself [consider Eq. (13.51) with $n = 1$].

Another important thing to recognize about the successive divisions of the time series described above and shown, for example, in Figure 13.9 is that the original sequence of numbers x_j, $j = 0, \ldots, n = 1$, is rearranged, in the final stage, in an order where the indices have been bit reversed. This is emphasized in Figure 13.9 by showing the binary representation of the index of each number in the original sequence and each number in the completely rearranged sequence. This rearrangement of the time series in bit-reversed order, plus recognition that the DFT of an individual number is the number itself, provides us with the information necessary to execute an FFT.

The first stage in the execution of an FFT is to rearrange the original time series in bit-reversed order, as shown, for example for $n = 8$ in Figure 13.9. The second stage in execution of the FFT involves the successive formation of the DFTs of pairs of numbers, then quadruples of numbers, and so on, until the DFT of the original time series is found.

For example, in Figure 13.9, the first step in the second stage forms the

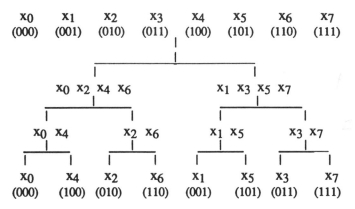

Figure 13.9 Successive divisions of time series x_j, $j = 0, \ldots, 7$, into subgroups for execution of the fast Fourier transform.

DFTs involving the pairs (x_0, x_4), (x_2, x_6), (x_1, x_5), and (x_3, x_7). The second step forms the DFTs involving the quadruples (x_0, x_2, x_4, x_6) and (x_1, x_3, x_5, x_7). The third, and final, step forms the DFTs involving all terms. Formulas like Eq. (13.55) are used at each step.

Figure 13.10 clarifies how the FFT computations are performed in the $n =$

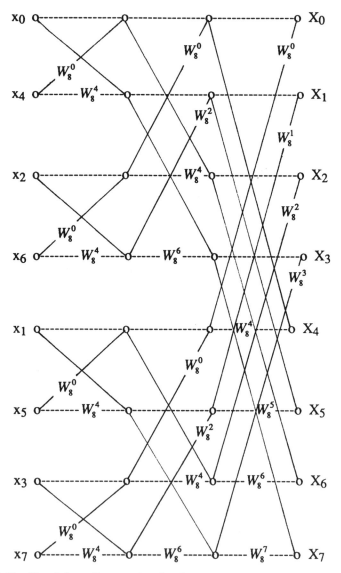

Figure 13.10 Signal flow diagram showing how computations in fast Fourier transform are executed for $n = 8$ case.

8 case using a signal flow diagram. The flow of computations is from left to right. The signal flow diagram shows, in the left column, how we start the computation with the values of the bit-reversed time series at the first set of nodes (circles in the diagram). In the first step of the computation, we form DFTs corresponding to the pairs of values (x_0, x_4), (x_2, x_6), ... At each node, we sum the terms corresponding to the lines leading into the node. The values corresponding to a particular line is the value at the node at the start of the line (left end) times the power of W on the line, if there is one. If there is no power of W, then the value corresponding to the line is simply the value at the starting node.

The values represented in the second column of nodes are the two-point DFTs. In the second step of computation, these values are used to generate the four-point DFTs in the third column of nodes. Finally, these values are used in the third computation step to generate the desired eight values of the DFT in the fourth column of nodes.

Figure 13.10 shows that the computation of an 8-point FFT involves, at each step of the computation, eight complex multiplications, and there are three such steps. If we were to perform a 16-point FFT, then the height of the flow diagram would double and one step would be added. In fact, the number of steps in the execution of an FFT equals $\log_2 n$, and n complex multiplications are executed at each step. Therefore, the total number of complex multiplications in an FFT is $n \log_2 n$. When n is large, this number of operations is substantially lower than n^2, the number of complex multiplications required to perform a direct DFT evaluation with Eq. (13.51). The FFT is very efficient, and the savings accrued with its use can be summarized in the speed ratio:

$$\text{Speed ratio} = \frac{n^2}{n \log_2 n} = \frac{n}{\log_2 n} \tag{13.57}$$

We note, finally, that the procedures described in Figure 13.10 can be used to develop computer subroutines and functions for computation of the FFT. A FORTRAN subroutine and a C language function for computation of the FFT are given in Appendix D.

13.6 GENERATION OF REAL-VALUED, FINITE-DURATION, SAMPLED REALIZATIONS OF STATIONARY, NORMAL RANDOM PROCESSES

There are many situations when we are required to generate realizations of random processes. Among these are the Monte Carlo analysis of system response and the laboratory random vibration testing of mechanical systems. In Monte Carlo analysis, we generate random realizations that simulate system inputs, compute the responses to these inputs, and use the methods of statistics to characterize the response random processes. In laboratory random vibration testing applications, typical test control systems continuously generate random inputs that excite motion in the test system with a desired spectral density.

We can generate finite-duration sampled realizations of stationary normal random processes using the formula developed to estimate the spectral density (see Chapters 5 and 14). Recall that the spectral density is approximately related to the expected value of the modulus squared of the DFT of the sampled, weakly stationary random process X_j, $j = 0, \ldots, n - 1$, via the formula

$$G_X(f_k) = \frac{2\Delta t}{n} E[|\overline{X}_k|^2] \qquad k = 0, \ldots, \tfrac{1}{2}n \tag{13.58}$$

where $G_X(f_k)$, $k = 0, \ldots, \tfrac{1}{2}n$, is the one-sided spectral density of the random process at frequencies $f_k = k\,\Delta f$, $k = 0, \ldots, \tfrac{1}{2}n$; $\Delta f = 1/n\,\Delta t$ is the frequency resolution; Δt is the sampling time increment; and $\overline{X}_k$, $k = 0, \ldots, \tfrac{1}{2}n$, is the DFT random process defined in Section 14.1.

We seek to generate a signal realization x_j, $j = 0, \ldots, n - 1$, from the random process X_j, $j = 0, \ldots, n - 1$, whose DFT satisfies the above equation, and we can do this by choosing the DFT of the realization with a modulus that satisfies

$$|\overline{x}_k| = \sqrt{\frac{n}{2\,\Delta t}\,G_X(f_k)} \qquad k = 0, \ldots, \tfrac{1}{2}n \tag{13.59}$$

The DFT of the random process realization has a phase, and the complete DFT can be expressed as

$$\overline{x}_k = |\overline{x}_k|\,e^{i\varphi_k} \qquad k = 0, \ldots, \tfrac{1}{2}n \tag{13.60}$$

The modulus of the DFT is deterministic because the spectral density is deterministic. Therefore, in order to randomize the realization, we select the phase ϕ_k of the DFT so that the phase values at various indices k are realizations of uniform $[-\pi, \pi)$ random variables.

We generate the DFT values so that they have conjugate symmetry, as discussed in Section 13.2; that is,

$$\overline{x}_{n/2+k}^* = \overline{x}_{n/2-k} \qquad k = 1, \ldots, \tfrac{1}{2}n - 1 \tag{13.61}$$

and

$$\mathrm{Im}(\overline{x}_0) = \mathrm{Im}(\overline{x}_{n/2}) = 0$$

These conditions guarantee that the time-domain signal will be real valued.

The time-domain random process realization is

$$x_j = \frac{1}{n} \sum_{k=0}^{n-1} \bar{x}_k \, e^{i2\pi jk/n} \qquad j = 0, \ldots, n-1 \qquad (13.62)$$

It is real valued and has the desired spectral density. When we require a realization whose mean is zero, we choose

$$\mathrm{Re}(\bar{x}_0) = 0 \qquad (13.63)$$

When the present signal is generated outside the index range $j \in [0, n-1]$, it is a periodic image of the signal inside the index range $j \in [0, n-1]$.

The approach presented here generates a digital signal with duration $n \, \Delta t = T$. When we require a continuous signal of finite duration, we can use a digital-to-analog (D/A) converter to make the signal continuous. When a continuous signal of arbitrary duration is required, an approach called overlap processing is employed. This technique is described in Smallwood and Paez (1993). This paper also shows how multiple, partially coherent, normal random process realizations can be generated. The technique presented there has its roots in a very robust version of the approach suggested by Dodds and Robson (1975). A classic paper on the simulation of multivariate and multidimensional random processes is the one by Shinozuka (1971).

The reason that the signal generated using Eq. (13.62) has a normal distribution is the central limit theorem (see Appendix A). The PDF of a single sinusoid is bimodal. Yet, when a sufficient number of independently phased sinusoids are added, their sum has a PDF that approaches the normal distribution. In fact, a relatively small number of elements is required in Eq. (13.62) in order for the sum to approach a normal random process as long as the spectral density is broad band. When the spectral density is narrow band, a few terms dominate the magnitude of the sum, and the requirements of the central limit theorem are violated. In view of this, when we seek to simulate a narrow-band random process or a random process composed of a few narrow-band components, we must ensure that the frequency resolution $\Delta f = 1/T$ used in the simulation is small enough so that several randomized harmonic components are included in each narrow-band frequency range. This may, under some circumstances, make it more efficient to directly use the DFT formulas for signal generation, rather than the FFT. When a signal source has narrow-band components superimposed on a broadband random signal, then realizations might be efficiently generated using a combination of DFT and FFT.

There are many applications when it is desirable to simulate nonstationary random processes, for example, when we wish to simulate earthquake motions or blast induced shock. Nonstationary random processes are often defined, for purposes of simulation, as the product of a deterministic function and a stationary random process. A generalization of this idea expresses the nonstationary random process $X(t)$, $t \in T$, as the sum of nonstationary components

$$X(t) = \sum_{m=1}^{M} g_m(t) \, Y_m(t) \qquad t \in T \qquad (13.64)$$

where each of the functions $g_m(t)$, $m = 1, \ldots , M$, is a deterministic function of time and the $Y_m(t)$, $t \in T$, are stationary random processes. More information on these types of representations is given in Priestly (1965).

PROBLEMS

13.1 Show how Eq. (13.37) is obtained from Eq. (13.35) via Fourier transformation.

13.2 Show that a decay of $|fT|^{-1}$ in Eq. (13.40) corresponds to a decay of 6 dB/octave.

13.3 Use Eq. (13.11) to find the DFT of

$$x_j = \sin\left(\frac{2\pi j}{n}\right) \qquad j = 0, \ldots , n - 1$$

and

$$y_j = \cos\left(\frac{2\pi(2)j}{n}\right) \qquad j = 0, \ldots , n - 1$$

where $n \geq 4$. Plot the results in terms of the real and imaginary parts and the modulus and phase of the DFT.

13.4 Suppose that, instead of the requirements of Eqs. (13.44), we require

$$\sum_{k=0}^{K} (-1)^k a_k = 0 \qquad \sum_{k=0}^{K} (-1)^k k^2 a_k = 0 \qquad \sum_{k=0}^{K} (-1)^k k^4 a_k \neq 0$$

What kind of decay must the spectral window have? What does this imply about the derivatives of $w(t)$ at $|t| = \frac{1}{2}T$?

13.5 Write a computer subroutine or function to execute the DFT (forward and inverse) as given in Eqs. (13.1) and (13.11). Use it to execute problem 13.3 for $n = 32$, and plot the results, as described there. Let

$$x_j = \sin\left(\frac{2\pi(\frac{3}{2})j}{n}\right) \qquad j = 0, \ldots , 32$$

$$y_j = \cos\left(\frac{2\pi(\frac{5}{2})j}{n}\right) \qquad j = 0, \ldots , 32$$

Find the DFTs and plot the results in terms of modulus and phase.

13.6 Program the FFT subroutine or function in Appendix D. Use it to compute the DFT of

$$x_j = 10 \cos \left(\frac{2\pi(\frac{15}{2})j}{n} \right) \quad j = 0, \ldots, 32$$

Plot the DFT modulus for $k = 0, \ldots, 16$. Let

$$y_j = w_j x_j \quad j = 0, \ldots, 32$$

where w_j is a temporal weighting. Using each of the three windows described in Eqs. (13.48), (13.49), and (13.50) with Eq. (13.35), compute the DFT of y_j and plot the resulting DFT modulus.

13.7 Use the FFT subroutine or function developed in problem 13.6 to write a subroutine or function to generate stationary random signal segments, as described in Eqs. (13.59)–(13.63). For $n = 128$ and $\Delta t = 0.1$ sec, generate a stationary random process realization that has the spectral density described by

$$G_X(f) = \frac{1}{f} \quad 0.05 \leq |f| \leq 5 \text{ Hz}$$

Plot the realization.

14

FREQUENCY-DOMAIN ESTIMATION OF RANDOM PROCESSES

The most important measures of random process behavior in engineering practice are the frequency-domain measures, namely the auto- and cross-spectral densities. The autospectral density is important because it provides a clean and simple means to interpret the physical and signal components that appear in a random signal, on the average. The cross-spectral density is important because it is related to the system frequency response function, and this entity completely characterizes a linear system. We develop, in this chapter, the formulas required to estimate the autospectral density of a weakly stationary random process as well as the cross-spectral density between random process pairs. Both quantities are then used to estimate the system frequency response function when the cross-spectral density is between excitation and response random processes. In the process, the estimator for ordinary coherence is developed. Bias and consistency are analyzed in most cases, and the sampling distributions required to create confidence intervals are developed. We assume, in this chapter, that the sources of signals to be analyzed have zero mean. If they do not, then the means must be estimated and removed from signals prior to analysis.

14.1 FUNDAMENTAL CONCEPTS IN ESTIMATION OF AUTOSPECTRAL DENSITY

The autospectral density is the fundamental measure of the frequency-domain behavior of a weakly stationary random process. During laboratory and field characterizations, it is certainly the most widely used random process measure. Therefore, we will develop the methods for its estimation with detail and precision. In the following sections, we develop the formulas for spectral density estimation that are Fourier transform based. These are most widely used in practice. In addition, however, there are other models and methods, including the method of maximum entropy, autoregressive (AR) models, moving-average (MA) models, mixed autoregressive–moving-average (ARMA) models, and others. Some of these methods are described in Lim and Oppenheim (1988).

The formulas developed for estimation of spectral density in the following sections are quite efficient because of the potential for use of the FFT to approximate Fourier transforms. For this reason, persons wishing to estimate the autocorrelation function of a stationary random process often first estimate the spectral density, then numerically approximate the Fourier transform to obtain an autocorrelation function estimate.

14.1.1 Direct Estimation of Autospectral Density

Many methods are available for developing the formulas for autospectral density estimation. To provide the reader with ample opportunity to intuitively understand spectral density estimation, we will cover two approaches. [A third approach that was widely popular for developing an intuitive understanding of the meaning of spectral density is based on the use of analog signals and their bandpass filtered components. This model is described in Chapter 5 and Bendat and Piersol (1986).] First, we present a direct approach related to the definition of spectral density.

Let $X(t)$, $-\infty < t < \infty$, be a weakly stationary random process with mean zero, two-sided spectral density $S_X(f)$, $-\infty < f < \infty$, and autocorrelation function $R_X(\tau)$, $-\infty < \tau < \infty$. Chapter 5 showed that the spectral density and autocorrelation functions are related via the expression [Eq. (5.52)]

$$S_X(f) = \int_{-\infty}^{\infty} R_X(\tau)\, e^{-i2\pi f\tau}\, d\tau \qquad -\infty < f < \infty \qquad (14.1)$$

where the cyclic frequency f is related to the circular frequency ω via $f = \omega/2\pi$. (The definition involving ω was used in Chapter 5.) The form of spectral density that is most frequently estimated in field and laboratory applications is the one-sided spectral density, denoted $G_X(f)$, $0 \le f < \infty$, and defined

$$G_X(f) = 2S_X(f) \qquad 0 \le f < \infty \qquad (14.2)$$

This is the form that we will investigate in this chapter. Note that although the letter W is often used in the literature and earlier in this book to denote one-sided spectral density, the letter G is used almost exclusively in the signal analysis literature and will be used in this chapter.

Let X_j, $j = 0, \ldots, n - 1$, be a discrete parameter random process that is a finite-duration, sampled version of $X(t)$, $-\infty < t < \infty$, where the sampling occurs at times $t_j = j\,\Delta t$, $j = 0, \ldots, n - 1$, and $X(j\,\Delta t)$ is denoted X_j, $j = 0, \ldots, n - 1$. The sampling occurs over a time interval of $T = n\,\Delta t$.

We can define a random process that is a transformation of X_j, $j = 0, \ldots,$ $n - 1$. It is the DFT, defined as

$$\overline{X}_k = \sum_{j=0}^{n-1} X_j e^{-i2\pi jk/n} \qquad k = 0, \ldots, n - 1 \qquad (14.3)$$

The DFT random process $\overline{X}_k$, $k = 0, \ldots, n - 1$, is a complex-valued, discrete-parameter random process that exists in a probabilitistic sense when the original random process X_j, $j = 0, \ldots, n - 1$, satisfies appropriate conditions. It is assumed that $\overline{X}_k$, $k = 0, \ldots, n - 1$, exists in a well-defined probabilistic sense. The mean of the random process $\overline{X}_k$, $k = 0, \ldots,$ $n - 1$, equals zero because X_j, $j = 0, \ldots, n - 1$, is a zero-mean random process.

The modulus squared of the random variable $\overline{X}_k$ is

$$|\overline{X}_k|^2 = \left(\sum_{j=0}^{n-1} X_j e^{-i2\pi jk/n} \right) \left(\sum_{m=0}^{n-1} X_m e^{i2\pi mk/n} \right)$$

$$= \sum_{j=0}^{n-1} \sum_{m=0}^{n-1} X_j X_m e^{-i2\pi(j-m)k/n} \qquad k = 0, \ldots, n - 1 \qquad (14.4)$$

The expected value of this quantity is

$$E[|\overline{X}_k|^2] = \sum_{j=0}^{n-1} \sum_{m=0}^{n-1} R_X(j - m) e^{-i2\pi(j-m)k/n} \qquad k = 0, \ldots, n - 1$$

$$(14.5)$$

where $R_X(j - m)$ is used to denote $R_X[(j - m)\, \Delta t]$. The double sum can be simplified using the change of variables.

$$j - m = r \qquad j = s \qquad (14.6)$$

This change of variables modifies the domain of the sum from $j \in [0, n - 1]$ and $m \in [0, n - 1]$ to the one shown in Figure 14.1. Using the change of variables, Eq. (14.4) becomes

$$E[|\overline{X}_k|^2] = \left(\sum_{r=-(n-1)}^{0} \sum_{s=0}^{(n-1)+r} + \sum_{r=1}^{n-1} \sum_{s=r}^{n-1} \right) R_X(r) e^{-i2\pi rk/n}$$

$$k = 0, \ldots, n - 1 \qquad (14.7)$$

The sum indexed with s must be executed first because it is expressed in terms of r. The sum is

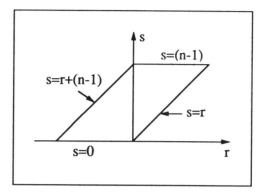

Figure 14.1 Domain of sum in Eq. (14.7).

$$E[|\overline{X}_k|^2] = n \sum_{r=-(n-1)}^{n-1} \left(1 - \frac{|r|}{n}\right) R_X(r)e^{-i2\pi rk/n} \qquad (14.8)$$

where it has been recognized that the sums in Eq. (14.7) differ only by an absolute value in r.

When the time lag $t_r = r\,\Delta t$, at which the autocorrelation function dies out, is small compared to $t_n = n\,\Delta t$,

$$\sum_{r=-(n-1)}^{n-1} \frac{|r|}{n} R_X(r)e^{-i2\pi rk/n} \cong 0 \qquad (14.9)$$

In view of this,

$$E[|\overline{X}_k|^2] \cong n \sum_{r=-(n-1)}^{n-1} R_X(r)e^{-i2\pi rk/n}$$

$$= \frac{n}{\Delta t}\left(\sum_{r=-(n-1)}^{n-1} R_X(r)e^{-i2\pi rk/n}\,\Delta t\right) \qquad k = 0, \ldots, n-1$$

$$(14.10)$$

The sum in the second expression is a zeroth-order approximation to the integral defining spectral density in Eq. (14.1). Therefore,

$$E[|\overline{X}_k|^2] \cong \frac{n}{\Delta t} S_X\left(\frac{k}{T}\right)$$

$$k = 0, \ldots, n-1 \qquad (14.11)$$

Based on this, the spectral density can clearly be estimated using the formula

$$\hat{G}_X(f_k) = \frac{2\,\Delta t}{n}\,\hat{E}[|\overline{X}_k|^2] \quad k = 0,\ldots,n/2 \quad (14.12)$$

where $f_k = k/T$, $k = 0,\ldots,n/2$, and $\hat{E}[|\overline{X}_k|^2]$ is the estimator for $E[|\overline{X}_k|^2]$. This latter estimate is simply the estimate for the expected value of a random variable, $|\overline{X}_k|^2$, that must be obtained using measured data, and one procedure is given in the following. (It is simply the approach of Chapter 11 applied to the random variable $|\overline{X}_k|^2$.) $\hat{G}_X(f_k)$ is only defined for index values $k = 0,\ldots,n/2$, because it is the one-sided spectral density.

Assume that the random process $X(t)$, $-\infty < t < \infty$, is both weakly stationary and ergodic, and that ξ_j, $j = 0,\ldots,(n \times M - 1)$, is a realization measured from the ensemble of $X(t)$, $-\infty < t < \infty$. The measurement includes $n \times M$ points. Divide the measured signal, ξ_j, $j = 0,\ldots,(n \times M - 1)$, into M segments of equal length n and denote these x_{mj}, $m = 1,\ldots,M$, $j = 0,\ldots,n-1$. In order to estimate the expected value of the modulus squared of the DFT of a sampled version of $X(t)$, $-\infty < t < \infty$, we first compute the DFT of each of the signals x_{mj}, $m = 1,\ldots,M$, $j = 0,\ldots,n-1$. These are denoted

$$\overline{x}_{mk} = \sum_{j=0}^{n-1} x_{mj}e^{-i2\pi jk/n} \quad m = 1,\ldots,M \quad k = 0,\ldots,n-1$$

$$(14.13)$$

Next, we compute the modulus squared of each DFT as follows

$$|\overline{x}_{mk}|^2 = \overline{x}_{mk}\overline{x}_{mk}^* \quad m = 1,\ldots,M \quad k = 0,\ldots,n-1 \quad (14.14)$$

The estimator for the expected value of $|\overline{X}_k|^2$ is simply the average of the sample moduli squared. Based on Eq. (11.1), this is

$$\hat{E}[|\overline{X}_k|^2] = \frac{1}{M}\sum_{m=1}^{M}|\overline{x}_{mk}|^2 \quad k = 0,\ldots,n/2 \quad (14.15)$$

When this formula is combined with Eq. (14.12), we obtain a spectral density estimator. While it would be all right to use these formulas with this approach to estimate the spectral density of a random process, this is not usually done because there are modifications that improve the accuracy of the estimate and make better use of available data. Nevertheless, these formulas and this approach form the basis for spectral density estimation used in the majority of modern applications.

14.1.2 Maximum Likelihood Estimation

A second approach for establishing a formula to estimate spectral density is the method of maximum likelihood. In order to establish an estimator formula for the spectral density, we must first establish the spectral density as a parameter in the PDF of some measure of a random process. As before, let $X(t)$, $-\infty < t < \infty$, be a weakly stationary random process with zero mean and two-sided spectral density $S_X(f)$, $-\infty < f < \infty$. Let X_j, $j = 0, \ldots$, $n - 1$, be a discrete-parameter random process that is a finite-duration, sampled version of $X(t)$, $-\infty < t < \infty$, where sampling occurs at times $t_j = j \, \Delta t$, $j = 0, \ldots, n - 1$, and $X(j \, \Delta t)$ is denoted, X_j, $j = 0, \ldots, n - 1$. Under the appropriate conditions (which we shall assume to exist), the random process X_j, $j = 0, \ldots, n - 1$, has a DFT that is also a random process denoted $\overline{X}_k$, $k = 0, \ldots, n - 1$. The random process $\overline{X}_k$, $k = 0, \ldots, n - 1$, is defined in Eq. (14.3). Each random variable $\overline{X}_k$ in the random process $\overline{X}_k$, $k = 0, \ldots, n - 1$, is, in general, complex valued, and when the random process X_j, $j = 0, \ldots, n - 1$, is real valued, the following relations hold:

$$\text{Re } \overline{X}_{n-k} = \text{Re } \overline{X}_k \qquad k = 1, \ldots, (\tfrac{1}{2}n - 1)$$

$$\text{Im } \overline{X}_{n-k} = -\text{Im } \overline{X}_k \qquad k = 1, \ldots, (\tfrac{1}{2}n - 1) \qquad (14.16)$$

$$\text{Im } \overline{X}_0 = \text{Im } \overline{X}_{n/2} = 0$$

That is, only $\tfrac{1}{2}n$ of the complex-valued random variables are independently defined. Each random variable $\overline{X}_k$ can be expressed in terms of real and imaginary parts,

$$\overline{X}_k = \text{Re } \overline{X}_k + i \text{ Im } \overline{X}_k \qquad k = 0, \ldots, n - 1 \qquad (14.17)$$

The real and imaginary parts can be defined using Eq. (14.3):

$$\text{Re } \overline{X}_k = \sum_{j=0}^{n-1} X_j \cos\left(\frac{2\pi jk}{n}\right) \qquad k = 0, \ldots, n - 1$$

$$(14.18)$$

$$\text{Im } \overline{X}_k = -\sum_{j=0}^{n-1} X_j \sin\left(\frac{2\pi jk}{n}\right) \qquad k = 0, \ldots, n - 1$$

Because of the central limit theorem, the random variables defined above are usually normally distributed when n is large. Because X_j, $j = 0, \ldots, n - 1$, is a zero-mean random process,

$$E[\text{Re } \overline{X}_k] = E[\text{Im } \overline{X}_k] = 0, \qquad k = 0, \ldots, n - 1 \qquad (14.19)$$

that is, the random variable $\overline{X}_k$ has real and imaginary parts whose expected values are zero. It can be shown (see problem 14.1) that

$$V[\text{Re } \overline{X}_k] = V[\text{Im } \overline{X}_k] = \frac{n}{2\,\Delta t} \, S_X(f_k) \qquad k = 1, \ldots, \tfrac{1}{2}n - 1$$

$$V[\text{Re } \overline{X}_0] = \frac{n}{\Delta t} \, S_X(f_0) \qquad V[\text{Im } \overline{X}_0] = 0 \tag{14.20}$$

$$V[\text{Re } \overline{X}_{n/2}] = \frac{n}{\Delta t} \, S_X(f_{n/2}) \qquad V[\text{Im } \overline{X}_{n/2}] = 0$$

[See Eq. (14.11), and note the relation between Re $\overline{X}_k$, Im $\overline{X}_k$, and $|\overline{X}_k|^2$.] Further, it can be shown (see problem 14.2) that the random variables Re $\overline{X}_k$ and Im $\overline{X}_k$, $k = 1, \ldots, \tfrac{1}{2}n - 1$, are uncorrelated. Because the random variables are normally distributed, this implies independence. In view of these facts, we can establish the following PDF specification. Let $\{\overline{X}\}$ denote the vector of random variables (Re $\overline{X}_0$ Re $\overline{X}_1$ Im $\overline{X}_1 \cdots$ Re $\overline{X}_{n/2-1}$ Im $\overline{X}_{n/2-1}$ Re $\overline{X}_{n/2})^T$, and let $\{\overline{x}\}$ denote its corresponding variate vector $(\overline{x}_{r0} \ \overline{x}_{r1} \ \overline{x}_{i1} \cdots \overline{x}_{r,n/2-1} \ \overline{x}_{i,n/2-1} \ \overline{x}_{r,n/2})^T$. Then the joint PDF of $\{\overline{X}\}$ is

$$f_{\{\overline{X}\}}(\{\overline{x}\}) = \frac{1}{(2\pi)^{n/2}|S|^{1/2}} \exp\left(-\tfrac{1}{2}\{\overline{x}\}^T\,[S]^{-1}\,\{\overline{x}\}\right) \qquad \{-\infty\} < \{\overline{x}\} < \{\infty\} \tag{14.21}$$

where $[S]$ is the diagonal covariance matrix that can be constructed based on Eq. (14.20),

$$[S] = \frac{n}{2\,\Delta t} \begin{bmatrix} 2S_0 & 0 & 0 & \cdots & 0 \\ 0 & S_1 & 0 & \cdots & 0 \\ 0 & 0 & S_1 & \cdots & 0 \\ \vdots & \vdots & \vdots & \ddots & \vdots \\ 0 & 0 & 0 & \cdots & 2S_{n/2} \end{bmatrix} \tag{14.21a}$$

where $S_k = S(f_k)$, $k = 0, \ldots, \tfrac{1}{2}n$. Its inverse is

$$[S]^{-1} = \frac{2\,\Delta t}{n} \begin{bmatrix} \tfrac{1}{2}S_0^{-1} & 0 & 0 & \cdots & 0 \\ 0 & S_1^{-1} & 0 & \cdots & 0 \\ 0 & 0 & S_1^{-1} & \cdots & 0 \\ \vdots & \vdots & \vdots & \ddots & \vdots \\ 0 & 0 & 0 & \cdots & \tfrac{1}{2}S_{n/2}^{-1} \end{bmatrix} \tag{14.21b}$$

and its determinant is

$$|S| = 4 \left(\frac{n}{2 \, \Delta t}\right)^n S_0 S_{n/2} \prod_{k=1}^{n/2-1} S_k^2 \qquad (14.21c)$$

When we make M independent measurements from the source that yields the random process X_j, $j = 0, \ldots , n - 1$, and use each measurement in the DFT formula to establish a random process $\overline{X}_k$, $k = 0, \ldots , n - 1$, we can use the results to develop M independent images of the vector $\{\overline{X}\}$ defined above. These can be denoted $\{\overline{X}_m\}$, $m = 1, \ldots , M$. The joint PDF of the $\{\overline{X}_m\}$ is the product from $m = 1$ through $m = M$ of Eq. (14.21) evaluated at $\{\overline{x}\} = \{\overline{x}_m\}$. When this function is taken as a function of the parameters $S_k = S(f_k)$, $k = 0, \ldots , \frac{1}{2}n$, it is the likelihood function of the spectral density:

$$L(\{S\}) = \frac{1}{(2\pi)^{Mn/2} \, |S|^{M/2}} \exp\left(-\frac{1}{2} \sum_{m=1}^{M} \{\overline{x}_m\}^T \, [S]^{-1} \, \{\overline{x}_m\}\right) \qquad (14.22)$$

where, formally, $\{\overline{x}_m\}$ is the variate vector corresponding to $\{\overline{X}_m\}$. In fact, the elements of $\{\overline{x}_m\}$ are the elements of the DFT of the mth measured random signal. These elements are Re $\overline{x}_k$ and Im $\overline{x}_k$, $k = 0, \ldots , n - 1$. The vector $\{S\}$ is defined $(S_0 \, S_1 \cdots S_{n/2})^T$.

The log-likelihood function is the natural logarithm of Eq. (14.22):

$$\mathcal{L}(\{S\}) = \frac{-Mn}{2} \ln(2\pi) - \frac{M}{2} \ln(|S|) - \frac{1}{2} \sum_{m=1}^{M} \{\overline{x}_m\}^T \, [S]^{-1} \, \{\overline{x}_m\} \qquad (14.23)$$

When we maximize the log-likelihood function with respect to the elements in $\{S\}$, we obtain the estimator formula for the spectral density. The log-likelihood function is a differentiable function of the elements of $\{S\}$, so the maximum occurs where

$$\frac{\partial}{\partial S_k} \mathcal{L}(\{S\}) = 0 \qquad k = 0, \ldots , \tfrac{1}{2}n \qquad (14.24)$$

The left-hand side of Eq. (14.24) is (for $k \neq 0$, $k \neq \frac{1}{2}n$)

$$-\frac{M}{2} \frac{2}{\hat{S}_k} + \frac{1}{2} \sum_{m=1}^{M} \frac{2 \, \Delta t}{n} (\overline{x}_{rmk}^2 + \overline{x}_{imk}^2) \frac{1}{\hat{S}_k^2} = 0 \qquad k = 0, \ldots , \tfrac{1}{2}n - 1$$

$$(14.25)$$

Note that $\overline{x}_{rmk}$ and $\overline{x}_{imk}$ are the real and imaginary parts of the kth term in the DFT of the mth measured signal. Therefore, the solution to Eq. (14.25) is

$$\hat{S}_k = \frac{\Delta t}{Mn} \sum_{m=1}^{M} |\bar{x}_{mk}|^2 \qquad k = 0, \ldots, \tfrac{1}{2}n \qquad (14.26)$$

Evaluation of Eq. (14.24) for $k = 0$ and $k = \tfrac{1}{2}n$ yields the same results. In view of these results, we can write

$$\hat{G}_X(f_k) = \frac{2\,\Delta t}{Mn} \sum_{m=1}^{M} |\bar{x}_{mk}|^2 \qquad k = 0, \ldots, \tfrac{1}{2}n \qquad (14.27)$$

This result is identical to that obtained using Eq. (14.15) in Eq. (14.12), and indeed, this type of result should be expected since the method of maximum likelihood is simply another way of obtaining the estimator formula.

The important thing to remember in noting that the spectral density estimator can be obtained by the method of maximum likelihood is that several qualities are implied in the estimate. One of these is that the estimate has minimum variance, that is, it makes optimum use of the data.

14.1.3 Bias, Consistency, and Sampling Distribution of Autospectral Density Estimator

Let us now consider bias in the spectral density estimator. To evaluate bias, we need to evaluate the mean value of the random variable $\hat{G}_X(f_k)$ that is formed when the variates $\bar{x}_{mk}$, $m = 1, \ldots, M$, $k = 0, \ldots, \tfrac{1}{2}n$, in Eq. (14.27) are replaced with the random variables $\overline{X}_{mk}$, $m = 1, \ldots, M$, $k = 0, \ldots, \tfrac{1}{2}n$. The random variable representing the one-sided spectral density at frequency $f_k = k/(n\,\Delta t)$ is

$$\hat{G}_X(f_k) = \frac{2\,\Delta t}{Mn} \sum_{m=1}^{M} |\overline{X}_{mk}|^2 \qquad k = 0, \ldots, \tfrac{1}{2}n \qquad (14.28)$$

The mean of the random variable is

$$E[\hat{G}_X(f_k)] = \frac{2\,\Delta t}{Mn} \sum_{m=1}^{M} E[|\overline{X}_{mk}|^2] = \frac{2\,\Delta t}{Mn} \sum_{m=1}^{M} E[|\overline{X}_k|^2]$$

$$= \frac{2\,\Delta t}{n} E[|\overline{X}_k|^2] \qquad k = 0, \ldots, \tfrac{1}{2}n \qquad (14.29)$$

where the fact that each random process $\overline{X}_{mk}$, $k = 0, \ldots, n - 1$, $m = 1, \ldots, M$, is simply an image of the random process $\overline{X}_k$, $k = 0, \ldots, n - 1$, has been used.

Sum all the elements on both sides of Eq. (14.29),

$$\sum_{k=0}^{n/2} E[\hat{G}_X(f_k)] = \frac{2\,\Delta t}{n} \sum_{k=0}^{n/2} E[|\overline{X}_k|^2] \qquad (14.30)$$

Recall from Section 13.2 that the DFT random process $\overline{X}_k$, $k = 0, \ldots ,$ $n - 1$, has a real part that is an even function and an imaginary part that is an odd function about $k = \frac{1}{2}n$. Further, recall from Section 13.1 that

$$\sum_{k=0}^{n-1} |\overline{X}_k|^2 = n \sum_{j=0}^{n-1} X_j^2 \qquad (14.31)$$

where the X_j, $j = 0, \ldots , n - 1$, are the random variables in the random process being characterized. In view of this, the expected values of both sides are equal and

$$\sum_{k=0}^{n/2} E[\hat{G}_X(f_k)] = \Delta t \sum_{j=0}^{n-1} E[X_j^2] \qquad (14.32)$$

Now, from Chapter 3, recall that the mean square of a zero-mean, stationary random process is simply the autocorrelation function evaluated at a lag of zero, that is,

$$E[X^2(t)] = R_X(0) \qquad -\infty < t < \infty \qquad (14.33)$$

Further, from Chapter 5, the autocorrelation function evaluated at a lag of zero is simply the area under the spectral density function curve. Therefore,

$$R_X(0) = \int_{-\infty}^{\infty} S_X(f) \, df \qquad (14.34)$$

and this is independent of time. Now note that X_j in Eq. (14.32) is simply a random variable in the random process X_j, $j = 0, \ldots , n - 1$, that represents a random variable in the random process $X(t)$, $-\infty < t < \infty$, evaluated at $t = j \, \Delta t$. Use Eq. (14.34) in Eq. (14.33) and the result in Eq. (14.32), and note that $\Delta f = 1/(n \, \Delta t)$ and that $S_X(f)$, $-\infty < f < \infty$, is a symmetric function. The result is

$$\sum_{k=0}^{n/2} E[\hat{G}_X(f_k)] = \frac{2}{\Delta f} \int_0^{\infty} S_X(f) \, df \qquad (14.35)$$

Finally, based on the developments of Sections 13.1, note that the kth term in the DFT $\overline{X}_k$, $k = 0, \ldots , n - 1$, is related to the kth frequency component in the signal represented. Assume that $S_X(f) = 0$ for $|f| > \frac{1}{2}n \, \Delta f$. Then

$$E[\hat{G}_X(f_k)] = \frac{2}{\Delta f} \int_{f_k - \Delta f/2}^{f_k + \Delta f/2} S_X(f) \, df \qquad k = 0, \ldots , \tfrac{1}{2}n \qquad (14.36)$$

This formula permits us to establish the bias in the one-sided spectral density estimator $\hat{G}_X(f_k)$, $k = 0, \ldots, \frac{1}{2}n$. To accomplish this, expand the spectral density $S_X(f)$ in a Taylor series about the frequency f_k:

$$S_X(f) = \sum_{m=0}^{\infty} \frac{(f - f_k)^m}{m!} S_X^{(m)}(f_k) \qquad f_k - \frac{1}{2}\Delta f \le f \le f_k + \frac{1}{2}\Delta f \quad (14.37)$$

where $S_X^{(m)}(f_k)$ is the mth derivative of the spectral density at frequency f_k. When we use Eq. (14.37) in Eq. (14.36), we show that

$$E[\hat{G}_X(f_k)] = 2S_X(f_k) + \sum_{m=1}^{\infty} \frac{2}{(2m + 1)(2m)!} \left(\frac{\Delta f}{2}\right)^{2m} S_X^{(2m)}(f_k)$$

$$\cong 2S_X(f_k) + \frac{(\Delta f)^2}{24} 2S_X''(f_k) \qquad k = 0, \ldots, \frac{1}{2}n \qquad (14.38)$$

where $S_X''(f_k)$ is the second derivative of the spectral density function. The second line simply takes the first term in the series as dominant over all the others. The second term on the second line of Eq. (14.38) is an approximation for the bias in the spectral density estimator. It is clearly only accurate when the higher order derivatives of the spectral density at f_k are small compared to their coefficients and when the spectral density is estimated with Eq. (14.27). When either of these conditions is violated, Eq. (14.38) does not reflect the true bias in the spectral density estimate.

The bias indicated by Eq. (14.38) is negative when $S_X''(f_k)$ is negative (i.e., there is a peak in the spectral density) and is positive when $S_X''(f_k)$ is positive (i.e., there is a trough in the spectral density). This means that the bias in estimated spectral density tends to make the estimate look "flatter" than it should. The autospectral density estimation bias defined in Eq. (14.38) is known as resolution bias error because is arises in connection with our choice of Δf.

The variance in the spectral density estimator can be evaluated directly using Eq. (14.28):

$$V[\hat{G}_X(f_k)] = \frac{4(\Delta t)^2}{(Mn)^2} V\left[\sum_{m=1}^{M} |\overline{X}_{mk}|^2\right] \qquad k = 0, \ldots, \frac{1}{2}n \qquad (14.39)$$

Recall that the real and imaginary parts of the DFT random process $\overline{X}_k$, $k = 0, \ldots, n - 1$, were assumed to be normally distributed and that their means and variances are given in Eqs. (14.19) and (14.20). Further, recall that the real and imaginary parts are uncorrelated, the variance of the square of a normally distributed random variable is given in Chapter 2, and the modulus square of a complex number is simply the sum of squares of the real and imaginary parts. Finally, the DFTs that come from different samples are uncorrelated. In view of these facts, the variance can be simplified to

$$V[\hat{G}_X(f_k)] = \frac{[2S_X(f_k)]^2}{M} \qquad k = 0, \ldots, \tfrac{1}{2}n \qquad (14.40)$$

This result makes it clear that the variance of the spectral density estimator approaches zero as the number of averages, M, used to obtain it increases. Therefore, the estimator is consistent.

The sampling distribution of the spectral density estimator can be developed based on Eq. (14.28). Consider first the summation portion of Eq. (14.28). The sum $\sum_{m=1}^{M} |\overline{X}_{mk}|^2$ is the sum of the squares of $2M$, uncorrelated, normally distributed random variables whose moments are given in Eqs. (14.19) and (14.20). When we divide each random variable squared by its variance, $(n/2 \, \Delta t) \, S_X(f_k)$, the sum becomes the sum of the squares of $2M$, uncorrelated, standard normal random variables. This is a chi-square random variable with $2M$ degrees of freedom. Now, when the sum in question is multiplied and divided by the quantity $2 \, \Delta t/Mn$, the numerator is the one-sided spectral density estimator and the denominator is $S_X(f_k)/M$. In view of this, the quantity

$$\frac{2M\hat{G}_X(f_k)}{2S_X(f_k)} = 2M \frac{\hat{G}_X(f_k)}{G_X(f_k)} = \chi^2_{2M} \qquad k = 0, \ldots, \tfrac{1}{2}n \qquad (14.41)$$

is a chi-square distributed random variable with $2M$ degrees of freedom.

This information can be used to establish confidence intervals on the spectral density. For example, the one-sided upper bound confidence interval on the one-sided spectral density at the $(1 - \alpha) \times 100\%$ level is

$$(0, 2M \, \hat{G}_X(f_k)/x_{1-\alpha}) \qquad (14.42)$$

where the quantity $x_{1-\alpha}$ satisfies the expression

$$P(\chi^2_{2M} > x_{1-\alpha}) = 1 - \alpha \qquad (14.42a)$$

One-sided lower bound confidence intervals and two-sided confidence intervals can also be easily obtained.

14.2 PRACTICAL ASPECTS OF ESTIMATION OF AUTOSPECTRAL DENSITY

The approach described in the previous section can be, and sometimes is, used to estimate the autospectral density of a stationary and ergodic random process. However, practical issues usually lead us to estimate autospectral density using a modified approach. Two issues lead us to the modifications. The first is the desire to improve upon the estimate of the Fourier transform of each segment of measured signal in the spectral density estimate obtained using the rectangular data window of the previous section. The second is the desire to make optimum use of limited data. We accomplish these improvements by (1) using data windows like the ones described in Chapter 13 and (2) using the available data in a scheme called overlap averaging.

The operations used in obtaining a spectral density estimate are shown in Figure 14.2. The measured signal, based upon which the spectral density is to be estimated, has the time duration T_{tot} and is separated into M overlapping segments, each with duration T. The fraction of overlap is q, and the overlap time is qT. The relation among these variables is

$$T_{tot} = [(M - 1)(1 - q) + 1]T \qquad (14.43)$$

We denote the original, continuous signal $s(t)$, $0 \le t \le T_{tot}$ and its mth overlapped segment as

$$s_m(\tau) = s[\tau + (m - 1)(1 - q)T] \qquad 0 \le \tau \le T \qquad m = 1, \ldots, M \qquad (14.44)$$

That is, each segment is defined over a time interval of $[0, T]$, and the mth time interval is taken from the original signal starting at time $(m - 1)(1 - q)T$. Now window and sample each of the overlapping segments as follows:

$$x_{mj} = w(j \, \Delta t) \times s_m(j \, \Delta t) \qquad m = 1, \ldots, M, \qquad j = 0, \ldots, n - 1 \qquad (14.45)$$

where $w(t)$, $0 \le t \le T$, is a temporal weighting like the ones described in Chaper 13 and $T = n \, \Delta t$. This yields a sampled ensemble of signals that can be used to estimate the spectral density of the underlying random process.

In fact, when the sampled signals are used as in Eqs. (14.12)–(14.15) and then scaled, we obtain an estimate of the autospectral density of the underlying random process. The scaling is required because of the fact that the window $w(t)$ modifies the amplitudes of the original measured signal. By analyzing the effect that the window has on the original measured signal and requiring that the estimated spectral density reflect the same mean-square value as the underlying random process, we find that the spectral density of a windowed signal must be corrected by a factor

$$Q = \frac{T}{\displaystyle\int_{-T/2}^{T/2} w^2(t) \, dt} = \frac{T}{\left[\displaystyle\int_{-T/2}^{T/2} w(t) \, dt\right]^2 B_e} = \frac{T}{a_0^2 B_e} \qquad (14.46)$$

where B_e is the effective bandwidth of the window, defined in Section 13.5, and a_0 is the leading term in the temporal weightings defined in the Nuttall

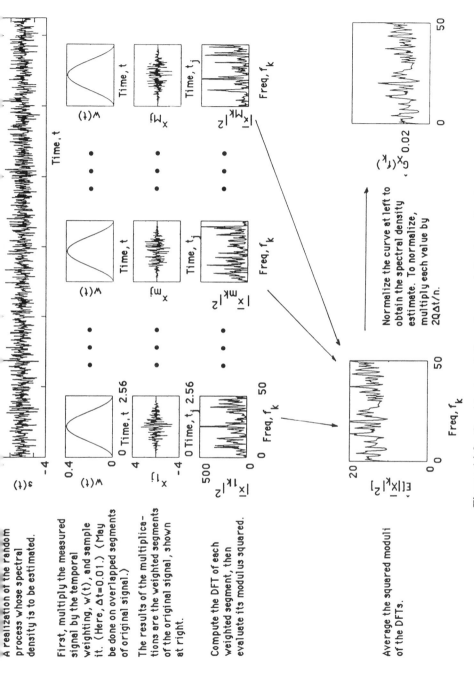

Figure 14.2 Graphic description of spectral density estimation.

A realization of the random process whose spectral density is to be estimated.

First, multiply the measured signal by the temporal weighting, $w(t)$, and sample it. (Here, $\Delta t=0.01$.) (May be done on overlapped segments of original signal.)

The results of the multiplications are the weighted segments of the original signal, shown at right.

Compute the DFT of each weighted segment, then evaluate its modulus squared.

Average the squared moduli of the DFTs.

Normalize the curve at left to obtain the spectral density estimate. To normalize, multiply each value by $2Q\Delta t/n$.

windows in Chapter 13. Based on this, the spectral density estimate is given by

$$\hat{G}_X(f_k) = \frac{2Q\,\Delta t}{Mn} \sum_{m=1}^{M} |\bar{x}_{mk}|^2 \qquad k = 0,\ldots,\tfrac{1}{2}n \qquad (14.47)$$

where the $\bar{x}_{mk}$, $m = 1,\ldots,M$, $k = 0,\ldots,\tfrac{1}{2}n$, are the DFTs of the windowed time-domain signals in Eq. (14.45).

The characteristics of the spectral density estimated using the above procedure are usually superior to those obtained without windowing and overlap averaging. The windowing improves the accuracy of the DFT computed Fourier transform approximation; the overlapping increases the number of averages that are used to form the spectral density estimate. Note, however, that these improvements are not unconditional. The improvement in approximation of the Fourier transform comes at the cost of widening the central lobe in the spectral window. The increase in the number of averages comes at the cost of sacrificing independence between averaged samples in Eq. (14.47). In view of the latter limitation, it is important to establish the effective number of averages in a spectral density estimate.

Recall from Eqs. (14.40) and (14.41) that it is important to know the number of averages in a spectral density estimate because the variance of the estimate and the sampling distribution of the estimate depend on this quantity. The references by Nuttall (1971) and Blackman and Tukey (1959) derive the effective number of averages in a spectral density estimate as

$$M_{\text{eff}} = \frac{E^2[\hat{G}(f)]}{V[\hat{G}(f)]} \qquad (14.48)$$

[This is also clear from Eq. (14.40).] Nuttall shows that this equals

$$M_{\text{eff}} = \frac{M}{\displaystyle\sum_{k=-(M-1)}^{M-1} \left(1 - \frac{|k|}{M}\right) \left|\frac{\phi_W(k(1-q)T)}{\phi_W(0)}\right|^2} \qquad (14.49)$$

where $\phi_W(\cdot)$ is the autocorrelation function of the temporal weighting $w(t)$. This is defined

$$\phi_W(\tau) = \int w(t)w(t-\tau)\,dt \qquad -\infty < \tau < \infty \qquad (14.50)$$

where the integral is taken over nonzero values of the integrand. This function obviously depends on the specific form of $w(t)$, and so too does the effective number of averages in a spectral density estimate. Figure 14.3 plots a lower bound on the ratio M_{eff}/M. The curves in the figure were computed using Eq.

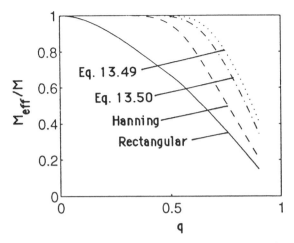

Figure 14.3 Ratio of effective number of averages to actual number of averages as function of overlap factor for overlap and window estimation of spectral density.

(14.49), but omitting the first factor in the sum in the denominator. The curves accurately reflect M_{eff}/M for large M. The curves corresponding to the rectangular window and those described in Section 13.4 are plotted. The curves make it clear that the temporal weightings of Section 13.4 are all superior to the rectangular temporal weighting and that the advantage of using an overlap factor starts to diminish for q greater than about 0.5. When we take into account the fact that data processing involves computation of the DFT of each signal segment, the increase in M_{eff} does not, at some point, effectively balance the increase in computations. Still, when the measured data are very limited and we need a low variance estimate of autospectral density, we can use overlap fractions up to about 0.75. Smallwood (1994) considers this issue of the optimum overlap fraction in great detail.

It is important to note that while the ideas of windowing and overlap averaging usually yield superior autospectral density estimates, they tend to diminish the accuracy of the approximation of the distribution of the normalized autospectral density estimator as a chi-square random variable. This is especially true when the overlap is great and the elements averaged to obtain the autospectral density estimate are correlated. In spite of this, we still usually assume that the quantity in Eq. (14.41) is chi-square distributed. The effect that windowing has on the distribution of the autospectral density estimator is discussed in Koopmans (1974).

Finally, we consider the number of averages that must be used in estimation of the spectral density. Obviously, as the number of averages used to estimate the spectral density increases, the variance of the estimate decreases; therefore, M_{eff} needs to be maximized. When there are no limitations on the duration of measurable realizations, storage space for measured signals, the cost of obtaining measurements, or the time required to analyze the signals, M_{eff} can be chosen very large. Values for M_{eff} on the order of 100 or 200 are not unreasonable.

However, when the total time of the measured signal, T_{tot}, is limited, for practical reasons, then we are not free to choose M_{eff} arbitrarily. Figure 14.2 shows that as M is increased, for a fixed overlap, the duration of the signal segment used for the DFT T must decrease. As T decreases, the frequency resolution $\Delta f = 1/T$ of the spectral density estimate grows poorer. When the signal we are measuring comes from a lightly damped mechanical structure, we run the real risk of missing some feature of the spectral density if Δf is too large. (Motion spectral densities of lightly damped structures tend to have very narrow peaks.) In view of this, trade-offs between estimation variance and frequency resolution must often be made in practical tests and analyses. In critical situations, it is prudent to try various parameter combinations.

By all means, at least a few averages must be used to estimate the spectral density, that is, the complete measured signal must be divided into at least a few segments. The spectral density estimate at each frequency is an average, and an average should never be based on one data value because estimation error will render the results practically useless. Surprisingly, some texts recommend the "one-average" approach; this should always be avoided.

Example 14.1. We demonstrate the use of the formulas and techniques developed in this section through the analysis of some random signals. In this illustration, one excitation and one response random signal were measured on a structure. The input was a force, and the response was an acceleration. Both signals were sampled at 16,384 times, and the sampling time interval was 0.001 sec. A short segment of the response signal is plotted in Figure 14.4a. The spectral density of the response was estimated using Eq. (14.47). The analysis was performed using 31 blocks of data, each with duration 1.024 sec ($=1024 \times 0.001$), and an overlap factor of 0.50. The temporal weighting described with Eq. (13.32) with the parameters in Eq. (13.46) was used to window the data. The estimated spectral density is shown in Figure 14.4b. The spectral density of the input forcing excitation was also estimated using the same formulas. It is shown in Figure 14.4c.

14.3 REAL-TIME ESTIMATION OF SPECTRAL DENSITY

In the practice of random signal analysis, we are often interested in monitoring the spectral density of an excitation or response over a long period of time. Of course, if the spectral density changes, then the source being monitored is not stationary. However, no real signal is truly stationary, and when a signal source varies slowly in time, we often seek to establish a real-time, or "running," estimate of spectral density. This section shows how a real-time estimate of spectral density is made and how we establish the effective number of averages,

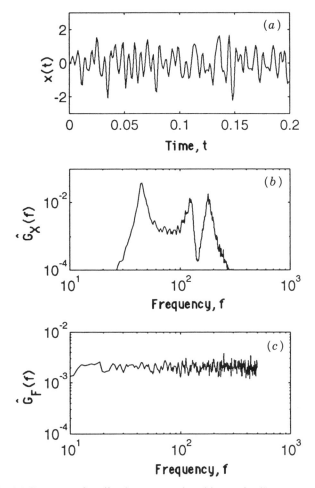

Figure 14.4 (a) Segment of realized response time history for linear structure excited by stationary random input. (b) Estimated spectral density of random process that is source of realization shown in part (a). (c) Estimated spectral density of random process that is source of excitation that drives response shown in part (a).

M_{ss}, on which it is based. For present purposes, we will assume that the signal source is in a steady state if its characteristics do not vary rapidly.

When a random signal source is in a steady state, we denote its estimated spectral density as $\hat{G}_{ss}(f_k)$. It has $2M_{ss}$ degrees of freedom [in chi-square terminology, see Eq. (14.41)] when based on M_{ss} effective averages. Its variance is [see Eq. (14.40)]

$$V[\hat{G}_{ss}(f_k)] = \frac{G_X^2(f_k)}{M_{ss}} \tag{14.51}$$

where $G_X(f_k)$ is the one-sided spectral density of the source random process.

We establish the "new" real-time estimate of the source spectral density, $\hat{G}_{new}(f_k)$, using the formula

$$\hat{G}_{new}(f_k) = \alpha\hat{G}_{up}(f_k) + (1 - \alpha)\hat{G}_{old}(f_k) \qquad (14.52)$$

where $\hat{G}_{old}(f_k)$ is the "old" estimate of spectral density, based on data measured in the past, and $\hat{G}_{up}(f_k)$ is the "update" estimate of spectral density, based on the data most recently measured. The number of averages used to obtain the update spectral density estimate is M_{up}, and this is often called the *number of averages per loop*. Here, $1/\alpha$ is often called the average weighting factor.

We assume that the update spectral density estimate is based on data distinct from the old spectral density estimate. Therefore, the variance of the new spectral density estimate is

$$V[\hat{G}_{new}(f_k)] = \alpha^2 V[\hat{G}_{up}(f_k)] + (1 - \alpha)^2\ V[\hat{G}_{old}(f_k)] \qquad (14.53)$$

But when the signal source is in a steady state, the new spectral density estimate must equal the old spectral density estimate, and these both equal $\hat{G}_{ss}(f_k)$. In view of this, the above equation becomes

$$V[\hat{G}_{ss}(f_k)] = \alpha^2 V[\hat{G}_{up}(f_k)] + (1 - \alpha)^2\ V[\hat{G}_{ss}(f_k)] \qquad (14.54)$$

Expand the square of $(1 - \alpha)$ on the right side, solve for the variance of the steady-state spectral density estimate, and note that the variance of the update spectral density is $G_X^2(f_k)/M_{up}$:

$$V[\hat{G}_{ss}(f_k)] = \frac{\alpha}{2 - \alpha}\ \frac{G_X^2(f_k)}{M_{up}} \qquad (14.55)$$

Compare this to the variance expression in Eq. (14.51). The effective number of averages in the steady-state spectral density estimate is

$$M_{ss} = \frac{2 - \alpha}{\alpha}\ M_{up} \qquad (14.56)$$

The number of degrees of freedom in the chi-square distribution for the estimate of the steady-state spectral density is twice the number of effective averages.

Figure 14.5 shows the ratio of the effective number of averages in the real-time spectral density estimate to the number of averages per loop, M_{ss}/M_{up}, as a function of α.

When the update spectral density estimate is established using windowing and overlap averaging, M_{up} is the actual number of averages as long as an overlap no greater than 50% is used. When an overlap fraction greater than 0.5 is used, M_{up} must be established using Figure 14.3.

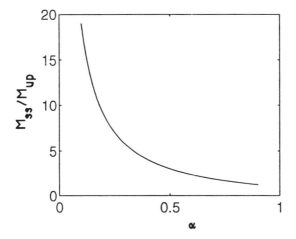

Figure 14.5 Ratio of effective number of averages to averages per loop versus α in real-time estimation of spectral density.

14.4 ESTIMATION OF CROSS-SPECTRAL DENSITY AND ORDINARY COHERENCE

The cross-spectral density between two random processes provides a fundamental measure of the degree of linear correlation between the random processes in the frequency domain. The cross-spectral density is, in general, complex valued. Its magnitude at a particular frequency indicates the average normalized magnitude of the product between the Fourier transform of one random process at that frequency and the complex conjugate of the Fourier transform of the other random process at the same frequency. Its phase at a particular frequency indicates the average phase difference between the two random processes at that frequency. The coherence is a real, normalized measure of the degree of linear relation between a pair of random processes in the frequency domain.

14.4.1 Cross-Spectral Density Estimation

A formula can be developed for the estimator of the cross-spectral density using either of the approaches applied to autospectral density estimation in Section 14.1. The steps required in the development are practically the same as the ones used there, and the entire development will not be presented here.

Let $X(t)$, $-\infty < t < \infty$, and $Y(t)$, $-\infty < t < \infty$, be a pair of random processes that can be simultaneously realized. Realizations of these random processes are sampled for analysis, and the sampled versions of these random processes possess corresponding DFT random processes as defined in Eq. (14.3). We can show that a relation similar to Eq. (14.11) exists for the cross-spectral density:

$$E[\overline{X}_k \overline{Y}_k^*] \cong \frac{n}{\Delta t} S_{XY}\left(\frac{k}{T}\right) \qquad k = 0, \ldots, \tfrac{1}{2}n \qquad (14.57)$$

This expression can be used to develop an estimator for the one-sided cross-spectral density:

$$\hat{G}_{XY}(f_k) = \frac{2\,\Delta t}{Mn} \sum_{m=1}^{M} \overline{x}_{mk} \overline{y}_{mk}^*, \qquad k = 0, \ldots, \tfrac{1}{2}n \qquad (14.58)$$

where $\overline{x}_{mk}$ and $\overline{y}_{mk}$, $m = 1, \ldots, M$, $k = 0, \ldots, \tfrac{1}{2}n$, are the DFTs of the mth measured realizations (or the mth segments in individual, measured realizations) of the sampled random processes $X(t)$, $-\infty < t < \infty$, and $Y(t)$, $-\infty < t < \infty$, respectively. This formula is based on the assumption that the measured realizations are not windowed [or, in other words, are windowed with the rectangular weighting, Eq. (13.26)].

An equivalent formula is defined for the case in which a window is applied to the samples measured from the two random processes. (The same window must be applied to both measured realizations.) It is

$$\hat{G}_{XY}(f_k) = Q \frac{2\,\Delta t}{Mn} \sum_{m=1}^{M} \overline{x}_{mk}\,\overline{y}_{mk}^* \qquad k = 0, \ldots, \tfrac{1}{2}n \qquad (14.59)$$

where Q is defined in Eq. (14.46) and it is assumed that $\overline{x}_{mk}$ and $\overline{y}_{mk}$, $m = 1, \ldots, M$, $k = 0, \ldots, \tfrac{1}{2}n$, are the DFTs of the measured signals that have been windowed and sampled.

In summary, the practical estimation of cross-spectral density between two random processes is nearly identical to the approach used in practical estimation of autospectral density. There is, however, one very important practical issue that must be addressed in cross-spectral density estimation. If the analyst requires that the phase of the cross-spectral density be accurately estimated, then she or he must ensure that the measured realizations of $X(t)$, $-\infty < t < \infty$, and $Y(t)$, $-\infty < t < \infty$, accurately reflect simultaneous signal activity. That is, there can be no lag in the measurement of the two random process realizations. This is generally accomplished, in laboratory and field measurements, by measuring the two realizations with one instrument using simultaneous sample and hold hardware. (For more information on this subject, consult commercial hardware specifications.) If care is not taken in this matter and a constant time lag between the two signal measurements is allowed, then an error in the phase estimate that varies linearly with frequency will result.

14.4.2 Coherence Estimation

The coherence is a normalized measure of the degree of linear relation between two weakly stationary random processes $X(t)$, $-\infty < t < \infty$, and $Y(t)$, $-\infty < t < \infty$. It can be defined in terms of the auto- and cross-spectral densities (see Section 5.2.2):

$$\gamma_{XY}^2(f) = \frac{|S_{XY}(f)|^2}{S_X(f)S_Y(f)} = \frac{|G_{XY}(f)|^2}{G_X(f)G_Y(f)} \qquad -\infty < f < \infty \quad (14.60)$$

It can be shown that the coherence must lie in the interval [0, 1]. When the random processes $X(t)$, $-\infty < t < \infty$, and $Y(t)$, $-\infty < t < \infty$, are completely uncorrelated at frequency f (in a linear sense), $\gamma_{XY}^2(f) = 0$. When they are linearly related to one another, $\gamma_{XY}^2(f) = 1$.

We estimate the coherence between two random processes using estimates of the quantities on the right-hand side in Eq. (14.60). Therefore,

$$\hat{\gamma}_{XY}^2(f_k) = \frac{|\hat{G}_{XY}(f_k)|^2}{\hat{G}_X(f_k)\hat{G}_Y(f_k)} \qquad k = 0, \ldots, \tfrac{1}{2}n \qquad (14.61)$$

where $f_k = k\,\Delta f$, $k = 0, \ldots, \tfrac{1}{2}n$, with $\Delta f = 1/T$, and T is the time duration of a single analysis block (for example) in Figure 14.2. The estimate in the numerator is obtained using Eq. (14.59); the estimates in the denominator are obtained using Eq. (14.47).

14.4.3 Bias and Consistency of Cross-Spectral Density Estimator

Bias can occur in estimation of the cross-spectral density, just as it does in estimation of the autospectral density, and the sources are the same. Most prominent among these is resolution bias that arises from our choice of the frequency resolution Δf. We assume in the following that the bias error is small.

Because the cross-spectral density estimator is a function of random variables, it is also a random variable, and it possesses a variance. The variance of the magnitude of the cross-spectral density estimator can be evaluated in the same general manner as the variance of the autospectral density estimator. The results for the autospectral density estimator are given in Eq. (14.40). To develop an analogous result for the cross-spectral density estimator, we perform the following operations:

1. Divide the cross-spectral density estimator into its real and imaginary parts.
2. Assume that the underlying random processes are normal random processes.
3. Compute the expected values of these parts in terms of the moments of the real and imaginary parts of the DFTs of the sampled forms of the random processes $X(t)$, $-\infty < t < \infty$, and $Y(t)$, $-\infty < t < \infty$. These moments involve the auto- and cross-spectral densities and the coherence.
4. Assume, for this computation, that $E[G_{XY}(f)] = G_{XY}(f)$.

The variance of the magnitude of the cross-spectral density estimator is approximately

$$V[|\hat{G}_{XY}(f_k)|] = \frac{|G_{XY}(f_k)|^2}{M\gamma_{XY}^2(f_k)} \qquad k = 0, \ldots, \tfrac{1}{2}n \qquad (14.62)$$

where $f_k = k\,\Delta f$, $k = 0, \ldots, \tfrac{1}{2}n$. This result can be used to infer the behavior of the frequency response function.

Though it can be done, we typically do not develop the sampling probability distribution of the cross-spectral density. It is difficult to develop because of the potential correlation between the random processes $X(t)$, $-\infty < t < \infty$, and $Y(t)$, $-\infty < t < \infty$. It can be shown that a form of the cross-spectral density estimate is governed by the Wishart distribution. This development is outlined in Koopmans (1974). Note that by developing the sampling distribution of the cross-spectral density estimator, we can obtain confidence intervals on the cross-spectral density real and imaginary parts, magnitude, and phase and on such functions of the cross-spectral density as coherence.

14.5 ESTIMATION OF FREQUENCY RESPONSE FUNCTION

The fundamental experimental measures of the dynamic behavior of a linear system are its frequency response functions. These establish, in the frequency domain, the ratios between the Fourier transforms of the responses at individual points on the system and the Fourier transform(s) of the input excitation(s). This section develops formulas for the frequency response functions of linear systems for the single-input, single-output (SISO) case and for the multiple-input, single-output (MISO) case. Then we consider the sampling distribution for a frequency response function magnitude squared and the variance of this quantitity.

Another input–output model for linear systems is also available, and it emphasizes the separate influences of the inputs on system response. It requires the use of multiple and partial spectral densities and coherence. This model is developed in Bendat and Piersol (1986).

14.5.1 Frequency Response in SISO Case

Consider the generic structure shown in Figure 14.6. The problem of estimation of the SISO frequency response function seeks to develop a formula for the frequency response function for the weakly stationary random process $X(t)$, $-\infty < t < \infty$, where only one of the input excitation random processes is nonzero. For present purposes, assume that the only input is the weakly stationary random process $F_1(t)$, $-\infty < t < \infty$, and because it is the only input, we denote it $F(t)$, $-\infty < t < \infty$. In the following development, it is assumed that the frequency response function is a deterministic but unknown characteristic of a linear physical system.

Based on the development in Chapter 9, it is clear that when a linear mechanical system is excited by a single input, the finite Fourier transform of the

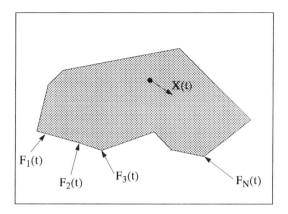

Figure 14.6 Linear structure with N input forces and one measured response.

response can be expressed as

$$\overline{X}(f) = H_{XF}(f)\overline{F}(f) \qquad -\infty < f < \infty \qquad (14.63)$$

where $\overline{F}(f)$, $-\infty < f < \infty$, is the finite Fourier transform of the input excitation, and $H_{XF}(f)$, $-\infty < f < \infty$, is the frequency response function of the
motion $\overline{X}(f)$ excited by the force $\overline{F}(f)$. We can evaluate Eq. (14.63) at each
of the discrete frequencies associated with the DFT to obtain

$$\overline{X}(f_k) = H_{XF}(f_k)\overline{F}(f_k) \qquad k = 0, \ldots, \tfrac{1}{2}n \qquad (14.64)$$

Next, we can multiply each side of the equation by the complex conjugate of
$\overline{F}(f_k)$, then take the expected value on both sides. The result is

$$E[\overline{X}(f_k)\overline{F}*(f_k)] = H_{XF}(f_k)E[|\overline{F}(f_k)|^2] \qquad k = 0, \ldots, \tfrac{1}{2}n \quad (14.65)$$

The frequency response function is not included in the expectation because it
is a deterministic quantity. Based on Eqs. (14.11) and (14.57), it is clear that
when Eq (14.65) is appropriately normalized, we obtain

$$S_{XF}(f_k) = H_{XF}(f_k)S_F(f_k) \qquad k = 0, \ldots, \tfrac{1}{2}n \qquad (14.66)$$

This relation makes it clear that the estimator for the frequency response function is

$$\hat{H}_{XF}(f_k) = \frac{\hat{G}_{XF}(f_k)}{\hat{G}_F(f_k)} \qquad k = 0, \ldots, \tfrac{1}{2}n \qquad (14.67)$$

Because the cross-spectral density estimator is, in general, complex valued, the frequency response function estimator must also be complex valued. Its magnitude at a given frequency governs the relation between the magnitudes of the system response and input at that frequency. Its phase at a given frequency governs the difference between the phases of the input and response. It is obvious that the quality of the estimate of the frequency response function depends heavily on the quality of the estimates of the input excitation auto-spectral density and the cross-spectral density between the input excitation and the response. Another restriction that may not be as clear is that the frequency response function estimate can only be accurate at frequencies where the input autospectral density has values *substantially different from zero*. In other words, at frequencies where the input is small, the response is small, and it is difficult to establish the relation between input and response. This statement cannot be made much more precise, without a great deal of difficulty, because the effects of random and bias errors on the auto- and cross-spectral densities have a great influence on the frequency response function estimate accuracy.

An estimator for the magnitude squared of the frequency response function can be developed by computing the complex magnitude squared on both sides of Eq. (14.64), then taking the expected value on both sides. The result is

$$E[|\overline{X}(f_k)|^2] = |H_{XF}(f_k)|^2 \, E[|\overline{F}(f_k)|^2] \qquad k = 0, \ldots, \tfrac{1}{2}n \qquad (14.68)$$

Based on Eq. (14.11), when normalized appropriately, this expression yields a formula involving the input and response autospectral densities:

$$S_X(f_k) = |H_{XF}(f_k)|^2 \, S_F(f_k) \qquad k = 0, \ldots, \tfrac{1}{2}n \qquad (14.69)$$

and this clearly gives rise to the estimator formula for the frequency response function magnitude squared:

$$|\hat{H}_{XF}(f_k)|^2 = \frac{\hat{G}_X(f_k)}{\hat{G}_F(f_k)} \qquad k = 0, \ldots, \tfrac{1}{2}n \qquad (14.70)$$

This estimator of the frequency response function magnitude squared does not yield results identical to those obtained with Eq. (14.67). Of course, the present estimator yields no information on the phase of the frequency response function.

Example 14.2. The example started in Section 14.2 is continued here. In this illustration, we first estimate the cross-spectral density between the two measured signals using Eq. (14.59) with the same parameters used to obtain the previous estimate. The magnitude and phase of the cross-spectral density estimate are shown in Figure 14.7a. Equation (14.67) was used to estimate the SISO frequency response function. The results are shown in Figure 14.7b, where the magnitude and phase of the estimated frequency response function

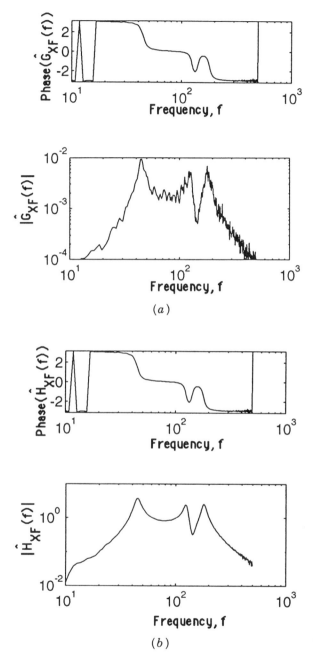

Figure 14.7 (a) Estimated phase and magnitude of cross-spectral density between random processes that are sources of response and excitation characterized in Figure 14.4. (b) Estimated phase and magnitude of SISO frequency response function for system response to excitation characterized in Figure 14.4. (c) Estimated coherence between response and excitation random processs characterized in Figure 14.4.

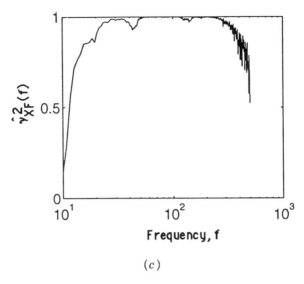

(c)

Figure 14.7 (*Continued*)

are plotted. Finally, Eq. (14.61) was used to estimate the coherence between the response and input. The estimate of coherence is shown in Figure 14.7c.

14.5.2 Sampling Distribution and Variance of Frequency Response Function Estimator

Because the frequency response function plays a crucial role in the characterization of linear systems, it is important to establish its accuracy. We do so in this section for random errors in estimation of the frequency response function of linear systems. Errors also occur due to bias, but these will not be developed in detail here. The text by Bendat and Piersol (1986) covers the sources of bias and the means to avoid bias errors in frequency response function estimation. Two major sources of bias are that which results from bias errors in the auto- and cross-spectral density estimates of the excitation and response and that which occurs due to nonlinearity in real system behavior. The former source is generally avoided through reasonable choice of analysis parameters in auto- and cross-spectral density estimation. The latter source cannot be avoided except through the use of data in estimation procedures that is taken during linear response of the systems to be analyzed. Of course, this is not always a realistic alternative. When a frequency response function is estimated based on data from a nonlinear system, it reflects an estimate that is best in a least squares sense.

Several approaches are available for approximating the variance of the estimator for the frequency response function. We choose, here, to consider one that is based on Eq. (14.67). There are several individual and joint measures of the frequency response function whose variance we might analyze, including real and imaginary parts, modulus, phase, modulus squared, and so forth. Here we approximate the variance of the modulus of the frequency response function. It is

$$|\hat{H}_{XF}(f_k)| = \frac{|\hat{G}_{XF}(f_k)|}{\hat{G}_F(f_k)} \qquad k = 0, \ldots, \tfrac{1}{2}n \qquad (14.71)$$

where $\hat{G}_{XF}(f_k)$ is defined in Eq. (14.59), and $\hat{G}_F(f_k)$ is defined in Eq. (14.47). At each frequency f_k, this is a ratio of two correlated random variables. They are correlated because the response is dependent upon the input excitation.

To show how we can easily approximate the variance of such a ratio, consider a random variable Z that is defined as a ratio of two other random variables $g(X, Y) = X/Y$. We seek a first order approximation to the variance of Z. To obtain one, we write the Taylor series approximation of the ratio and truncate it following the linear terms

$$Z = g(X, Y) \cong g(\mu_X, \mu_Y), + (X - \mu_X)\left.\frac{\partial g}{\partial x}\right|_\mu + (Y - \mu_Y)\left.\frac{\partial g}{\partial y}\right|_\mu \qquad (14.72)$$

where μ_X and μ_Y are the means of the random variables X and Y. The partial derivatives are evaluated at the means of the random variables X and Y. The mean and the variance of this approximate expression for Z are easily evaluated using the principles developed in Chapter 2. They are

$$E[Z] \cong g(\mu_X, \mu_Y) = \frac{\mu_X}{\mu_Y}$$

$$V[Z] \cong V[X]\left(\left.\frac{\partial g}{\partial x}\right|_\mu\right)^2 + V[Y]\left(\left.\frac{\partial g}{\partial y}\right|_\mu\right)^2 + 2\,Cov(X,Y)\left(\left.\frac{\partial g}{\partial x}\right|_\mu\right)\left(\left.\frac{\partial g}{\partial y}\right|_\mu\right)$$

$$= \frac{\mu_X^2}{\mu_Y^2}\left[\frac{\sigma_X^2}{\mu_X^2} + \frac{\sigma_Y^2}{\mu_Y^2} - 2\,\frac{Cov(X,Y)}{\mu_X\mu_Y}\right] \qquad (14.73)$$

where σ_X, σ_Y, $Cov(X, Y)$, are the standard deviations of X and Y and the covariance between X and Y, respectively.

To apply this variance formula to the modulus of the frequency response function, we assume that the estimates of $|\hat{G}_{XF}(f_k)|$ and $\hat{G}_F(f_k)$ are unbiased, and we take the covariance between $|\hat{G}_{XF}(f_k)|$ and $\hat{G}_F(f_k)$ to equal $|G_{XF}(f_k)|G_F(f_k)$ (see Bendat and Piersol, 1986). The variances of $|\hat{G}_{XF}(f_k)|$ and $\hat{G}_F(f_k)$ are given in Eqs. (14.40) and (14.62). The approximation for the variance of the modulus of the frequency response function is

$$V[|\hat{H}_{XF}(f_k)|] = \frac{(1 - \gamma_{XF}^2(f_k))}{M\gamma_{XF}^2(f_k)}|H_{XF}(f_k)|^2 \qquad k = 0, \ldots, \tfrac{1}{2}n \quad (14.74)$$

where $\gamma_{XF}^2(f_k)$ is the coherence between the response and the input at frequency f_k. This is a measure of the degree of variation of the frequency response function estimate.

Because the response of a physical system is correlated with the input excitation, the sampling distribution of the frequency response function is complicated. We can, however, establish approximate confidence intervals on measures of the frequency response function by assuming a normal distribution. If we make this assumption for the modulus of the frequency response function, assume that the estimate is unbiased, and use the approximate variance of Eq. (14.74), then the techniques of Chapter 11 can be used to establish confidence intervals on $|H_{XF}(f_k)|$.

14.5.3 Frequency Response Function in MISO Case

Consider again the generic structure shown in Figure 14.6. The problem of estimation of the MISO frequency response function seeks to develop formulas for the frequency response functions of the weakly stationary random process $X(t)$, $-\infty < t < \infty$, where all the input excitation random processes $F_j(t)$, $-\infty < t < \infty, j = 1, \ldots, N$, are nonzero. As in the previous development, it is assumed that the frequency response functions are characteristics of a linear physical system that are deterministic but unknown.

Chapter 9 develops the matrix equations governing the response of a multi-degree-of-freedom system in the frequency domain. We can specialize these to show that the Fourier transform of the response at a single point on a structure excited by a collection of force inputs is

$$\overline{X}(f) = \{H_{XF}(f)\}^T\{\overline{F}(f)\} \qquad -\infty < f < \infty \qquad (14.75)$$

where $\{\overline{F}(f)\}$, $-\infty < f < \infty$, is the vector of Fourier transforms of the exciting forces (like the ones shown, for example, in Figure 14.6) and $\{H_{XF}(f)\}$, $-\infty < f < \infty$, is the vector of frequency response functions relating the separate inputs to the response; the T superscript denotes the matrix operation of transposition. We can evaluate Eq. (14.75) at each of the discrete frequencies associated with the DFT to obtain

$$\overline{X}(f_k) = \{H_{XF}(f_k)\}^T \{\overline{F}(f_k)\} \qquad k = 0, \ldots, \tfrac{1}{2}n \qquad (14.76)$$

Now postmultiply both sides of the equation by the vector transpose of the complex conjugate of $\{\overline{F}(f_k)\}$, then take the expected value on both sides. The result is

$$E[\overline{X}(f_k)\{\overline{F}*(f_k)\}^T] = \{H_{XF}(f_k)\}^T E[\{\overline{F}(f_k)\}\{\overline{F}*(f_k)\}^T]$$

$$k = 0, \ldots, \tfrac{1}{2}n \qquad (14.77)$$

As we did in Section 14.5.1, we now recognize that the vector elements on the left side and the matrix elements on the right side are related to auto- and cross-spectral densities of the input forces and the response. The vector elements on the left are related to the cross-spectral densities between the response random process $X(t)$, $-\infty < t < \infty$, and the input random processes $F_j(t)$, $-\infty < t < \infty$, $j = 1, \ldots, N$. The matrix elements on the right side are related to the auto- and cross-spectral densities of the random processes $F_j(t)$, $-\infty < t < \infty$, $j = 1, \ldots, N$. When we normalize Eq. (14.77), we obtain

$$\{S_{XF}(f_k)\}^T = \{H_{XF}(f_k)\}^T[S_F(f_k)] \qquad k = 0, \ldots, \tfrac{1}{2}n \qquad (14.78)$$

Finally, when we compute the inverse of the force spectral density matrix on the right side, then postmultiply both sides of the equation by this quantity, we obtain

$$\{S_{XF}(f_k)\}^T[S_F(f_k)]^{-1} = \{H_{XF}(f_k)\}^T \qquad k = 0, \ldots, \tfrac{1}{2}n \qquad (14.79)$$

This is the expression upon which the estimate to the vector of frequency response functions is based. Specifically,

$$\{\hat{H}_{XF}(f_k)\}^T = \{\hat{G}_{XF}(f_k)\}^T[\hat{G}_F(f_k)]^{-1} \qquad k = 0, \ldots, \tfrac{1}{2}n \qquad (14.80)$$

where $\{\hat{H}_{XF}(f_k)\}$, $k = 0, \ldots, \tfrac{1}{2}n$, contains elements that estimate the frequency response functions that relate the response to the input forces; $\{\hat{G}_{XF}(f_k)\}$, $k = 0, \ldots, \tfrac{1}{2}n$, is the vector whose elements are the estimators of the one-sided cross-spectral densities between the response random process and the input random processes; and $[\hat{G}_F(f_k)]$, $k = 0, \ldots, \tfrac{1}{2}n$, is the matrix whose elements are the estimators of the input force auto- and cross-spectral densities.

There are obviously a number of difficulties that can occur in the use of the above equation. Foremost among these is ill-conditioning in one or more of the matrices $[\hat{G}_F(f_k)]$, $k = 0, \ldots, \tfrac{1}{2}n$, causing difficulties in their inversions. This is almost certain to occur in practical applications for one of two reasons. First, some of the input force random processes may have little signal content at some frequencies. This causes the corresponding row and column of $[\hat{G}_F(f_k)]$ to be small, including a very small value on the diagonal. Second, two of the input force random processes may be highly correlated at a frequency, and this causes the corresponding rows in $[\hat{G}_F(f_k)]$ to be nearly equal. To circumvent this problem of ill conditioning, we usually perform the inversion in Eq. (14.80) using an approach called singular value decomposition. This has been shown to work very well in a wide variety of real applications. This is discussed in Press et al. (1988).

To establish the frequency response functions at other points on the system, we simply use Eq. (14.80) with the appropriate response random process realizations.

Example 14.3. We now consider the system used in the previous examples, except that now two inputs are applied and responses are measured at two locations. Each measured signal is sampled 16,384 times, and the sampling interval is 0.001 sec. The sampled response signals resemble the one shown in Figure 14.4a. The 2×2 spectral density matrix of the input forces was estimated using the two measured input forces. The two 2×1, response-to-

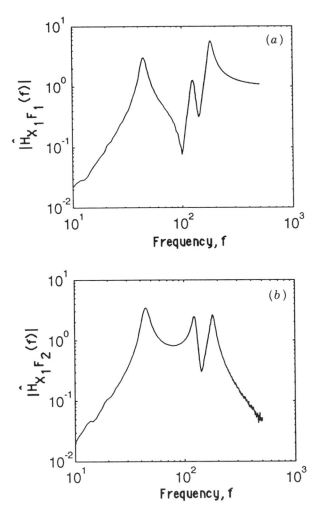

Figure 14.8 (a) Estimated frequency response function of response measured at location 1 to excitation measured at location 1 for linear system. (b) Estimated frequency response function of response measured at location 1 to excitation measured at location 2 for linear system. (c) Estimated frequency response function of response measured at location 2 to excitation measured at location 1 for linear system. (d) Estimated frequency response function of response measured at location 2 to excitation measured at location 2 for linear system.

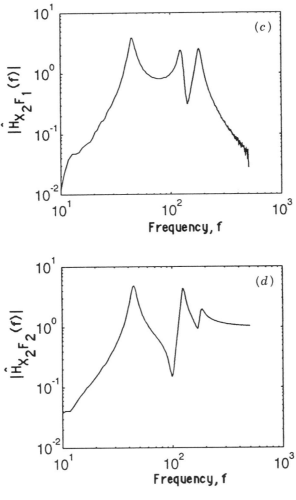

Figure 14.8 (*Continued*)

input cross-spectral density vectors were estimated using the measured inputs and responses. The spectral density matrix and vectors were used in Eq. (14.80) to estimate the four response-to-input frequency response functions. These are shown for response location 1 in Figures 14.8*a* and *b* and for response location 2 in Figures 14.8*c* and *d*.

PROBLEMS

14.1 Show that if X_j, $j = 0, \ldots, n - 1$, is a discrete-time sampled portion of a continuous, stationary random process, and if $\bar{X}_k$, $k = 0$,

. . . , $n - 1$, is defined as its DFT, then the moments of the real and imaginary parts of $\bar{X}_k$ are given by Eqs. (14.20). Assume that

$$E\,[\bar{X}_k \bar{X}_m^*] = \begin{cases} \dfrac{n}{\Delta t}\, S_X(f_k) & m = k \\ 0 & m \neq k \end{cases}$$

$$E[\bar{X}_k \bar{X}_m] = 0 \qquad \text{all } m,\, k$$

(*Hint:* For the second-order moments, write $\bar{X}_k$ and $\bar{X}_m$ in terms of their real and imaginary parts and compute $E[\bar{X}_k \bar{X}_m^*]$ and $E[\bar{X}_k \bar{X}_m]$. Use the resulting expressions to establish the desired moments.)

14.2 For the random process framework described in the previous problem, show that when the random process under consideration is a white noise,

$$E\,[\bar{X}_k \bar{X}_m^*] = \begin{cases} \dfrac{n}{\Delta t}\, S_X(f_k) & m = k \\ 0 & m \neq k \end{cases}$$

$$E[\bar{X}_k \bar{X}_m] = 0 \qquad \text{all } m,\, k$$

This is also approximately true when the random process under consideration is wide band.

14.3 Develop expressions for the one-sided lower bound confidence interval and the two-sided confidence interval for the spectral density of a stationary random process.

14.4 Develop the one-sided upper bound confidence interval and the two-sided confidence interval for the magnitude of the frequency response function in the SISO case.

14.5 Draw a flow chart of the operations required to estimate the spectral density based on Figure 14.2.

14.6 Draw a flow chart of the operations required to estimate the cross-spectral density.

14.7 Draw a flow chart of the operations required to estimate the frequency response function in the SISO case.

14.8 Draw a flow chart of the operations required to estimate the frequency response function vector in the MISO case.

14.9 Write computer functions or subroutines to compute estimates of autospectral density, cross-spectral density, and frequency response functions for the SISO and MISO cases based on the flow charts in problems 14.5–14.8. Use the Hanning window, where applicable.

14.10 Following the procedures outlined in the Problems section of Chapter 12 or Chapter 13, generate a realization of a zero-mean-, white-noise random process. Generate 8192 values, with a source variance of 1.0. Interpret these points as a sample input signal with $\Delta t = 0.001$ sec. Denote the signal q_j, $j = 0, \ldots, 8191$. Generate a "response" using the formula

$$x_{j+1} = q_j + 1.90 \, x_j - 0.97 \, x_{j-1} \qquad j = 1, \ldots, 8190$$

$$x_0 = x_1 = 0.0$$

Use the computer programs developed in the previous problem to estimate the autospectral densities of q_j and x_j, $j = 0, \ldots, 8191$. Estimate their cross-spectral density and SISO frequency response function. Use the Hanning window with 512 points in each DFT. Use a 50% overlap. This will yield 31 averages. Plot the results.

APPENDIX A

CONVERGENCE OF RANDOM PROCESSES

This appendix addresses some important topics in the calculus of random processes, namely, convergence, continuity, derivatives, and integrals. It briefly discusses the concept of ergodicity and introduces the central limit theorem. Readers who wish to pursue these points further should refer to Papoulis (1965) or Lin (1967).

A.1 INTRODUCTORY COMMENTS

We saw in Chapter 3, where random processes were first discussed, and Chapters 7–9, where the responses to stochastic inputs were considered, that (1) we need to write differential equations that involve stochastic responses and their derivatives and (2) we often write convolution integrals involving input stochastic processes to express response stochastic processes. Specifically, the linear differential equations that govern the responses of linear systems involve the response displacement and its first two derivatives as well as the input random process. Further, the convolution integral that expresses any measure of the linear response of a structure involves the input stochastic process and the deterministic impulse response function of the structure. In addition, we often know the probabilistic specification of the one measure of the response of a structure—displacement, velocity, or acceleration—and wish to establish another measure of response through differentiation or integration.

In view of the above it is necessary to be able to differentiate and integrate random processes in some sense. But a random process is not a single deterministic function, so we need to consider what the derivative and integral operations mean for random processes. (Note that this is of primary importance in the specification of theoretical random vibration models. When we make measurements of excitation and response random process realizations in the laboratory or in the field, the sources from which we measure are, by necessity, continuous, differentiable, and integrable. That is, all real measured signals have finite magnitude, finite slope, and so on. It is only when we seek to develop a realistic model of a physical system that we need to worry about continuity, differentiability, and integrability.)

Consider first a deterministic function. A deterministic function may be differentiable if it is finite and continuous. Its derivative is defined at a point if it exists as a unique limit. A deterministic function has a Riemann integral if it is finite everywhere and has at most a finite number of discontinuities.

These definitions serve as a good starting point for consideration of differentiability and integrability of random processes. Every random process is an ensemble consisting of all the realizations of a random process. If we know the specific mathematical characteristics of all the members of a random process ensemble, then we may be able to assess its differentiability and integrability. For example, if we know that all the elements in the ensemble of the random process are finite and continuous, then the random process is said to be integrable. If, in addition, all the elements in the ensemble are differentiable, then the random process is said to be differentiable. Usually, though, we do not know the characteristics of all the elements in the ensemble, and we must consider other criteria for differentiability and integrability.

A.2 MODES OF CONVERGENCE

In order to develop the capability to consider these issues, we introduce the idea of stochastic convergence. Let $X_j, j = 1, 2, \ldots$, be a sequence of random variables. We say that the sequence of random variables converges to the random variable X as $j \to \infty$ if the sequence satisfies certain requirements. These requirements can be cast in different frameworks, and the different frameworks reflect how "strongly" the random variable sequence converges to the limit X.

Consider that each object in the sequence $X_j, j = 1, 2, \ldots$, is a random variable, and every random variable has a probability distribution with probability density function, cumulative distribution function, mean, variance, and so on. The random variable X is characterized by the same functions and parameters.

One way in which the sequence $X_j, j = 1, 2, \ldots$, can converge to X is characterized by the statement

$$\lim_{j \to \infty} P(X_j = X) = 1 \tag{A.1}$$

This is called convergence with probability 1, for obvious reasons, and is one of the strongest modes of stochastic convergence.

Another mode of convergence is characterized by the statement

$$\lim_{j \to \infty} P(|X_j - X| \geq \epsilon) = 0 \qquad (A.2)$$

for every $\epsilon > 0$. This is known as convergence in probability and is a weaker form of convergence than convergence with probability 1. This form of convergence is implied by convergence with probability 1 but does not necessarily imply convergence with probability 1.

A third mode of convergence is implied by the statement

$$\lim_{j \to \infty} F_{X_j}(x) = F_X(x) \qquad (A.3)$$

This is known as convergence in distribution and is weaker than the previous forms defined. The central limit theorem, derived for a special case in Section A.7, shows that sums of independent, identically distributed random variables converge in distribution to the normal distribution.

The fact that we often are interested in the mean-square characteristics of a random process (like the autocorrelation function and spectral density) suggests a fourth very useful form of convergence. This form of convergence considers the existence of derivatives and integrals for random processes in what is known as a mean-square sense (or in quadratic mean).

Mean-square convergence is defined as follows. A sequence of random variables X_j, $j = 1, 2, \ldots$, is said to converge to the random variable X in a mean-square sense if

$$\lim_{j \to \infty} E[(X_j - X)^2] = 0 \qquad (A.4)$$

This does not necessarily imply that every realization of X_j approaches a realization of X as j approaches infinity, but only that the average value of the squares of the differences between realizations of X_j and X approaches zero as j approaches infinity.

In practical situations, we cannot always characterize the random variable X in Eq. (A.4), so we need another criterion for determining whether the sequence X_j, $j = 1, 2, \ldots$, converges to any limit in the mean-square sense. For this purpose we use the Cauchy convergence criterion. The Cauchy convergence criterion states that, given any $\epsilon > 0$, if there exists a value of N such that, when $j > m > N$,

$$E[(X_j - X_m)^2] < \epsilon \qquad (A.5)$$

then the sequence X_j, $j = 1, 2, \ldots$, converges in a mean-square sense to a limit as $j \to \infty$.

A.3 MEAN-SQUARE CONTINUITY

Consider how the mean-square convergence criterion is applied to differentiability of random processes. In order for a random process to be differentiable in a mean-square sense, it is first necessary that it be continuous in a mean-square sense. A random process is continuous at time t in a mean-square sense if

$$\lim_{h \to 0} E[(X(t + h) - X(t))^2] = 0 \qquad (A.6)$$

If the random process has the autocorrelation function $R_X(t_1, t_2)$, $t_1, t_2 \in T$, then

$$E[(X(t + h) - X(t))^2] = R_X(t + h, t + h) - 2R_X(t + h, t)$$
$$+ R_X(t, t) \qquad t, t + h \in T \qquad (A.7)$$

and the random process is mean-square continuous at time t if $R_X(t_1, t_2)$ is continuous in t_1 and t_2 at $t_1 = t_2 = t$. The random process is mean-square continuous everywhere it is defined if $R_X(t_1, t_2)$ is continuous for all $t_1 = t_2 \in T$. [It can be shown that if $R_X(t_1, t_2)$ is continuous for $t_1 = t_2 \in T$, then it is continuous for all $t_1, t_2 \in T$.]

For a stationary process, Eq. (A.6) becomes

$$\lim_{h \to 0} E[(X(t + h) - X(t))^2] = \lim_{h \to 0} 2[R_X(0) - R_X(h)] \qquad (A.8)$$

Hence, a stationary process is continuous in the mean square sense as long as the autocorrelation function is continuous at the origin:

$$\lim_{h \to 0} R_X(h) = R_X(0) \qquad (A.9)$$

This means that the autocorrelation function cannot have a delta function at the origin, as does white noise, for example.

The mean-square continuity results have two important implications. First, if a random process is mean-square continuous, then its mean value is continuous. Second, in developments involving the operations of expectation and limit, the order of the operations is interchangeable. This latter result is used extensively in the following.

A.4 MEAN-SQUARE DIFFERENTIABILITY

Consider the mean-square differentiability of the random process $X(t)$, $t \in T$. The derivative of the random process at time t is defined in the standard way:

$$\dot{X}(t) = \frac{d}{dt} X(t) = \lim_{h \to 0} \frac{X(t + h) - X(t)}{h} \qquad t \in T \qquad (A.10)$$

This derivative exists in a mean-square sense if there is a random process $\dot{X}(t)$, $t \in T$, such that

$$\lim_{h \to 0} E\left[\left(\frac{X(t + h) - X(t)}{h} - \dot{X}(t)\right)^2\right] = 0 \qquad (A.11)$$

However, we do not know the characteristics of the random variable $\dot{X}(t)$, so we must use Cauchy's convergence criterion to establish the mean-square differentiability of $X(t)$, $t \in T$. Cauchy's criterion requires that in order for $X(t)$, $t \in T$, to be mean-square differentiable at time t, the following limit must exist:

$$\lim_{\substack{h_1 \to 0 \\ h_2 \to 0}} E\left[\left(\frac{X(t + h_1) - X(t)}{h_1} - \frac{X(t + h_2) - X(t)}{h_2}\right)^2\right] = 0 \qquad (A.12)$$

This implies that when the quantity

$$\frac{\partial^2}{\partial t_1 \, \partial t_2} R_X(t_1, t_2) \qquad (A.13)$$

is finite and unique for $t_1 = t_2 = t$, then the random process $X(t)$, $t \in T$, is mean-square differentiable at time t.

A derivative random process is completely characterized by its joint probability density functions of all orders, and these are, in general, difficult to establish. However, all the probability density functions of a normal random process are described in terms of their second-order moments. Therefore, a derivative random process that is normal is completely characterized by its mean function and its autocorrelation function. Because of the regenerative characteristic of normal random variables under linear operations, and because of the definition of the derivative, Eq. (A.10), the derivative of any normal random process is also a normal random process. Therefore, the joint distribution of a normal random process and its derivative random process is completely characterized by the means and autocorrelation functions of the individual random processes and the cross-correlation function between the random process and its derivative. Using the definition of derivative and the interchangeability between the operations of limit and expectation, these can be shown to be

$$\mu_{\dot{X}}(t) = \frac{d}{dt} \mu_X(t) = \dot{\mu}_X(t) \qquad t \in T \qquad (A.14)$$

$$R_{\dot{X}\dot{X}}(t_1, t_2) = \frac{\partial^2}{\partial t_1 \, \partial t_2} R_{XX}(t_1, t_2) \qquad t_1, t_2 \in T \qquad (A.15)$$

$$R_{X\dot{X}}(t_1, t_2) = \frac{\partial}{\partial t_2} R_{XX}(t_1, t_2) \qquad t_1, t_2 \in T \qquad (A.16)$$

All the expressions developed above are simplified when the random processes under consideration are weakly stationary. Let $X(t)$, $-\infty < t < \infty$, be a weakly stationary random process with autocorrelation function $R_X(\tau)$, $-\infty < \tau < \infty$. When the random process is continuous in a mean-square sense, as defined by Eq. (A.9), then it is also mean-square differentiable when $R_X''(0)$ is finite and unique. Because the autocorrelation function of a stationary random process is symmetric in τ, this implies that $R_X'(0)$ must equal zero for the random process to be mean-square differentiable. The mean, auto-, and cross-correlation functions corresponding to those given in Eqs. (A.14)–(A.16) are given in Section 3.2.3.

A.5 MEAN-SQUARE INTEGRABILITY

Consider the mean-square integrability of a random process $X(t)$, $t \in T$. The mean-square Riemann integral of a random process, if it exists, is defined as the limit of its Riemann sum:

$$I = \lim_{n \to \infty} I_n = \lim_{\substack{n \to \infty \\ \Delta t_j \to 0}} \sum_{j=1}^{n} X(t_j) \, \Delta t_j = \int_a^b X(t) \, dt \qquad a, b \in T \quad (A.17)$$

where $a = t_1 < t_2 < \cdots < t_n = b$, and the Δt_j partition the interval $[a, b]$. The integral clearly only makes sense when $a, b \in T$. The integral exists in a mean-square sense if there is a random variable I such that

$$\lim_{\substack{n \to \infty \\ \Delta t_j \to 0}} E\left[\left(\sum_{j=1}^{n} X(t_j) \, \Delta t_j - I \right)^2 \right] = 0 \qquad (A.18)$$

However, as in the analysis of the derivative, we do not know the characteristics of the random variable I. Therefore, we must use Cauchy's criterion to determine the conditions that are required for $X(t)$, $t \in T$, to be mean-square integrable. That is,

$$\lim_{\substack{n_1 \to \infty \\ \Delta t_j \to 0 \\ n_2 \to \infty \\ \Delta t_k = 0}} E\left[\left(\sum_{j=1}^{n_1} X(t_j) \, \Delta t_j - \sum_{k=1}^{n_2} X(t_k) \, \Delta t_k \right)^2 \right] = 0 \qquad (A.19)$$

When the autocorrelation function of the random process $X(t)$, $t \in T$, is $R_X(t_1, t_2)$, $t_1, t_2 \in T$, then expansion of Eq. (A.19) shows that the condition required for mean-square integrability is that its autocorrelation function possess a double Riemann integral with respect to both variables t_1 and t_2 over the square in the Cartesian space defined by $a \leq t_1, t_2 \leq b$. That is, it is required that $R_X(t_1, t_2)$ be finite and have at most a finite number of discontinuities for $a \leq t_1, t_2 \leq b$.

If $R_X(t_1, t_2)$, $t_1, t_2 \in T$, is continuous on the diagonal $t_1 = t_2 \in T$, then it is continuous everywhere else. Therefore, a mean-square continuous random process is mean-square integrable.

The definition of the mean-square integral and the fact that the expectation and limit operations can be interchanged allow us to establish expressions for the mean, mean square, and variance of a mean-square integral. They are

$$\mu_I = \int_a^b \mu_X(t) \, dt \qquad\qquad a, b \in T \qquad\qquad (A.20)$$

$$E[I^2] = \int_a^b dt_1 \int_a^b dt_2 \, R_X(t_1, t_2) \qquad a, b \in T \qquad\qquad (A.21)$$

$$\sigma_I^2 = \int_a^b dt_1 \int_a^b dt_2 \, C_X(t_1, t_2) \qquad a, b \in T \qquad\qquad (A.22)$$

where $C_X(t_1, t_2)$, $t_1, t_2 \in T$, is the autocovariance function of the random process $X(t)$, $t \in T$.

The formulas developed above simplify when the random process under consideration is weakly stationary. Let $R_X(\tau)$, $-\infty < \tau < \infty$, be the auto-correlation function of the random process. Then in order for the random process to be mean-square integrable on the interval $[a, b]$, $R_X(\tau)$ must possess the integral

$$\int_a^b dt_1 \int_a^b dt_2 \, R_X(t_2 - t_1) = (b - a) \int_{-(b-a)}^{(b-a)} \left(1 - \frac{|\tau|}{b - a}\right) R_X(\tau) \, d\tau$$

$$-\infty < a < b < \infty \qquad\qquad (A.23)$$

The mean and variance of the mean-square integral of the stationary random process are

$$\mu_I = \int_a^b \mu_X \, dt = \mu_X(b - a) \qquad\qquad -\infty < a < b < \infty$$

$$(A.24)$$

$$\sigma_I^2 = (b - a) \int_{-(b-a)}^{(b-a)} \left(1 - \frac{|\tau|}{b - a}\right) C_X(\tau) \, d\tau \qquad -\infty < a < b < \infty$$

$$(A.25)$$

The probability distribution of the integral of a random process can, in principle, be obtained, but in general, it is very difficult to establish. In the special case where the random process is normal, its integral is also normal because of the linearity of the integration operation.

The concept of mean-square integrability of a random process can be generalized through a straightforward extension of the development presented above. Let us define the random process $Y(t)$, $t \in T$, as an integral transformation of the random process $X(t)$, $t \in T$:

$$Y(t) = \int_a^b g(t - \tau) X(\tau) \, d\tau \qquad a, b \in T \qquad \text{(A.26)}$$

where $g(t)$ may be a complex valued function. (That is, this may represent a convolution integral, Fourier or Laplace transform, etc.) The transformation exists in a mean-square sense when

$$\left| \int_a^b d\tau_1 \int_a^b d\tau_2 \, g(t - \tau_1) g^*(t - \tau_2) R_X(\tau_1, \tau_2) \right| < \infty \qquad \text{(A.27)}$$

The mean and variance of the mean-square transformation, if it exists, are

$$\mu_Y(t) = \int_a^b g(t - \tau) \mu_X(\tau) \, d\tau \qquad\qquad a, b \in T \quad \text{(A.28)}$$

$$\sigma_Y^2(t) = \int_a^b d\tau_1 \int_a^b d\tau_2 \, g(t - \tau_1) g^*(t - \tau_2) C_X(\tau_1, \tau_2) \qquad a, b \in T \quad \text{(A.29)}$$

Many forms of the autocorrelation and autocovariance functions of the random process, $Y(t)$, $t \in T$, and the cross-correlation and cross-covariance functions between the random process $Y(t)$, $t \in T$, and $X(t)$, $t \in T$, exist, depending on whether we take the complex conjugate of neither, one, or both elements and which elements we take the complex conjugate of. One form of the autocorrelation and autocovariance functions is

$$R_{YY^*}(t_1, t_2) = \int_a^b d\tau_1 \int_a^b d\tau_2 \, g(t_1 - \tau_1) g^*(t_2 - \tau_2) R_X(\tau_1, \tau_2) \qquad a, b \in T$$

$$\text{(A.30)}$$

$$C_{YY^*}(t_1, t_2) = \int_a^b d\tau_1 \int_a^b d\tau_2 \, g(t_1 - \tau_1) g^*(t_2 - \tau_2) C_X(\tau_1, \tau_2) \qquad a, b \in T$$

$$\text{(A.31)}$$

Formulas like these are useful when we seek to establish relations between responses and themselves, and excitations and responses, and when the responses are expressed as convolution integrals of the inputs, as in Chapters 8 and 9. They are also useful when we seek to characterize transforms of random processes, as in Chapter 14.

A.6 ERGODICITY

The concept of ergodicity is very important in the practical application of random vibrations and random signal analysis, because analysts frequently have very little data with which to characterize a random process. Indeed, a single realization must often be used to characterize the random process. Ergodicity concerns the problem of estimating the statistics of a random process from a single realization. A stationary random process is said to be ergodic with respect to a particular probability distribution parameter if the time average estimate of that parameter converges to the underlying value. For example, a process is ergodic in the mean if its time average mean value, defined by

$$\langle X(t) \rangle = \lim_{T \to \infty} \frac{1}{2T} \int_{-T}^{+T} X(t) \, dt \tag{3.16}$$

is equal to the ensemble average, defined by

$$\langle X(t) \rangle = E[X(t)] = \mu_X \tag{3.17}$$

The question of ergodicity is of fundamental importance in data acquisition and analysis. The purpose of this section is to briefly introduce the reader to the theory.

We recognize Eq. (3.16) as the limiting value of the integral of a random process. Hence, $M = \langle X(t) \rangle$ is a random variable. From Eqs. (3.16) and (A.24), the mean value μ_M is given by

$$\mu_M = \lim_{T \to \infty} \frac{1}{2T} (2T\mu_X) = \mu_X \tag{A.32}$$

That is, the expected value of the time average is equal to the ensemble average. The variance of M, from Eq. (A.25), is

$$\sigma_M^2 = \lim_{T \to \infty} \frac{1}{T} \int_0^{2T} \left(1 - \frac{\tau}{2T} \right) C_X(\tau) \, d\tau \tag{A.33}$$

In order for M to converge to M_X in the mean-square sense, its variance must vanish as T tends to infinity. The ergodic theorem for the mean value states this requirement as follows:

$$\langle X(t) \rangle = \lim_{T \to \infty} \frac{1}{2T} \int_{-T}^{+T} X(t) \, dt = E[X(t)] = \mu_X$$

if and only if

$$\lim_{T \to \infty} \frac{1}{T} \int_0^{2T} \left(1 - \frac{\tau}{2T}\right) C_X(\tau) \, d\tau = 0 \qquad \text{(A.34)}$$

There are ergodic theorems for all the averages that one would want to calculate from a single random process realization rather than from an ensemble, for example, the autocorrelation function, the probability distribution function, or the spectral density. All these theorems are based on the same principle, namely, that the variance of the time integral, which is a random variable, should tend to zero as the length of the integral tends to infinity.

A.7 CENTRAL LIMIT THEOREM

The central limit theorem is an example of convergence in distribution. It shows that under certain circumstances the distribution of a sum of random variables approaches the normal distribution. Consider the sum S_n defined in Eq. (2.133) in Section 2.5.2.

$$S_n = \sum_{i=1}^{n} X_i \qquad \text{(2.133)}$$

We will present the outline of a proof for the simple case in which the random variables X_i are independent and identically distributed with mean μ_X and variance σ_X^2. The mean, variance, and standardized variable are

$$\mu_{S_n} = n\mu_X \qquad \text{(A.35)}$$

$$\sigma_{S_n}^2 = n\sigma_X^2 \qquad \text{(A.36)}$$

$$Z_n = \frac{S_n - n\mu_X}{\sqrt{n\sigma_X^2}} = \sum_{i=1}^{n} \frac{X_i - \mu_X}{\sqrt{n\sigma_X^2}} \qquad \text{(A.37)}$$

Recall that the moment-generating function of the sum of independent random variables is equal to the product of the moment-generating functions of the addends. The strategy of the proof is to express the moment-generating function of Z_n as such a product and show that its limit is the moment-generating function of the standard normal distribution:

$$M_Z(t) = \exp\left(\tfrac{1}{2} t^2\right) \qquad (A.38)$$

By substituting the definition of Z_n [Eq. (A.37)] into the equation of the moment-generating function of the sum of independent and identically distributed random variables [Eq. (2.116)] and rearranging terms, we get

$$M_{Z_n}(t) = \exp\left(-\frac{\sqrt{n}\,\mu_X}{\sigma_X} t\right)\left[M_X\left(\frac{t}{\sqrt{n}\,\sigma_X}\right)\right]^n \qquad (A.39)$$

We take the natural logarithm to get

$$\ln M_{Z_n}(t) = \left(-\frac{\sqrt{n}\,\mu_X}{\sigma_X} t\right) + n \ln\left[M_X\left(\frac{t}{\sqrt{n}\,\sigma_X}\right)\right] \qquad (A.40)$$

We wish to replace $M_X(t)$ in this equation. The moment-generating function is approximated by the first three terms (up to t^2) of a Maclaurin series:

$$M_X(t) = 1 + \frac{d}{dt} M_X(t)\big|_{t=0}\, t + \frac{1}{2}\frac{d^2}{dt^2} M_X(t)\big|_{t=0}\, t^2 + \cdots \qquad (A.41)$$

From the moment-generating property, the coefficients of t and t^2 are the mean μ_X and the mean square $\mu_X^2 + \sigma_X^2$ of X, respectively:

$$M_X(t) = 1 + \mu_X t + \tfrac{1}{2}(\mu_X^2 + \sigma_X^2)t^2 + \cdots \qquad (A.42)$$

Put this back into the equation for the logarithm of M_{Z_n} to get

$$\ln M_{Z_n}(t) = \left(-\frac{\sqrt{n}\,\mu_X}{\sigma_X} t\right)$$
$$+ n \ln\left[1 + \mu_X \frac{t}{\sqrt{n}\,\sigma_X} + \frac{1}{2}(\mu_X^2 + \sigma_X^2)\left(\frac{t}{\sqrt{n}\,\sigma_X}\right)^2 + \cdots\right]$$

$$(A.43)$$

The logarithm is replaced by its power series expansion:

$$\ln(1 + \xi) = \xi - \tfrac{1}{2}\xi^2 + \tfrac{1}{3}\xi^3 - \cdots \qquad |\xi| < 1 \qquad (A.44)$$

Note that ξ is equal to the term in t plus the term in t^2. Dropping the higher order terms for convenience (they go to zero when the limit is taken), we get

$$\ln M_{Z_n}(t) \approx \left(-\frac{\sqrt{n}\,\mu_X}{\sigma_X}\,t\right) + n\left\{\left[\mu_X\frac{t}{\sqrt{n}\,\sigma_X} + \frac{1}{2}(\mu_X^2 + \sigma_X^2)\left(\frac{t}{\sqrt{n}\,\sigma_X}\right)^2\right]\right.$$

$$\left. -\frac{1}{2}\left[\mu_X\frac{t}{\sqrt{n}\,\sigma_X} + \frac{1}{2}(\mu_X^2 + \sigma_X^2)\left(\frac{t}{\sqrt{n}\,\sigma_X}\right)^2\right]^2\right\} \tag{A.45}$$

The reader may verify that upon expansion the only term without n appearing in the denominator is $\frac{1}{2}t^2$. So, in the limit, all terms except this one vanish, leaving

$$\lim_{n\to\infty} \ln M_{Z_n}(t) = \tfrac{1}{2}t^2 \tag{A.46}$$

Thus, in the limit, the moment-generating function is the moment-generating function for a standard normal variable and the theorem is proved.

APPENDIX B

INTEGRALS OF TRANSFER FUNCTIONS[1]

For stable linear systems with transfer functions $H_n(\omega)$ of the form

$$H_n(\omega) = \frac{B_0 + (i\omega)B_1 + (i\omega)^2 B_2 + \cdots + (i\omega)^{n-1} B_{n-1}}{A_0 + (i\omega)A_1 + (i\omega)^2 A_2 + \cdots + (i\omega)^n A_n}$$

the integral I_n of the complex modulus squared, $|H_n(\omega)|^2$,

$$I_n = \int_{-\infty}^{\infty} |H_n(\omega)|^2 \, d\omega$$

is given by

$$I_1 = \pi \frac{B_0^2}{A_0 A_1} \qquad n = 1$$

$$I_2 = \pi \frac{A_0 B_1^2 + A_2 B_0^2}{A_0 A_1 A_2} \qquad n = 2$$

$$I_3 = \pi \frac{A_0 A_3 (2 B_0 B_2 - B_1^2) - A_0 A_1 B_2^2 + A_2 A_3 B_0^2}{A_0 A_3 (A_0 A_3 - A_1 A_2)} \qquad n = 3$$

$$I_4 = \pi \left\{ \left[\frac{A_0 B_3^2 (A_0 A_3 - A_1 A_2) + A_0 A_1 A_4 (2 B_1 B_3 - B_2^2)}{A_0 A_4 (A_0 A_3^2 + A_1^2 A_4 - A_1 A_2 A_3)} \right] \right.$$

$$\left. + \left[\frac{-A_0 A_3 A_4 (B_1^2 - 2 B_0 B_2) + A_4 B_0^2 (A_1 A_4 - A_2 A_3)}{A_0 A_4 (A_0 A_3^2 + A_1^2 A_4 - A_1 A_2 A_3)} \right] \right\} \qquad n = 4$$

[1]From James, Nichols, and Phillips (1947).

416

APPENDIX C

FORMULAS FOR APPROXIMATE EVALUATION OF SOME INTEGRALS USEFUL IN PROBABILITY

There are numerous formulas available for the approximate evaluation of the probability density functions, cumulative distribution functions, and inverses of the cumulative distribution functions of the random variables of interest in random vibrations. A listing of many such functions is given in Abramowitz and Stegun (1964). Three of the simplest formulas are given here, for convenience.

First, we give an approximation for the cumulative distribution function of a standard normal random variable (tabulated in Section E.1):

$$0 \leq z \leq \infty$$

$$\Phi(z) = 1 - \tfrac{1}{2}(1 + c_1 z + c_2 z^2 + c_3 z^3 + c_4 z^4)^{-4} + \epsilon(z)$$

$$|\epsilon(z)| < 2.5 \times 10^{-4}$$

$$c_1 = 0.196854 \qquad c_2 = 0.115194$$

$$c_3 = 0.000344 \qquad c_4 = 0.019527$$

The second formula can be used to find the inverse of the cumulative distribution function of a standard normal random variable:

$$1 - \Phi(z_p) = p \qquad 0 < p \leq 0.5$$

$$z_p = t - \frac{a_0 + a_1 t}{1 + b_1 t + b_2 t^2} + \epsilon(p) \qquad t = \sqrt{\ln \frac{1}{p^2}}$$

$$|\epsilon(p)| < 3 \times 10^{-3}$$

$$a_0 = 2.30753 \qquad a_1 = 0.27061$$

$$b_1 = 0.99229 \qquad b_2 = 0.04481$$

The third formula is an approximation for the gamma function:

$$0 \leq x \leq 1$$

$$\Gamma(x + 1) = x! = 1 + a_1 x + a_2 x^2 + a_3 x^3 + a_4 x^4 + a_5 x^5 + \epsilon(x)$$

$$|\epsilon(x)| \leq 5 \times 10^{-5}$$

$$a_1 = -0.5748646 \qquad a_4 = 0.4245549$$

$$a_2 = 0.9512363 \qquad a_5 = -0.1010678$$

$$a_3 = -0.6998588$$

To evaluate the gamma function for arguments outside the interval [1, 2] use the formula

$$\Gamma(x + 1) = x\Gamma(x) = x! = x(x - 1)! \qquad x > 1$$

APPENDIX D

SOME FAST FOURIER TRANSFORM PROGRAMS

D.1 C Language

The following is a C language function for the computation of the discrete Fourier transform (DFT) of a complex signal via the fast Fourier transform (FFT) method. The function is based on the description given in Section 13.5 and computes the forward and inverse DFTs given by Eqs. (13.11) and (13.1), respectively. The function is named fft, must be declared void, and is designed to be called with the following four arguments:

xr   Pointer to the array of doubles that contains the real part of the signal whose FFT is to be computed.

xi   Pointer to the array of doubles that contains the imaginary part of the signal whose FFT is to be computed. (When the signal to be FFTed is purely real, the contents of this array is values that are all equal to zero.)

npt   An integer indicating the number of elements in each of the arrays xr and xi. This must be a power of 2; the funciton does not check for this.

isign An integer that indicates the direction of the FFT: -1 indicates a forward transform, in the terminology of Chapter 13; 1 or $+1$ indicates an inverse transform, in the terminology of Chapter 13.

```
/*          fft.c
            fft via bit reversal and dft.
*/

#include <math.h>
#define PI 3.1415926535898

void fft( xr, xi, npt, isign )
double *xr, *xi;
int np, isign;
{
        int j, jpow, jksum, j1, j2=0, k, log2_npt, nptd2, nptm1;
        double ur, ui, wr, wi, tr, ti, dnpt;
        /*  Set some limits. */
        nptm1 = npt - 1;
        nptd2 = npt / 2;
        /*      Compute log2(npt). */
        log2_npt = ((int)( (0.001+log( ((double)(npt)) ))/log( 2.0 ) ) );
        /*      Divide all elements by npt if this is an inverse. */
        if( isign == (+1) )
        {   dnpt = ((double)(npt));
                for( j=0; j<npt; j++ )
                {       xr[j] / = dnpt;
                        xi[j] /= dnpt;
                }
        }
        /* Perform a bit reversal on the input data. j1 indexes the original
        data; j2 indexes the bit reversed data. */
        for( j1=0; j1<nptm1; j1++ )
        {       if( j1<j2 )
                {       tr = xr[j2];        ti = xi[j2];
                        xr[j2] = xr[j1]; xi[j2] = xi[j1];
                        xr[j1] = tr;        xi[j1] = ti;
                }
                k = nptd2;
                while( k<(j2+1) )
                {       j2 -= k;
                        k /= 2;
                }
                j2 += k;
        }
        /*  Calculate the dft. */
        for( j=0; j<log2_npt; j++ )
        {       ur = 1.0; ui = 0.0;
                jpow = pow( 2.0, j+1 );
                k = jpow/2;
                wr = cos( PI/k); wi = isign*sin( PI/k );
                for( j1=0; j1<k; j1++)
                {       for( j2=j1; j2<npt; j2+=jpow )
                        {       jksum = j2 + k;
                                tr = xr[jksum]*ur-xi[jksum]*ui;
                                ti = xi[jksum]*ur+xr[jksum]*ui;
                                xr[jksum] = xr[j2]-tr; xi[jksum] = xi[j2]-ti;
                                xr[j2] += tr; xi[j2] += ti;
                        }
                        tr = ur; ti = ui;
                        ur = tr*wr-ti*wi; ui = tr*wi+ti*wr;
                }
        }
}
```

D.2 FORTRAN

The following is a FORTRAN subroutine for the computation of the DFT via the FFT. The subroutine is essentially a translation of the C language function given above. The subroutine arguments are the same as described above.

```
c         fft.f
c         fft via bit reversal and dft.
          subroutine fft( xr, xi, npt, isign )
          double precision xr(npt), xi(npt), ur, ui, wr, wi, tr, ti, PI
          PI = 3.1415926535898
c         Set some limits.
          nptm1 = npt - 1
          nptd2 = npt / 2
c         Compute log2(npt).
          log2_npt = 0.01 + alog10(real(npt))/alog10(2.0)
c         Divide all elements by npt is this is an inverse.
          if( isign.eq.(+1) )then
                  dnpt = real(npt)
                  do 101 j=1,npt
                          xr(j) = xr(j) / dnpt
                          xi(j) = xi(j) / dnpt
101               continue
          endif
c         Perform a bit reversal on the input data. j1 indexes the original
c         data; j2 indexes the bit reversed data.
          j2 = 1
          do 211 j1=1,nptm1
                  if( j1.lt.j2 )then
                          tr = xr(j2)
                          ti = xi(j2)
                          xr(j2) = xr(j1)
                          xi(j2) = xi(j1)
                          xr(j1) = tr
                          xi(j1) = ti
                  endif
                  k = nptd2
201               if( k.lt.j2 )then
                          j2 = j2-k
                          k = k/2
                          go to 201
                  endif
                  j2 = j2+k
211       continue
c         Calculate the dft.
          do 321 j=1,log2_npt
                  ur = 1.0
                  ui = 0.0
                  jpow = 2**j
                  k = jpow/2
                  wr = dcos( PI/real(k) )
                  wi = isign * dsin ( PI/real(k) )
                  do 311 j1=1,k
                          do 301 j2=j1,npt,jpow
                                  jksum = j2+k
                                  tr = xr(jksum)*ur-xi(jksum)*ui
                                  ti = xi(jksum)*ur+xr(jksum)*ui
                                  xr(jksum) = xr(j2)-tr
                                  xi(jksum) = xi(j2)-ti
                                  xr(j2) = xr(j2)+tr
                                  xi(j2) = xi(j2)+ti
```

```
301              continue
                 tr = ur
                 ti = ui
                 ur = tr*wr-ti*wi
                 ui = tr*wi+ti*wr
311          continue
321      continue
         return
         end
```

APPENDIX E

TABLES

E.1 TABLE OF CUMULATIVE DISTRIBUTION FUNCTION OF STANDARD NORMAL RANDOM VARIABLE

The values of the cumulative distribution function of a standard normal random variable $\Phi(z)$ are listed in the table, where

$$\Phi(z) = \frac{1}{\sqrt{2\pi}} \int_{-\infty}^{z} e^{-x^2/2} \, dx \qquad 0 \le z \le 4.99$$

Values of the cumulative distribution function for negative argument can be calculated using the relation

$$\Phi(-z) = 1 - \Phi(z), \qquad z \ge 0$$

z	.00	.01	.02	.03	.04	.05	.06	.07	.08	.09
0.0	.5000000	.5039894	.5079783	.5119665	.5159534	.5199388	.5239222	.5279032	.5318814	.5358564
0.1	.5398278	.5437953	.5477584	.5517168	.5556700	.5596177	.5635595	.5674949	.5714237	.5753454
0.2	.5792597	.5831662	.5870644	.5909541	.5948349	.5987063	.6025681	.6064199	.6102612	.6140919
0.3	.6179114	.6217195	.6255158	.6293000	.6330717	.6368307	.6405764	.6443088	.6480273	.6517317
0.4	.6554217	.6590970	.6627573	.6664022	.6700314	.6736448	.6772419	.6808225	.6843863	.6879331
0.5	.6914625	.6949743	.6984682	.7019440	.7054015	.7088403	.7122603	.7156612	.7190427	.7224047
0.6	.7257469	.7290691	.7323711	.7356527	.7389137	.7421539	.7453731	.7485711	.7517478	.7549029
0.7	.7580363	.7611479	.7642375	.7673049	.7703500	.7733726	.7763727	.7793501	.7823046	.7852361
0.8	.7881446	.7910299	.7938919	.7967306	.7995458	.8023375	.8051055	.8078498	.8105703	.8132671
0.9	.8159399	.8185887	.8212136	.8238145	.8263912	.8289439	.8314724	.8339768	.8364569	.8389129
1.0	.8413447	.8437524	.8461358	.8484950	.8508300	.8531409	.8554277	.8576903	.8599289	.8621434
1.1	.8643339	.8665005	.8686431	.8707619	.8728568	.8749281	.8769756	.8789995	.8809999	.8829768
1.2	.8849303	.8868606	.8887676	.8906514	.8925123	.8943502	.8961653	.8979577	.8997274	.9014747
1.3	.9031995	.9049021	.9065825	.9082409	.9098773	.9114920	.9130850	.9146565	.9162067	.9177356
1.4	.9192433	.9207302	.9221962	.9236415	.9250663	.9264707	.9278550	.9292191	.9305634	.9318879
1.5	.9331928	.9344783	.9357445	.9369916	.9382198	.9394292	.9406201	.9417924	.9429466	.9440826
1.6	.9452007	.9463011	.9473839	.9484493	.9494974	.9505285	.9515428	.9525403	.9535213	.9544860
1.7	.9554345	.9563671	.9572838	.9581849	.9590705	.9599408	.9607961	.9616364	.9624620	.9632730
1.8	.9640697	.9648521	.9656205	.9663750	.9671159	.9678432	.9685572	.9692581	.9699460	.9706210
1.9	.9712834	.9719334	.9725711	.9731966	.9738102	.9744119	.9750021	.9755808	.9761482	.9767045
2.0	.9772499	.9777844	.9783083	.9788217	.9793248	.9798178	.9803007	.9807738	.9812372	.9816911
2.1	.9821356	.9825708	.9829970	.9834142	.9838226	.9842224	.9846137	.9849966	.9853713	.9857379
2.2	.9860966	.9864474	.9867906	.9871263	.9874545	.9877755	.9880894	.9883962	.9886962	.9889893
2.3	.9892759	.9895559	.9898296	.9900969	.9903581	.9906133	.9908625	.9911060	.9913437	.9915758
2.4	.9918025	.9920237	.9922397	.9924506	.9926564	.9928572	.9930531	.9932443	.9934309	.9936128

	.00	.01	.02	.03	.04	.05	.06	.07	.08	.09
2.5	.9937903	.9939634	.9941323	.9942969	.9944574	.9946139	.9947664	.9949151	.9950600	.9952012
2.6	.9953388	.9954729	.9956035	.9957308	.9958547	.9959754	.9960930	.9962074	.9963189	.9964274
2.7	.9965330	.9966358	.9967359	.9968333	.9969280	.9970202	.9971099	.9971972	.9972821	.9973646
2.8	.9974449	.9975229	.9975988	.9976726	.9977443	.9978140	.9978818	.9979476	.9980116	.9980738
2.9	.9981342	.9981929	.9982498	.9983052	.9983589	.9984111	.9984618	.9985110	.9985588	.9986051
3.0	.9986501	.9986938	.9987361	.9987772	.9988171	.9988558	.9988933	.9989297	.9989650	.9989992
3.1	.9990324	.9990646	.9990957	.9991260	.9991553	.9991836	.9992112	.9992378	.9992636	.9992886
3.2	.9993129	.9993363	.9993590	.9993810	.9994024	.9994230	.9994429	.9994623	.9994810	.9994991
3.3	.9995166	.9995335	.9995499	.9995658	.9995811	.9995959	.9996103	.9996242	.9996376	.9996505
3.4	.9996631	.9996752	.9996869	.9996982	.9997091	.9997197	.9997299	.9997398	.9997493	.9997585
3.5	.9997674	.9997759	.9997842	.9997922	.9997999	.9998074	.9998146	.9998215	.9998282	.9998347
3.6	.9998409	.9998469	.9998527	.9998583	.9998637	.9998689	.9998739	.9998787	.9998834	.9998879
3.7	.9998922	.9998964	.9999004	.9999043	.9999080	.9999116	.9999150	.9999184	.9999216	.9999247
3.8	.9999277	.9999305	.9999333	.9999359	.9999385	.9999409	.9999433	.9999456	.9999478	.9999499
3.9	.9999519	.9999539	.9999557	.9999575	.9999593	.9999609	.9999625	.9999641	.9999655	.9999670
4.0	.9999683	.9999696	.9999709	.9999721	.9999733	.9999744	.9999755	.9999765	.9999775	.9999784
4.1	.9999793	.9999802	.9999811	.9999819	.9999826	.9999834	.9999841	.9999848	.9999854	.9999861
4.2	.9999867	.9999872	.9999878	.9999883	.9999888	.9999893	.9999898	.9999902	.9999907	.9999911
4.3	.9999915	.9999918	.9999922	.9999925	.9999929	.9999932	.9999935	.9999938	.9999941	.9999943
4.4	.9999946	.9999948	.9999951	.9999953	.9999955	.9999957	.9999959	.9999961	.9999963	.9999964
4.5	.9999966	.9999968	.9999969	.9999971	.9999972	.9999973	.9999974	.9999976	.9999977	.9999978
4.6	.9999979	.9999980	.9999981	.9999982	.9999982	.9999983	.9999984	.9999985	.9999986	.9999986
4.7	.9999987	.9999988	.9999988	.9999989	.9999989	.9999990	.9999990	.9999991	.9999991	.9999991
4.8	.9999992	.9999992	.9999993	.9999993	.9999994	.9999994	.9999994	.9999994	.9999995	.9999995
4.9	.9999995	.9999995	.9999996	.9999996	.9999996	.9999996	.9999996	.9999997	.9999997	.9999997

E.2   TABLE OF PERCENTAGE POINTS OF χ^2 DISTRIBUTION

The percentage points of the χ^2 distribution are listed in the table. These are the values of a, tabulated as a function of v and p, such that

$$\frac{1}{2^{v/2}\Gamma(v/2)} \int_0^a s^{v/2-1} e^{-s/2}\, ds = p$$

For large values of v, one can use the formula

$$a \cong v\left(1 - \frac{2}{9v} + z_p\sqrt{\frac{2}{9v}}\right)^3$$

where z_p is to be obtained from the table of the standard normal cumulative distribution function, Section E.1, and satisfies $\Phi(z_p) = p$.

v \ p	0.005	0.01	0.025	0.05	0.10	0.20	0.80	0.90	0.95	0.975	0.99	0.995
1	0.0000	0.0002	0.0010	0.0039	0.0158	0.0642	1.6424	2.7055	3.8415	5.0239	6.6349	7.8794
2	0.0100	0.0201	0.0506	0.1026	0.2107	0.4463	3.2189	4.6052	5.9915	7.3778	9.2103	10.5966
3	0.0717	0.1148	0.2158	0.3518	0.5844	1.0052	4.6416	6.2514	7.8147	9.3484	11.3449	12.8382
4	0.2070	0.2971	0.4844	0.7107	1.0636	1.6488	5.9886	7.7794	9.4877	11.1433	13.2767	14.8603
5	0.4117	0.5543	0.8312	1.1455	1.6103	2.3425	7.2893	9.2364	11.0705	12.8325	15.0863	16.7496
6	0.6757	0.8721	1.2373	1.6354	2.2041	3.0701	8.5581	10.6446	12.5916	14.4494	16.8119	18.5476
7	0.9893	1.2390	1.6899	2.1673	2.8331	3.8223	9.8033	12.0170	14.0671	16.0128	18.4753	20.2777
8	1.3444	1.6465	2.1797	2.7326	3.4895	4.5936	11.0301	13.3616	15.5073	17.5345	20.0902	21.9550
9	1.7349	2.0879	2.7004	3.3251	4.1682	5.3801	12.2421	14.6837	16.9190	19.0228	21.6660	23.5894
10	2.1559	2.5582	3.2470	3.9403	4.8652	6.1791	13.4420	15.9872	18.3070	20.4832	23.2093	25.1882
11	2.6032	3.0535	3.8157	4.5748	5.5778	6.9887	14.6314	17.2750	19.6751	21.9201	24.7250	26.7568
12	3.0738	3.5706	4.4038	5.2260	6.3038	7.8073	15.8120	18.5493	21.0261	23.3367	26.2170	28.2995
13	3.5650	4.1069	5.0088	5.8919	7.0415	8.6339	16.9848	19.8119	22.3620	24.7356	27.6883	29.8195
14	4.0747	4.6604	5.6287	6.5706	7.7895	9.4673	18.1508	21.0641	23.6848	26.1189	29.1412	31.3194
15	4.6009	5.2293	6.2621	7.2609	8.5468	10.3070	19.3107	22.3071	24.9958	27.4884	30.5779	32.8013
16	5.1422	5.8122	6.9077	7.9616	9.3122	11.1521	20.4651	23.5418	26.2962	28.8453	31.9999	34.2672
17	5.6972	6.4078	7.5642	8.6718	10.0852	12.0023	21.6146	24.7690	27.5871	30.1910	33.4087	35.7185
18	6.2648	7.0149	8.2307	9.3905	10.8649	12.8570	22.7595	25.9894	28.8693	31.5264	34.8053	37.1564
19	6.8440	7.6327	8.9065	10.1170	11.6509	13.7158	23.9004	27.2036	30.1435	32.8523	36.1909	38.5823
20	7.4338	8.2604	9.5908	10.8508	12.4426	14.5784	25.0375	28.4120	31.4104	34.1696	37.5662	39.9968
21	8.0337	8.8972	10.2829	11.5913	13.2396	15.4446	26.1711	29.6151	32.6706	35.4789	38.9322	41.4011
22	8.6427	9.5425	10.9823	12.3380	14.0415	16.3140	27.3015	30.8133	33.9244	36.7807	40.2894	42.7957
23	9.2604	10.1957	11.6885	13.0905	14.8480	17.1865	28.4288	32.0069	35.1725	38.0756	41.6384	44.1813
24	9.8862	10.8564	12.4012	13.8484	15.6587	18.0618	29.5533	33.1962	36.4150	39.3641	42.9798	45.5585
25	10.5196	11.5240	13.1197	14.6114	16.4734	18.9398	30.6752	34.3816	37.6525	40.6465	44.3141	46.9279
26	11.1602	12.1981	13.8439	15.3792	17.2919	19.8202	31.7946	35.5632	38.8851	41.9232	45.6417	48.2899
27	11.8076	12.8785	14.5734	16.1514	18.1139	20.7030	32.9117	36.7412	40.1133	43.1945	46.9629	49.6449
28	12.4613	13.5647	15.3079	16.9279	18.9392	21.5880	34.0266	37.9159	41.3371	44.4608	48.2782	50.9934
29	13.1211	14.2565	16.0471	17.7084	19.7677	22.4751	35.1394	39.0875	42.5570	45.7223	49.5879	52.3356
30	13.7867	14.9535	16.7908	18.4927	20.5992	23.3641	36.2502	40.2560	43.7730	46.9792	50.8922	53.6720
31	14.4578	15.6555	17.5387	19.2806	21.4336	24.2551	37.3591	41.4217	44.9853	48.2319	52.1914	55.0027

v \ p	0.005	0.01	0.025	0.05	0.10	0.20	0.80	0.90	0.95	0.975	0.99	0.995
32	15.1340	16.3622	18.2908	20.0719	22.2706	25.1478	38.4663	42.5847	46.1943	49.4804	53.4858	56.3281
33	15.8153	17.0735	19.0467	20.8665	23.1102	26.0422	39.5718	43.7452	47.3999	50.7251	54.7755	57.6484
34	16.5013	17.7891	19.8063	21.6643	23.9523	26.9383	40.6756	44.9032	48.6024	51.9660	56.0609	58.9639
35	17.1918	18.5089	20.5694	22.4650	24.7967	27.8359	41.7780	46.0588	49.8018	53.2034	57.3421	60.2748
36	17.8867	19.2327	21.3359	23.2686	25.6433	28.7350	42.8788	47.2122	50.9985	54.4373	58.6192	61.5812
37	18.5858	19.9602	22.1056	24.0749	26.4921	29.6355	43.9782	48.3634	52.1923	55.6680	59.8925	62.8833
38	19.2889	20.6914	22.8785	24.8839	27.3430	30.5373	45.0763	49.5126	53.3835	56.8955	61.1621	64.1814
39	19.9959	21.4262	23.6543	25.6954	28.1958	31.4405	46.1730	50.6598	54.5722	58.1201	62.4281	65.4756
40	20.7065	22.1643	24.4330	26.5093	29.0505	32.3450	47.2685	51.8051	55.7585	59.3417	63.6907	66.7660

E.3 TABLE OF PERCENTAGE POINTS OF t DISTRIBUTION

The percentage points of the t distribution are listed in the table. These are the values of a, tabulated as a function of v and p, such that

$$\frac{\Gamma[(v + 1/2]}{\sqrt{\pi v}\,\Gamma(v/2)} \int_{-a}^{a} \left(1 + \frac{s^2}{v}\right)^{-(v + 1)/2} ds = p$$

v \ p	0.6	0.8	0.9	0.95	0.98	0.99
1	1.376	3.078	6.314	12.706	31.821	63.657
2	1.061	1.886	2.920	4.303	6.965	9.925
3	0.978	1.638	2.353	3.182	4.541	5.841
4	0.941	1.533	2.132	2.776	3.747	4.604
5	0.920	1.476	2.015	2.571	3.365	4.032
6	0.906	1.440	1.943	2.447	3.143	3.707
7	0.896	1.415	1.895	2.365	2.998	3.499
8	0.889	1.397	1.860	2.306	2.896	3.355
9	0.883	1.383	1.833	2.262	2.821	3.250
10	0.879	1.372	1.812	2.228	2.764	3.169
11	0.876	1.363	1.796	2.201	2.718	3.106
12	0.873	1.356	1.782	2.179	2.681	3.055
13	0.870	1.350	1.771	2.160	2.650	3.012
14	0.868	1.345	1.761	2.145	2.624	2.977
15	0.866	1.341	1.753	2.131	2.602	2.947
16	0.865	1.337	1.746	2.120	2.583	2.921
17	0.863	1.333	1.740	2.110	2.567	2.898
18	0.862	1.330	1.734	2.101	2.552	2.878
19	0.861	1.328	1.729	2.093	2.539	2.861
20	0.860	1.325	1.725	2.086	2.528	2.845
21	0.859	1.323	1.721	2.080	2.518	2.831
22	0.858	1.321	1.717	2.074	2.508	2.819
23	0.858	1.319	1.714	2.069	2.500	2.807
24	0.857	1.318	1.711	2.064	2.492	2.797
25	0.856	1.316	1.708	2.060	2.485	2.787
26	0.856	1.315	1.706	2.056	2.479	2.779
27	0.855	1.314	1.703	2.052	2.473	2.771
28	0.855	1.313	1.701	2.048	2.467	2.763
29	0.854	1.311	1.699	2.045	2.462	2.756
30	0.854	1.310	1.697	2.042	2.457	2.750
40	0.851	1.303	1.684	2.021	2.423	2.704
60	0.848	1.296	1.671	2.000	2.390	2.660
80	0.846	1.292	1.664	1.990	2.374	2.639
100	0.845	1.290	1.660	1.984	2.364	2.626
150	0.844	1.287	1.655	1.976	2.351	2.609
200	0.843	1.286	1.653	1.972	2.345	2.601

E.4 TABLE OF PERCENTAGE POINTS OF F DISTRIBUTION

Three sets of percentage points of the F distribution are listed, where these are the values of a, tabulated as a function of v_1 and v_2, such that

$$\frac{\Gamma[(v_1 + v_2)/2]}{\Gamma(v_1/2)\Gamma(v_2/2)} \left(\frac{v_1}{v_2}\right)^{v_1/2} \int_0^a \frac{s^{(v_1 - 2)/2}}{\left[1 + (v_1/v_2)s\right]^{(v_1 + v_2)/2}} \, ds = p$$

Table with v_1 across columns and v_2 down rows.

$p = 0.90$

v_2	1	2	3	4	6	8	10	15	20	30	40	60	80	100	150	200
1	39.86	49.50	53.59	55.83	58.20	59.44	60.19	61.22	61.74	62.26	62.53	62.79	62.93	63.01	63.11	63.17
2	8.53	9.00	9.16	9.24	9.33	9.37	9.39	9.42	9.44	9.46	9.47	9.47	9.48	9.48	9.48	9.49
3	5.54	5.46	5.39	5.34	5.28	5.25	5.23	5.20	5.18	5.17	5.16	5.15	5.15	5.14	5.14	5.14
4	4.54	4.32	4.19	4.11	4.01	3.95	3.92	3.87	3.84	3.82	3.80	3.79	3.78	3.78	3.77	3.77
6	3.78	3.46	3.29	3.18	3.05	2.98	2.94	2.87	2.84	2.80	2.78	2.76	2.75	2.75	2.74	2.73
8	3.46	3.11	2.92	2.81	2.67	2.59	2.54	2.46	2.42	2.38	2.36	2.34	2.33	2.32	2.31	2.31
10	3.29	2.92	2.73	2.61	2.46	2.38	2.32	2.24	2.20	2.16	2.13	2.11	2.09	2.09	2.08	2.07
15	3.07	2.70	2.49	2.36	2.21	2.12	2.06	1.97	1.92	1.87	1.85	1.82	1.80	1.79	1.78	1.77
20	2.97	2.59	2.38	2.25	2.09	2.00	1.94	1.84	1.79	1.74	1.71	1.68	1.66	1.65	1.64	1.63
30	2.88	2.49	2.28	2.14	1.98	1.88	1.82	1.72	1.67	1.61	1.57	1.54	1.52	1.51	1.49	1.48
40	2.84	2.44	2.23	2.09	1.93	1.83	1.76	1.66	1.61	1.54	1.51	1.47	1.45	1.43	1.42	1.41
60	2.79	2.39	2.18	2.04	1.87	1.77	1.71	1.60	1.54	1.48	1.44	1.40	1.37	1.36	1.34	1.33
80	2.77	2.37	2.15	2.02	1.85	1.75	1.68	1.57	1.51	1.44	1.40	1.36	1.33	1.32	1.30	1.28
100	2.76	2.36	2.14	2.00	1.83	1.73	1.66	1.56	1.49	1.42	1.38	1.34	1.31	1.29	1.27	1.26
150	2.74	2.34	2.12	1.98	1.81	1.71	1.64	1.53	1.47	1.40	1.35	1.30	1.28	1.26	1.23	1.22
200	2.73	2.33	2.11	1.97	1.80	1.70	1.63	1.52	1.46	1.38	1.34	1.29	1.26	1.24	1.21	1.20

$p = 0.95$

v_2	1	2	3	4	6	8	10	15	20	30	40	60	80	100	150	200
1	161.45	199.50	215.71	224.58	233.99	238.88	241.88	245.95	248.01	250.10	251.14	252.20	252.72	253.04	253.46	253.68
2	18.51	19.00	19.16	19.25	19.33	19.37	19.40	19.43	19.45	19.46	19.47	19.48	19.48	19.49	19.49	19.49
3	10.13	9.55	9.28	9.12	8.94	8.85	8.79	8.70	8.66	8.62	8.59	8.57	8.56	8.55	8.54	8.54
4	7.71	6.94	6.59	6.39	6.16	6.04	5.96	5.86	5.80	5.75	5.72	5.69	5.67	5.66	5.65	5.65
6	5.99	5.14	4.76	4.53	4.28	4.15	4.06	3.94	3.87	3.81	3.77	3.74	3.72	3.71	3.70	3.69
8	5.32	4.46	4.07	3.84	3.58	3.44	3.35	3.22	3.15	3.08	3.04	3.01	2.99	2.97	2.96	2.95
10	4.96	4.10	3.71	3.48	3.22	3.07	2.98	2.85	2.77	2.70	2.66	2.62	2.60	2.59	2.57	2.56
15	4.54	3.68	3.29	3.06	2.79	2.64	2.54	2.40	2.33	2.25	2.20	2.16	2.14	2.12	2.10	2.10

	v_1	1	2	3	4	6	8	10	15	20	30	40	60	80	100	150	200
v_2																	

$p = 0.95$

v_2	1	2	3	4	6	8	10	15	20	30	40	60	80	100	150	200
20	4.35	3.49	3.10	2.87	2.60	2.45	2.35	2.20	2.12	2.04	1.99	1.95	1.92	1.91	1.89	1.88
30	4.17	3.32	2.92	2.69	2.42	2.27	2.16	2.01	1.93	1.84	1.79	1.74	1.71	1.70	1.67	1.66
40	4.08	3.23	2.84	2.61	2.34	2.18	2.08	1.92	1.84	1.74	1.69	1.64	1.61	1.59	1.56	1.55
60	4.00	3.15	2.76	2.53	2.25	2.10	1.99	1.84	1.75	1.65	1.59	1.53	1.50	1.48	1.45	1.44
80	3.96	3.11	2.72	2.49	2.21	2.06	1.95	1.79	1.70	1.60	1.54	1.48	1.45	1.43	1.39	1.38
100	3.94	3.09	2.70	2.46	2.19	2.03	1.93	1.77	1.68	1.57	1.52	1.45	1.41	1.39	1.36	1.34
150	3.90	3.06	2.66	2.43	2.16	2.00	1.89	1.73	1.64	1.54	1.48	1.41	1.37	1.34	1.31	1.29
200	3.89	3.04	2.65	2.42	2.14	1.98	1.88	1.72	1.62	1.52	1.46	1.39	1.35	1.32	1.28	1.26

$p = 0.99$

v_2	1	2	3	4	6	8	10	15	20	30	40	60	80	100	150	200
2	98.50	99.00	99.17	99.25	99.33	99.37	99.40	99.43	99.45	99.47	99.47	99.48	99.49	99.49	99.49	99.49
3	34.12	30.82	29.46	28.71	27.91	27.49	27.23	26.87	26.69	26.50	26.41	26.32	26.27	26.24	26.20	26.18
4	21.20	18.00	16.69	15.98	15.21	14.80	14.55	14.20	14.02	13.84	13.75	13.65	13.61	13.58	13.54	13.52
6	13.75	10.92	9.78	9.15	8.47	8.10	7.87	7.56	7.40	7.23	7.14	7.06	7.01	6.99	6.95	6.93
8	11.26	8.65	7.59	7.01	6.37	6.03	5.81	5.52	5.36	5.20	5.12	5.03	4.99	4.96	4.93	4.91
10	10.04	7.56	6.55	5.99	5.39	5.06	4.85	4.56	4.41	4.25	4.17	4.08	4.04	4.01	3.98	3.96
15	8.68	6.36	5.42	4.89	4.32	4.00	3.80	3.52	3.37	3.21	3.13	3.05	3.00	2.98	2.94	2.92
20	8.10	5.85	4.94	4.43	3.87	3.56	3.37	3.09	2.94	2.78	2.69	2.61	2.56	2.54	2.50	2.48
30	7.56	5.39	4.51	4.02	3.47	3.17	2.98	2.70	2.55	2.39	2.30	2.21	2.16	2.13	2.09	2.07
40	7.31	5.18	4.31	3.83	3.29	2.99	2.80	2.52	2.37	2.20	2.11	2.02	1.97	1.94	1.90	1.87
60	7.08	4.98	4.13	3.65	3.12	2.82	2.63	2.35	2.20	2.03	1.94	1.84	1.78	1.75	1.70	1.68
80	6.96	4.88	4.04	3.56	3.04	2.74	2.55	2.27	2.12	1.94	1.85	1.75	1.69	1.65	1.61	1.58
100	6.90	4.82	3.98	3.51	2.99	2.69	2.50	2.22	2.07	1.89	1.80	1.69	1.63	1.60	1.55	1.52
150	6.81	4.75	3.91	3.45	2.92	2.63	2.44	2.16	2.00	1.83	1.73	1.62	1.56	1.52	1.46	1.43
200	6.76	4.71	3.88	3.41	2.89	2.60	2.41	2.13	1.97	1.79	1.69	1.58	1.52	1.48	1.42	1.39

REFERENCES

Abramowitz, M., and Stegun, L. A. (1964), *Handbook of Mathematical Functions*, Applied Math Series, No. 55, National Bureau of Standards, Washington, D.C.

Almar-Naess, A. (Ed.) (1985), *Fatigue Handbook: Offshore Structures*, Tapir, Trondheim, Norway.

Ang, A. H. S., and Tang, W. H. (1975), *Probability and Concepts in Engineering Planning and Design*, Vol. 1, Wiley, New York.

Ang, A. H. S., and Tang, W. (1984), *Probability Concepts in Engineering Planning and Design*, Vol. II: *Decision, Risk, and Reliability*, Wiley, New York.

Augusti, G., Baratta, A., Casciati, F. (1984), *Probability Methods in Structural Engineering*, Chapman & Hall, New York.

Baca, T. J. (1982), "Characterization of Conservatism in Mechanical Shock Testing of Structures," Ph.D. Dissertation, Stanford University, Stanford, CA.

Baca, T. J. (1984), "Alternative Shock Characterizations for Consistent Shock Test Specifications," *Shock Vibrat. Bull.*, **54**, (Pt. 2), pp. 109–130.

Baca, T. J., and Paez, T. L. (1987), "Mechanical Shock Test Specification Using Nonstationary Random Excitation Models," *Proceedings of the 1987 Society for Experimental Mechanics Fall Conference*, October, Savannah, GA, Society for Experimental Mechanics, pp. 193–198.

Bendat, J. S., and Piersol, A. G. (1966), *Measurement and Analysis of Random Data*, Wiley, New York.

Bendat, J. S., and Piersol, A. G. (1971), *Random Data, Analysis and Measurement Procedures*, Wiley, New York.

Bendat, J. S., and Piersol, A. G. (1986), *Random Data, Analysis and Measurement Procedures*, 2nd ed., Wiley, Interscience, New York.

Bendat, J. S., and Piersol, A. G. (1980), *Engineering Applications of Correlation and Spectral Analysis*, Wiley-Interscience, New York.

433

Benjamin, J. R., and Cornell, C. A. (1970), *Probability, Statistics, and Decision for Civil Engineers*, McGraw-Hill, New York.

Beyer, W. H. (1978), *CRC Standard Mathematical Tables*, 25th ed., CRC Press, Boca Raton, FL.

Biot, M. A. (1941), "A Mechanical Analyzer for the Prediction of Earthquake Stresses," *Bull. Seismol. Soc. Am.*, **31**, 151–171.

Blackman, R. B., and Tukey, J. W. (1959), *The Measurement of Power Spectra from the Point of View of Communications Engineering*, Dover Publications, New York.

Bochner, S. (1933), "Monotone Funcktionen Stieljessche Integrale und Harmonische Analyse," *Math. Ann.*, **108**, 376–385.

Bochner, S. (1959), *Lectures on Fourier Integrals*, Tenenbaum and H. Pollard (transl.), Princeton University Press, Princeton, NJ.

Bolotin, V. V. (1984), *Random Vibration of Elastic Systems*, Martinus Nijhoff, The Hague, The Netherlands.

Box, G. E. P., and Jenkins, G. M. (1976), *Time Series Analysis*, Holden-Day, San Francisco.

Box, G. E. P., Muller, N. E., and Marsaglia, G. (1958), "Polar Methods for Normal Deviates," *Ann. Math. Stat.*, **28**, 610.

Bracewell, R. N. (1978), *The Fourier Transform and Its Applications*, McGraw-Hill, New York.

Broek, D. (1984), *Elementary Engineering Fracture Mechanics*, Martinus Nijhoff, The Hague, The Netherlands.

Brownlee, K. A. (1960), *Statistical Theory and Methodology in Science and Engineering*, Wiley, New York.

Cacoullos, T. (1966), "Estimation of Multivariate Density," *Ann. Inst. Stat. Math.*, **18**(2), 179–189.

Campbell, G. A., and Foster, R. M. (1948), *Fourier Integrals for Practical Applications*, Van Nostrand, New York.

Champeney, D. C. (1973), *Fourier Transforms and Their Physical Applications*, Academic, New York.

Clough, R. W., and Penzien, J. (1975), *Dynamics of Structures*, McGraw-Hill, New York.

Collins, J. A. (1981), *Failure of Materials in Mechanical Design*, McGraw-Hill, New York.

Cooley, J. W., and Tukey, J. W. (1965), "An Algorithm for the Machine Calculation of Complex Fourier Series," *Math. Computat.*, **19**, 297.

Crandall, S. H. (Ed.) (1958), *Random Vibration*, Technology Press of MIT and John Wiley, New York.

Crandall, S. H. (Ed.) (1963), *Random Vibration*, MIT Press, Cambridge, MA.

Crandall, S. H., and Mark, W. D. (1963), *Random Vibration in Mechanical Systems*, Academic, New York.

Crandall, S. H., Chandiramani, K. L., and Cook, R. G. (1966), "Some First Passage Problems in Random Vibration," *J. Appl. Mech.*, **33**(3), 532–538.

Der Kiureghian, A. (1980), "Structural Response to Stationary Excitation," *ASCE J. Eng. Mech. Div.*, **106**(6), 1195–1213.

Dodds, C. J., and Robson, J. D. (1975), "Partial Coherence in Multivariate Random Processes," *J. Sound Vibrat.*, **42**, 243–247.

Doob, J. L. (1953), *Stochastic Processes*, Wiley, New York.

Dowling, N. E. (1972), "Fatigue Failure Predictions of Complicated Stress-Strain Histories," *ASTM J. Mater.*, **7**(1), 71–87.

Dowling, N. E. (1993), *Mechanical Behavior of Materials*, Prentice-Hall, Englewood Cliffs, NJ.

Drenick, R. F. (1970), "Model Free Design of Seismic Structures," *J. Eng. Mech. Div. ASCE*, **96**, 483–493.

Einstein, A. (1905), "Investigations on the Theory of Brownian Movement," *Ann. d. Phys.*, **17**, 549. (Republished along with other papers on the same subject by Einstein under the title, *Investigations on the Theory of Brownian Movement*, by Dover Publications, New York, in 1956. Ed. R. Furth, trans. A. D. Cowper.)

Elishakoff, I. (1983), *Probabilistic Methods in the Theory of Structures*, Wiley, New York.

Feller, W. (1957), *An Introduction to Probability Theory and Its Applications*, Vol. I, Wiley, New York.

Feller, W. (1966), *An Introduction to Probability Theory and Its Applications*, Vol. 2, Wiley, New York.

Forristall, G. Z. (1984), "The Distribution of Measured and Simulated Wave Heights as a Function of Spectral Shape," *J. Geophys. Res.*, **89**(C6), 10547–10552.

Fuchs, H. O., and Stephens, R. K. (1980), *Metal Fatigue in Engineering*, Wiley, New York.

Gumbel, E. J. (1958), *Statistics of Extremes*, Columbia University Press, New York.

Gurney, T. R. (1979), *Fatigue of Welded Structures*, 2nd ed., Cambridge University Press, Cambridge, United Kingdom.

Gurney, T. R. (1983), "Fatigue Tests Under Variable Amplitude Loading," The Welding Institute, Abington Hall, Cambridge, England, Report No. 220/83.

Harris, F. (1978), "On the Use of Windows for Harmonic Analysis with the Discrete Fourier Transform," *Proc. IEEE*, **66**(1), 51–83.

Hertzberg, R. W. (1989), *Deformation and Fracture Mechanics of Engineering Materials*, Wiley, New York.

Himelblau, H., and Piersol, A. G. (1989), "Evaluation of a Procedure for the Analysis of Nonstationary Vibroacoustic Data," *J. Environ. Sci.*, Vol. XXXII, No. 2, 35–42.

Hines, W. W., and Montgomery, D. C. (1990), *Probability and Statistics in Engineering and Management Science*, 3rd ed., Wiley, New York.

Hoel, P. G., Port, S. C., and Stone, C. J. (1972), *Introduction to Stochastic Processes*, Houghton Mifflin, Boston.

Hurty, W. C., and Rubinstein, M. F. (1964), *Dynamics of Structures*, Prentice-Hall, Englewood Cliffs, NJ.

Ibrahim, R. A. (1985), *Parametric Random Vibration*, Wiley, New York.

Inman, D. J. (1994), *Engineering Vibration*, Prentice-Hall, Englewood Cliffs, NJ.

Jacobsen, L. S., and Ayre, R. S. (1958), *Engineering Vibrations with Applications to Structures and Machinery*, McGraw-Hill, New York.

James, H. M., Nichols, N. B., and R. S. Phillips (1947), *Theory of Servomechanisms*, McGraw-Hill, New York.

James, M. L., Smith, G. M., Wolford, J. C., and Whaley, P. W. (1994), *Vibration of Mechanical and Structural Systems*, Harper Collins, New York.

Jenkins, G. M., and Watts, D. G. (1969), *Spectral Analysis and Its Applications*, Holden-Day, San Francisco.

Karlin, S., and Taylor, H. M. (1975), *A First Course in Stochastic Processes*, Academic, New York.

Karlin, S., and Taylor, H. M. (1981), *A Second Course in Stochastic Processes*, Academic, New York.

Khinchine, A. (1934), "Korrelationstheorie der stationären stochastischen Prozesse," *Math. Ann.*, **109**, 604–615.

Koopmans, L. (1974), *The Spectral Analysis of Time Series*, Academic Press, New York.

Krenk, S. (1978), "A Double Envelope for Stochastic Processes," Structural Research Laboratory, Technical University of Denmark, Lyngby, Denmark, Report No. 134.

Larsen, C. E., and Lutes, L. D. (1991), "Predicting the Fatigue Life of Offshore Structures by the Single-Moment Spectral Method," *Probabilistic Engineering Mechanics*, **6**(2), 96–108.

Lim, S. L., and Oppenheim, A. V. (1988), *Advanced Topics in Signal Processing*, Prentice-Hall, Englewood Cliffs, NJ.

Lin, Y. K. (1970), "First Excursion Failure of Randomly Excited Structures," *AIAA J.*, **8**(4), 720–728.

Lin, Y. K. (1967), *Probabilistic Theory of Structural Dynamics*, McGraw-Hill, New York. (Republished in 1976 by Krieger, Huntington, NY.)

Lutes, L. D., and Larsen, C. E. (1990), "Improved Spectral Method for Variable Amplitude Fatigue Prediction," *ASCE J. Struc. Eng.*, **116**(4), 1149–1164.

Lyon, R. H. (1961), "On the Vibration Statistics of a Randomly Excited Hard-Spring Oscillator," *J. Acoust. Soc. Am.*, **33**(10), 1395–1403.

Madsen, H. O., Krenk, S., and Lind, N. C. (1986), *Methods of Structural Safety*, Prentice-Hall, Englewood Cliffs, NJ.

Marley, M. J. (1991), *Time Variant Reliability Under Fatigue Degradation*, Norwegian Institute of Technology, Trondheim, Norway.

Melchers, R. E. (1987), *Structural Reliability Analysis and Prediction*, Wiley, New York.

MIL-HDBK-5 (1987), "Metallic Materials and Elements for Aerospace Vehicle Structures," MIL-HDBK-5E, Naval Publications and Forms Center, Philadelphia, PA.

Miller, I., and Freund, J. E. (1985), *Probability and Statistics for Engineers*, Prentice-Hall, Englewood Cliffs, NJ.

Miner, M. A. (1945), "Cumulative Damage in Fatigue," *ASME Trans.*, **67**, A159–164.

Mood, A. M., Graybill, F. A., and Boes, D. C. (1974), *Introduction to the Theory of Statistics*, McGraw-Hill, New York.

Nelson, D. V. (1978), "Cumulative Fatigue Damage in Metals," Ph.D. Dissertation, Stanford University, Stanford, CA.

Newland, D. E. (1984), *Random Vibrations and Spectral Analysis*, 2nd ed., Longman, New York.

Nigam, N. C. (1983), *Introduction to Random Vibrations*, MIT Press, Cambridge, MA.

Nuttall, A. (1971), "Spectral Estimation by Means of Overlapped Fast Fourier Transform Processing of Windowed Data," *NUSC Rept.* No. 4169, Naval Underwater Systems Center, New London, CT.

Nuttall, A. (1981), "Some Windows with Very Good Sidelobe Behavior," *IEEE Trans. Acoust., Speech Signal Proc.*, **ASSP-29**(1), 84–91.

Ortiz, K. (1985), "On the Stochastic Modeling of Fatigue Crack Growth," Ph.D. Dissertation, Stanford University, Stanford, CA.

Ortiz, K., and Chen, N. K. (1987), "Fatigue Damage Prediction for Stationary Wide-Band Stresses," ICASP 5, presented at the Fifth International Conference on the Applications of Statistics and Probability in Civil Engineering, Vancouver, Canada.

Ortiz, K., Kung, C. J., and Perng, H. L. (1988), "Modeling Fatigue Crack Growth Resistance, *dn/da*," in *Probabilistic Methods in Civil Engineering*, P. D. Spanos, (Ed.), American Society of Civil Engineers, New York, pp. 21–25.

Otnes, R. K., and Enochson, L. (1978), *Applied Time Series Analysis*, Vol. 1, Wiley, New York.

Paez, T. (1985), "Characterization of Nonstationary Random Processes," *Proceedings of the Institute of Environmental Sciences Annual Meeting*, May, Las Vegas, NV, Institute of Environmental Sciences, pp. 495–500.

Palmgren, A. (1924), "Die Lebensdauer von Kugellagern," *Zeits. Ver. Deuts. Ing.*, **68**(14), 339.

Papoulis, A. (1962), *The Fourier Integral and Its Applications*, McGraw-Hill, New York.

Papoulis, A. (1965), *Probability, Random Variables, and Stochastic Processes*, McGraw-Hill, New York.

Paris, P. C. (1964), "The Fracture Mechanics Approach to Fatigue," in *Fatigue, An Interdisciplinary Approach*, J. J. Burke, N. L. Reed, and V. Weiss (Ed.), Syracuse University Press, New York, pp. 107–132.

Parzen, E. (1962a), "On Estimation of a Probability Density Function and Mode," *Ann. Math. Stat.*, **33**, 1065–1076.

Parzen, E. (1962b), *Stochastic Processes*, Holden-Day, San Francisco.

Perng, H.-L. (1989), "Damage Accumulation in Random Loads," Ph.D. Dissertation, University of Arizona, Tucson, AZ.

Powell, A. (1958), "On the Fatigue Failure of Structures Due to Vibrations Excited by Random Pressure Fields," *J. Acoust. Soc. Am.*, **30**(12), 1130–1135.

Press, N. H., Flannery, B. P., Tenkolsky, S. A., and Vettering, W. T. (1988), *Numerical Recipes in C, The Art of Scientific Computing*, Cambridge University Press, Cambridge (also available in Fortran and Pascal).

Priestly, M. B. (1965), "Evolutionary Spectra and Nonstationary Random Processes," *J. Roy. Stat. Soc. B*, **27**, 204–229.

Priestly, M. B. (1967), "Power Spectral Analysis of Nonstationary Random Processes," *J. Sound Vibrat.*, **6**(1), 86–97.

Rao, S. S. (1990), *Mechanical Vibrations*, Addison-Wesley, New York.

Rayleigh, J. W. S. (1877), *The Theory of Sound*, Vols. I and II, The Macmillan Company, London. (The second revised and enlarged edition of 1894 was republished by Dover Publications, New York, in 1945.)

Rice, J. R. (1964), "Theoretical Prediction of Some Statistical Characteristics of Random Loadings Relevant to Fatigue and Fracture," Ph.D. Dissertation, Lehigh University, Bethlehem, PA.

Rice, J. R., and Beer, F. P. (1965), "On the Distribution of Rises and Falls in a Continuous Random Process," *J. Basic Eng. ASME*, June, **87**(2), 398–404.

Rice, S. O. (1944, 1945), "Mathematical Analysis of Random Noise," *Bell Syst. Tech. J.*, **23**, 282–332; **24**, 46–156; reprinted in N. Wax, *Selected Papers on Noise and Stochastic Processes*, Dover, New York, 1954.

Richard, R. M. (1990), "Damping and Vibration Considerations for the Design of Optical Systems in a Launch/Space Environment," Optical Sciences Center, University of Arizona, Tucson, AZ.

Richard, R. M., Cho, M., and Pollard, W. (1988), "Dynamic Analysis of the SIRTF One-Meter Mirror During Launch," *Proc. Int. Soc. Opt. Eng.*, **973**(11), 86–99.

Roberts, J. B., and Spanos, P.D. (1990), *Random Vibration and Statistical Linearization*, Wiley, New York.

Robson, J. D. (1964), *An Introduction to Random Vibration*, Elsevier, New York.

Rolfe, S. T., and Barsom, J. M. (1987), *Fracture and Fatigue Control in Structures*, Prentice-Hall, Englewood Cliffs, NJ.

Rona, T. P. (1958), "Instrumentation for Random Vibration Analysis," in *Random Vibration*, S. H. Crandall (Ed.), Technology Press of MIT and John Wiley, New York.

Roy, R. V., and Spanos, P. D. (1993), "Power Spectral Density of Nonlinear System Reponse: The Recursion Method," *ASME J. App. Mech.*, **60**, 358–365.

Schuëller, G. I., and Shinozuka, M. (Eds.) (1987), *Stochastic Methods in Structural Dyanmics*, Martinus Nijhjoff, Boston.

Schuster, A. (1898), "On the Investigation of Hidden Periodicities with Application to a Supposed 26 Day Period of Meterological Phenomena," *Terrestr. Magnet.*, **3**, 13–41.

Schuster, A. (1900), "The Periodogram of Magnetic Declination As Obtained from the Records of the Greenwich Observatory During the years 1871–1895," *Trans. Camb. Phil. Soc.*, **18**, 107–135.

Schuster, A. (1906), "The Periodogram and Its Optical Analogy," *Proc. Roy. Soc. Lond. Ser. A*, **77**, 136–140.

Schütz, W. (1993), "Fatigue Life Prediction: A Review of the State of the Art," in *Structural Failure, Product Liability and Technical Insurance*, Vol. IV, H. P. Rossmanith (Ed.), Elsevier, New York.

Schütz, W. (1979), "The Prediction of Fatigue in the Crack Initiation and Propagation Stages: A State of the Art Survey," *Eng. Fract. Mech.*, **11**(2), 405–421.

Shin, Y. S., and Lukens, R. W. (1983), "Probability Based High Cycle Fatigue Life Predictions," *Random Fatigue Life Prediction*, American Society of Mechanical Engineers, New York.

Shinozuka, M. (1970), "Maximum Structural Response to Seismic Excitations," *J. Eng. Mech. Div. ASCE*, **96**, 729–738.

Shinozuka, M. (1971), "Simulation of Multivariate and Multidimensional Random Processes," *J. Acoust. Soc. Am.*, **49**(1), 357–367.

Smallwood, D. O. (1981), "An Improved Recursive Formula for Calculating Shock Response Spectra," *Shock Vibrat. Bull.*, Shock and Vibration Information Center, **51**(2), 211.

Smallwood, D. O. (1989), "Characterizing Transient Vibrations Using Band Limited Moments," *Proceedings of the 60th Shock and Vibration Symposium*, Vol. III, November, Portsmouth, VA, Shock and Vibration Information Center, pp. 93–112.

Smallwood, D. O. (1992), "Variance of Temporal Moment Estimates of a Squared Time History," *Proceedings of the 63rd Shock and Vibration Symposium*, October, Arlington, VA, Shock and Vibration Information Center, pp. 389–399.

Smallwood, D. O. (1993), "Characterizing Transient Vibrations Using Band Limited Temporal Moments," *Proceedings of the Institute of Environmental Sciences Annual Meeting*, May, Chicago, p. 291–302, Institute of Environmental Sciences.

Smallwood, D. O., and Paez, T. L. (1993), "A Frequency Domain Method for the Generation of Partially Coherent, Normal Stationary Time Domain Signals," *Shock Vibrat.*, **1**(1), 45–53.

Smallwood, D. O. (1994), "Note on the Variance of Autospectral Density Processed Using Welch's Method," *Proceedings of the IES Annual Meeting*, V. 2, Institute of Environmental Sciences, May, Chicago, IL.

Sobczyk, K., and Spencer, B. F. (1991), *Random Fatigue; From Data to Theory*, Academic, New York.

Society of Automotive Engineers (SAE) (1977), *Fatigue Under Complex Loading*, AE-6, SAE, Warrendale, PA.

Society of Automotive Engineers (SAE) (1988), *Fatigue Design Handbook*, AE-10, SAE, Warrendale, PA.

Soong, T. T. (1981), *Probabilistic Modeling and Analysis in Science and Engineering*, Wiley, New York.

Soong, T. T., and Grigoriu, M. (1993), *Random Vibration of Mechanical and Structural Systems*, Prentice-Hall, Englewood Cliffs, NJ.

Spanos, P. D. (1980), "Formulation of Stochastic Linearization for Symmetric or Asymmetric Multidegree of Freedom Nonlinear Systems," *J. Appl. Mech.*, **47**, 209–211.

Spanos, P. D. (1983), "Spectral Moments in the Calculation of Linear System Output," *ASME J. App. Mech.*, **50**, 901–903.

Spanos, P. D. (1987), "An Approach to Calculating Random Vibration Integrals," *ASME J. App. Mech.*, **54**, 409–413.

Stearns, S. (1975), *Digital Signal Analysis*, Hayden, Rochelle Park, NJ.

Steidel, R. F. (1979), *An Introduction to Mechanical Vibrations*, Wiley, New York.

Sun, C. T., and Lu, Y. P. (1995), *Vibration Damping of Structural Elements*, Prentice-Hall, Englewood Cliffs, NJ.

Tapia, R. A., and Thompson, J. R. (1978), *Nonparametric Probability Density Estimation*, Johns Hopkins University Press, Baltimore, MD.

Tayfun, M. A. (1981), "Distribution of Crest-to-Trough Wave Heights," *J. Waterways Harbors Div. ASCE*, **107**, 149–158.

Taylor, G. I. (1920), "Diffusion by Continuous Movements," *Proc. Lond. Math. Soc. Ser. 2*, **20**, 196–212.

Thoft-Christiansen, P., and Baker, M. J. (1982), *Structural Reliability Theory and Its Applications*, Springer-Verlag, New York.

Thomson, W. T. (1981), *Theory of Vibrations with Applications*, Prentice-Hall, Englewood Cliffs, NJ.

Timoshenko, S. (1928), *Vibration Problems in Engineering*, Van Nostrand, New York.

U.S. Air Force, Material Laboratory (1988), *Damage Tolerant Design Handbook*, Material Laboratory, Wright Patterson AFB, Ohio.

Vanmarcke, E. (1984), *Random Fields, Analysis and Synthesis*, MIT Press, Cambridge, MA.

Vanmarcke, E. (1975), "On the Distribution of First Passage Time for Normal Stationary Random Processes," *J. Appl. Mech.*, **42**, 215–220.

Veers, P. S., Winterstein, S. R., Nelson, D. V., and Cornell, A. C. (1989), "Variable Amplitude Load Models for Fatigue Damage and Crack Growth," *Development of Fatigue Loading Spectra*, ASTM STP1006:172–197, American Society for Testing and Materials, Philadelphia, PA.

Wax, N. (Ed.) (1954) *Selected Papers on Noise and Stochastic Processes*, Dover, New York.

Weaver, W., Timoshenko, S., and Young, D. H. (1990), *Vibration Problems in Engineering*, Wiley, New York.

Wiener, N. (1930), "Generalized Harmonic Analysis," *Acta Math.*, **55**, 117–258.

Wiener, N. (1933), *The Fourier Integral and Certain of Its Applications*, Cambridge University Press, New York.

Wilson, E. L., Der Kiureghian, A., and Bayo, E. P. (1981), "A Replacement for the SRSS Method in Seismic Analysis," *Earthquake Eng. Struct. Dyn.*, **9**, 187–192.

Winterstein, S. R. (1988), "Nonlinear Vibration Models for Extremes and Fatigue," *J. Eng. Mech. ASCE*, **114**(10), pp. 1772–1790.

Wirsching, P. H. (1991), "Advanced Fatigue Reliability Analysis," *Int. J. Fatigue*, **13**(5), 389–394.

Wirsching, P. H., and Chen, Y. N. (1988), "Consideration of Probability Based Fatigue Design Criteria for Marine Structures," *Marine Struct.*, **1**, 23–45.

Wirsching, P. H., and Light, M. C. (1980), "Fatigue Under Wide Band Random Stresses," *ASCE J. Struct. Div.*, **106**, 1593–1607.

Yang, C. Y. (1986), *Random Vibration of Structures*, Wiley, New York.

Yang, J. N. (1974), "Statistics of Random Loading Relevant to Fatigue," *J. Eng. Mech. Div. ASCE*, **100**(EM3), 469–475.

Yang, J. N., and Shinozuka, M. (1971), "On the First Excursion Probability in Stationary Narrow Band Random Vibration," *J. Appl. Mech.*, **37**(4), 1017–1022.

Yao, J. T. P., Kozin, F., Wen, Y. K., Yang, J. N., Schuëller, G. I., and Ditlevsen, O. (1986), "Stochastic Fatigue Fracture and Damage Analysis," *Struct. Saf.*, **3**, 231–267.

INDEX